W0225758

New Techniques for Future Accelerators II

RF and Microwave Systems

ETTORE MAJORANA
INTERNATIONAL SCIENCE SERIES
Series Editor:
Antonino Zichichi
European Physical Society
Geneva, Switzerland

(PHYSICAL SCIENCES)

Recent volumes in the series:

Volume 25   **FRONTIERS IN NUCLEAR DYNAMICS**
Edited by R. A. Broglia and C. H. Dasso

Volume 26   **TOKAMAK START-UP: Problems and Scenarios
Related to the Transient Phases of a
Thermonuclear Fusion Reactor**
Edited by Heinz Knoepfel

Volume 27   **DATA ANALYSIS IN ASTRONOMY II**
Edited by V. Di Gesu, L. Scarsi, P. Crane,
J. H. Friedman, and S. Levialdi

Volume 28   **THE RESPONSE OF NUCLEI UNDER EXTREME
CONDITIONS**
Edited by R. A. Broglia and G. F. Bertsch

Volume 29   **NEW TECHNIQUES FOR FUTURE ACCELERATORS**
Edited by M. Puglisi, S. Stipcich, and G. Torelli

Volume 30   **SPECTROSCOPY OF SOLID-STATE LASER-TYPE
MATERIALS**
Edited by Baldassare Di Bartolo

Volume 31   **FUNDAMENTAL SYMMETRIES**
Edited by P. Bloch, P. Pavlopoulos, and R. Klapisch

Volume 32   **BIOELECTROCHEMISTRY II: Membrane Phenomena**
Edited by G. Milazzo and M. Blank

Volume 33   **MUON-CATALYZED FUSION AND FUSION WITH
POLARIZED NUCLEI**
Edited by B. Brunelli and G. G. Leotta

Volume 34   **VERTEX DETECTORS**
Edited by Francesco Villa

Volume 35   **LASER SCIENCE AND TECHNOLOGY**
Edited by A. N. Chester, V. S. Letokhov, and S. Martellucci

Volume 36   **NEW TECHNIQUES FOR FUTURE ACCELERATORS II:
RF and Microwave Systems**
Edited by M. Puglisi, S. Stipcich, and G. Torelli

A Continuation Order Plan is available for this series. A continuation order will bring delivery of
each new volume immediately upon publication. Volumes are billed only upon actual shipment.
For further information please contact the publisher.

New Techniques for Future Accelerators II

RF and Microwave Systems

Edited by

Mario Puglisi

Pavia University
Pavia, Italy

Stanislao Stipcich

INFN National Laboratory of Frascati
Frascati, Italy

and

Gabriele Torelli

Pisa University
Pisa, Italy

Springer Science+Business Media, LLC

Library of Congress Cataloging in Publication Data

Seminar on New Techniques for Future Accelerators II—RF and Microwave Systems
(1987: Erice, Sicily)
New techniques for future accelerators II: RF and microwave systems / edited by
Mario Puglisi, Stanislao Stipcich, and Gabriele Torelli.
    p.   cm—(Ettore Majorana international science series. Physical sciences;
v. 36)
  "Proceedings of the third workshop of the INFN Eloisatron Project: Seminar on New
Techniques for Future Accelerators II—RF and Microwave Systems, held May 23–30,
1987, in Erice, Italy"—T.p. verso.
  Includes bibliographical references and index.

ISBN 978-1-4612-8064-4   ISBN 978-1-4613-0751-8 **(eBook)**
DOI 10.1007/978-1-4613-0751-8

  1. Particle accelerators—Technique—Congresses. I. Puglisi, Mario. II. Stipcich,
Stanislao. III. Torelli, Gabriele. IV. Istituto nazionale di fisica nucleare. V. Title. VI.
Series.
QC787.P3S46 1987   88-39414
539.7′3—dc19   CIP

Proceedings of the Third Workshop of the INFN
Eloisatron Project: Seminar on New Techniques
for Future Accelerators II—RF and Microwave Systems,
held May 23–30, 1987, in Erice, Italy

© 1989 **Springer Science+Business Media New York**
Originally published by Plenum Press, New York in 1989
Softcover reprint of the hardcover 1st edition 1989

A Division of Plenum Publishing Corporation
233 Spring Street, New York, N.Y. 10013

All rights reserved

No part of this book may be reproduced, stored in a retrieval system, or transmitted
in any form or by any means, electronic, mechanical, photocopying, microfilming,
recording, or otherwise, without written permission from the Publisher

CONTENTS

Opening Address 1
   G. Torelli

Prospects for High-Energy Particle Physics and Future Colliders 3
   E. Picasso

Accelerator Amplitude and Phase Control 21
   J. Kummer

RF Structures for Linear Accelerators 37
   A. Schempp

RF Driven Linear Colliders 51
   W. Schnell

Frequency Scaling and Gyroklystron Sources for Linear Colliders
     with SLAC-Type Linac Structures 65
   M. Reiser, W. Lawson, A. Mondelli and D. Chernin

RF System of a Synchrotron for Protons and Heavy Ions 131
   D. Böhne

Basic Features of Superconducting RF Cavities 145
   H. Piel

One Dimensional Model for the Interaction of a High Charged
     Bunched Beam with a Superconducting Cavity 181
   R. Bonifacio, L. Ferrucci, C. Pagani and L. Serafini

RF-Structures: A Survey 189
   D.T. Tran

Superconducting Radio Frequency Cavities for Accelerators 225
   H. Lengeler

RF System Design for Control of Heavy Beam Loading in
     Circular Machines 245
   G.H. Rees

Components for High-Power RF Systems in Modern Accelerators 267
    G. Schaffer

Radio Frequency Systems for Present and Future Accelerators 309
    E.C. Raka

RF Cavity Primer for Cyclic Proton Accelerators 331
    J.E. Griffin

Concluding Remarks 359
    M. Puglisi

Participants 361

Index 365

OPENING ADDRESS

I am very glad to open this third Workshop of the "INFN Eloisatron Project" in this delightful small town of Erice, so conducive to peace and meditation. The excellent management of the "Ettore Majorana Centre" makes this an ideal place for a fruitful interchange of ideas and suggestions among scientists. We owe this atmosphere to the efforts of Prof. Zichichi and the staff of the Centre.

This Workshop is the second one devoted to accelerator science, and hence it will be focused on the best way to construct very-high-energy machines like the Eloisatron in order to obtain the best possible performance. We are obviously not in charge of a project; we are here only to discuss the present status of the technique and to teach to younger people what we know at the present time. Our purpose is also to indicate directions and areas which should be researched more deeply in order to advance our knowledge and our technical capabilities.

The first Workshop was devoted to very general problems and projects of accelerators. This one is much more specialized and is devoted to RF and microwave systems, which, as everyone knows, are important components of accelerating machines. In the near future, the aim of reaching higher and higher energies will force the accelerating system (based on RF, microwave, or laser) to become perhaps the most important part of such machines.

In spite of this, to my knowledge, RF specialists have never met in a conference exclusively devoted to the acceleration problem. This is the occasion, then, to profit from the opportunity we have of being together to exchange ideas in such a beautiful place as Erice.

I am therefore most happy to open this Workshop and to wish that it will be the first of a series of most enriching meetings.

Gabriele Torelli

PROSPECTS FOR HIGH-ENERGY PARTICLE PHYSICS

AND FUTURE COLLIDERS

E. Picasso

European Organization for Nuclear Research, Geneva
and Scuola Normale Superiore, Pisa

PART 1 - THE PHYSICS

1.1. An overview of past discoveries

Man's history has been shaped by the irresistible urge to find out
how Nature works and how we can use this knowledge to improve our way
of living.

Some of the most basic questions we have always asked are: What is
the ultimate structure of matter? What are the forces through which matter interacts? How did the Universe begin? Will it ever end?

All these questions fall in the domain of elementary particle physics,
upon which our knowledge of other sciences ultimately rests. In the same
way as Biology and Medicine are founded on Chemistry, and Chemistry on
Physics, so all Physics is based on the study of the elementary particles
and the forces which govern their behaviour.

Until very recently it could scarcely have been foreseen that some
of the old-age mysteries of astronomy might find their solution by looking at the elementary particles, where powerful forces give all matter
its shape and form. Particle physicists, together with nuclear and atomic
physicists, astrophysicists and cosmologists are beginning to understand
not simply what matter is, but where it comes from, when and how. Questions of this type are important, because everything else we want to know
more about, is deeply related to them.

Accelerators are to particle physics what telescopes are to astronomy,
or microscopes to biology. All these instruments reveal and shed light
onto worlds which otherwise would remain hidden from our view. They are
undoubtedly the very tools of scientific progress.

Fig. 1 shows the chart of the discoveries from the fifties to the
present day, in the various high-energy centers where accelerators and
colliders are found[1]. The greater part of our knowledge in subatomic and

subnuclear physics derives from the use of powerful accelerators which have given us the possibility of exploring ever higher energies with increased luminosities. In parallel with the accelerators, it is important to develop highly-performing particle detectors enabling the physicists to reconstruct with great precision and efficiency those events which are to be studied.

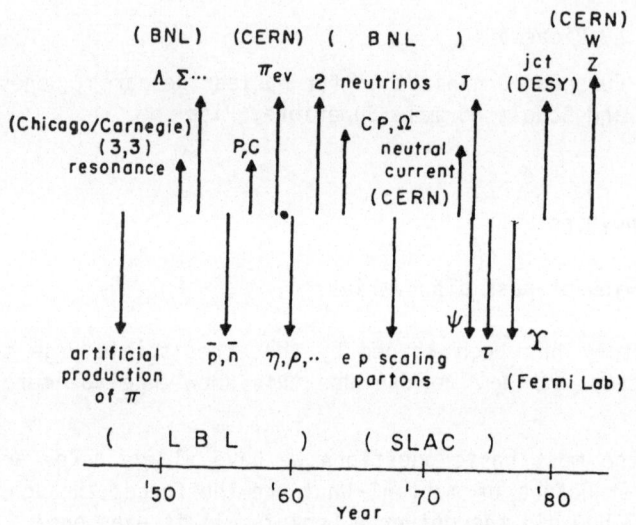

Fig. 1. The chart lists almost all the major discoveries made in particle physics for more than three decades (From T.D. Lee: Symmetries, asymmetries and world of particles).

The first particle accelerators were built approximately fifty years ago. These early machines had energies of the order of a few MeV and were destined to study a world that looked relativy simple. Matter was then known to be composed of four basic constituents: protons, neutrons, electrons and neutrinos. These constituents interact via four forces: the weak, the electromagnetic, the strong forces and gravitation. The weak force is responsible for a large number of physical processes: nuclear β-decay, numerous decays of elementary particles, reactions induced by neutrinos from accelerators and nuclear reactors, effects involving parity violation, those involving γ-decays of nuclei and atomic optical spectra. All known leptons and hadrons are subject to the weak interaction and the weak force plays an important role in such astrophysical phenomena as the burning of the Sun and the explosions of supernova.

The electromagnetic force accounts for the interaction between charges and currents. The description of the interaction between the electromagnetic field and the electron-positron field constitutes the main problem of Quantum ElectroDynamics (QED). The basic equation of QED gives the impression of being extremely simple. It is a theory which describes an enormous range of phenomena. Furthermore, it is an elegant theory, because all the observable quantities are expressed only in terms of coupling constants and lepton masses, which are both assumed to have been given[2]. Quantum electrodynamics is the relativistic quantum field theory developed in the 1930's and 1940's, which extends Maxwell's theory to atomic scales.

The strong forces accounts for the law which describes the dynamic behaviour of the nucleus. The nucleus is a tighly-bound collection of protons and neutrons, confined in a region about 10^{-12} cm across. The force that binds the protons and the neutrons together to form the nucleus, is much stronger and considerably shorter in range that the electromagnetic force. Particles that participate in the strong interaction are called hadrons. The mystery of the strong force and the structure of nuclei seemed very intractable as little as fifteen years ago. Experiments during the fifties showed two highly interesting facts. First, the strong forces makes no distinction between protons and neutrons, i.e. the proton and the neutron transform into each other under isospin rotations. Second, the structure of the protons and the neutrons is as rich as that of nuclei and furthermore, many new hadrons were discovered, just as elementary as the protons and the neutrons themselves.

The table of the elementary particles in the mid-60's displayed much of the same complexity and symmetry as the periodic table of the elements. In 1961 both Gell-Mann and Ne'eman[3] proposed that all hadrons could be classified in multiplets of the symmetry group called SU(3) (see Fig.2). The great triumph of this theory was the prediction and the subsequent discovery of a new hadron, the Ω^-, in 1964 at the Brookhaven AGS.

Finally there is a force which is well known to all of us, which has a far-reaching action, to which all particles are subjected, indipendent of their properties: gravitation. This gravitational force between two particles is extremely small and when the properties of the particles are studied, it is entirely ignored. It becomes important at the macroscopic level, when millions of millions of thousands of million atoms are present, and it is responsible for the large structure of the Universe.

In recent years, the energy of the accelerating machines has grown by five or six orders of magnitude to reach 100 GeV to 1 TeV. The view just described above, of what the elementary constituents of matter were, has turned out to be either incomplete or wrong. The simple picture of four constituents (protons, neutrons, electrons and neutrinos), became ever more complicated as the machines of higher energies were built. More and more mesons and isobars of the nucleon were discovered and in the early 1960's there were more than 100 of these "elementary particles". All this was swept away in the 1960's, in fact, in spite of the SU(3) classification scheme, the belief that all these so-called "elementary particles" were truly elementary became more and more untenable.

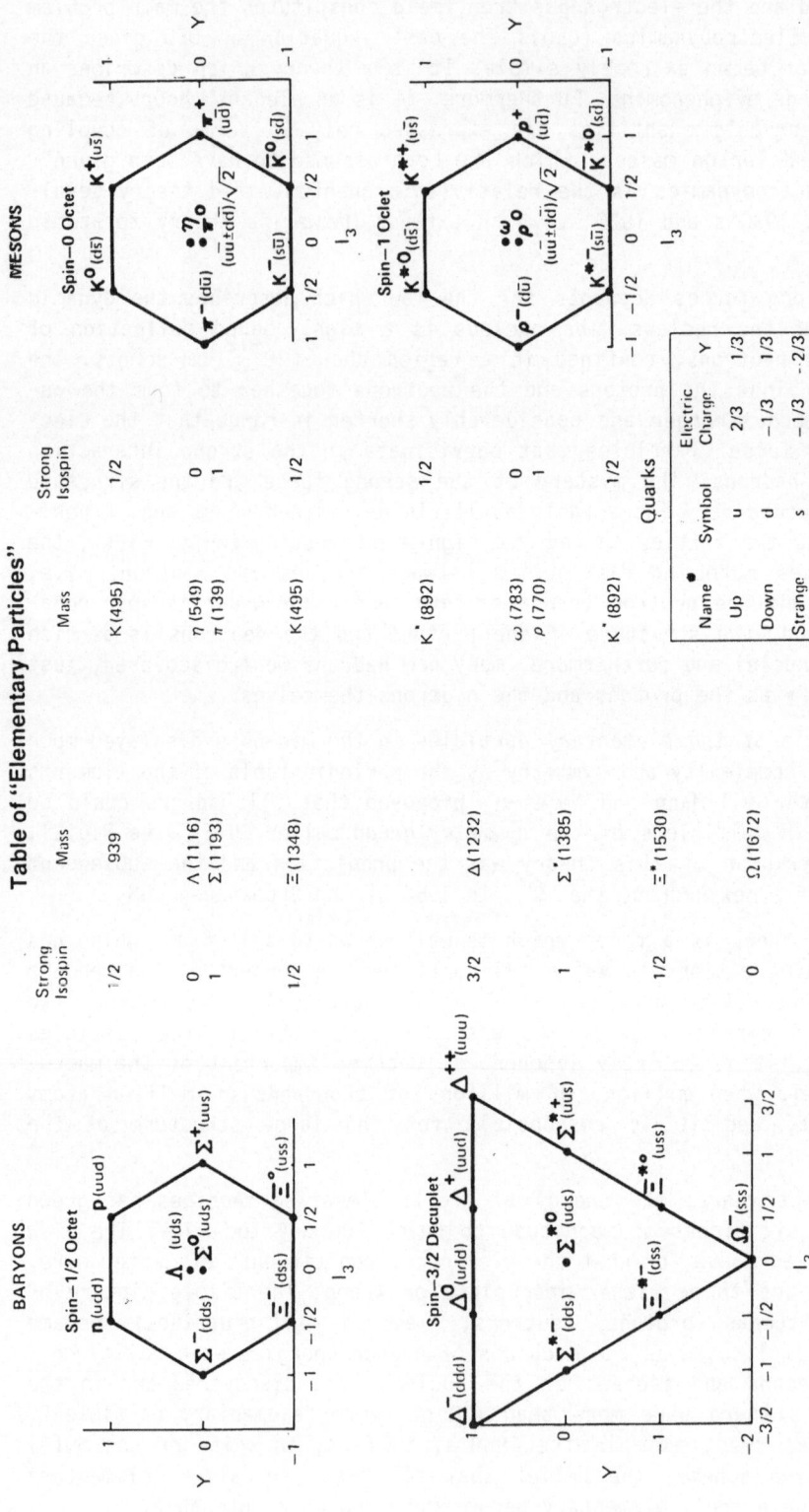

Fig. 2. The eight-fold way classified the hadrons into multiplets of the symmetry group SU(3). Particles of each SU(3) multiplet that lie on a horizontal line form strong-isospin SU(2) multiplets. Particles are plotted according to the quantum number I3 and strong hypercharge Y=S+B, where S is strangeness and B is the baryon number I3; Y corresponds to the two diagonal generators of SU(3). (From S.A. Raby, R.C. Slansky and G.B. West: Particle Physics and the Standard Model, Science N11, 1984).

The most contradictory evidence was the finite size of hadrons which contrasted with the point-like nature of leptons. The hadronic zoo was eventually tamed by postulating the existence of a small number of truly elementary point-like particles called quarks. In 1963, Gell-Mann and, indipendently Zweig[4] suggested that all hadrons could be constructed from three half spin fermions, designated u, d and s (up, down and strange) (Fig. 3). The SU(3) symmetry that manifested itself in the table of "elementary particles" arose from an invariance of the Lagrangian of the strong interaction to rotation among these objects. As is well-known, this global symmetry is exact only if the u, d and s quarks have identical masses, which implies that the particles states populating a given SU(3) multiplet also have the same mass. Since this is certainly not the case, SU(3) is a broken global symmetry, and the breaking mechanism arises from the difference in the masses of the u, d and s quarks. The

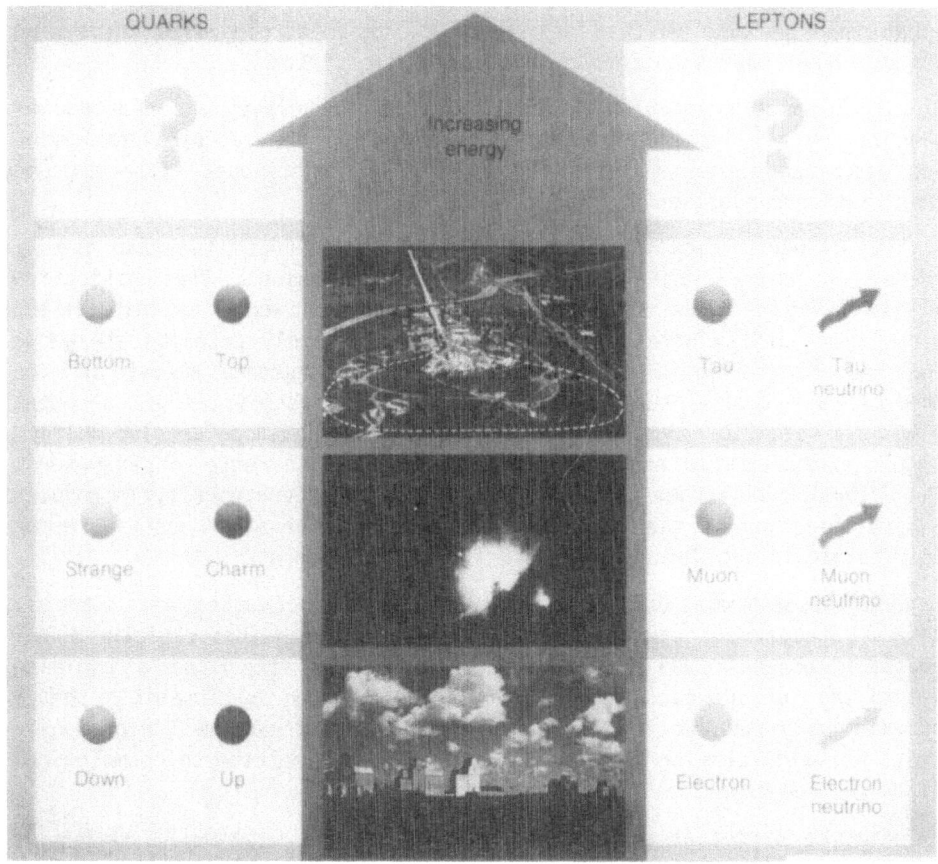

Fig. 3. The three families of quarks and leptons. The world in which we live our daily life, has to do with the family of the lowest energy (u,d) and e, ν_e. The second family of quarks and leptons is found in an interval of energy typical of the astrophysics processes. The third family existed very briefly at the moment of the Big Bang. All these kinds of quarks and leptons are produced in particle accelerators at varying energies. (From Elementary Particles: The Eternal Quest, SSC publ.,1985).

origin of these quark masses is one of the great questions that remain without reply. The SU(3) symmetry among the u, d and s quarks is preserved by the strong interaction and is known as flavour symmetry. Another and quite different SU(3) symmetry is verified by quarks, a local symmetry that is associated with the strong force and is known as the SU(3) colour. The three "colours" play the role of the electric charge and the three colour varieties of each quark form a triplet under the SU(3) local gauge symmetry. A local phase transformation of the quark field, can rotate the colour and thereby change a "red" quark into a "blue" one. The colour transformations of Quantum ChromoDynamics (QCD) change the particle, while the local transformations of quantum electrodynamics change the phase of an electron.

1.2. A brief survey of the present knowledge of the subatomic world[5]

In these past fifty years we have unified two forces in one, in fact our present picture is that the weak and the electromagnetic forces are but different manifestations of the same basic force.

The dynamic principles remain the same. Relativity and quantum mechanics are still the basic framework in which the electroweak phenomena are described. Space is still thought to be continuous, although some physicists are questioning that as well.

It is worth recalling the attention on the fact that two central themes of physics are "unification" and "extension". The Unification allows us to describe coherently those phenomena which at first sight might seem totally unrelated. A theory must not only describe the known phenomena, but also make predictions of new phenomena. Almost all the theories are incomplete because they provide a description of phenomena only in a specific range of parameters. A theory could be deeply modified if one extends it to explain phenomena over a wide range of parameters. The extension of a theory to a new range of parameters usually introduces a simplification in the theory itself and therefore introduces a greater unification in the description of different phenomena.

The best-known example of extension and unification is, perhaps, Newton's theory of gravity (1666), which unifies the description of macroscopic objects falling to earth with that of the planets revolving around the Sun. Newton's theory is superseded only by Einstein's theory of relativity when we try to describe phenomena at extremely high densities, or velocities or again when the events are related to cosmological distances and time scales.

Another outstanding example of unification is Maxwell's theory of electrodynamics which unifies electricity and magnetism. From Maxwell's unification of electricity and magnetism emerged the concept of electromagnetic waves. The speed of electromagnetic waves or light, is given by $(\varepsilon_0 \mu_0)^{-1/2}$ and is then determined in terms of purely static electric and magnetic measurements alone. It is interesting to emphasize that apart from the crucial change in Ampère's law, Maxwell's equations were well known to natural philosophers before the advent of Maxwell. The unification and the extension to the phenomena of electromagnetic waves become clear only when Maxwell expresses his known equations in term of the "right" set of variables, namely the fields of $\vec{E}$ and $\vec{B}$.

Now we know that matter is made up of quarks (and antiquarks) and leptons (and antileptons). These particles have one property in common: their intrinsic angular momentum (or spin) is a half-integer number in units of Planck's action, $s = 1/2\ \hbar$. Quarks have a fractionary electric charge, whereas leptons have unitary electric charge or have neutral charge (the neutrinos). The forces between the particles reveal themselves by the exchange of other particles or bosons. The electrically-charged particles interact between each other by the exchange of photons, these are the bosons of zero-gravitational mass and bear no electric charge.

Quarks interact with each other through the exchange of gluons which are the mediators of the strong forces. Quarks equally interact with each other through the exchange of photons, because they can also carry an electric charge. Contrary to the quarks, the electrons, or more generally the leptons, are not subject to the strong interactions, because they do not carry a strong charge, that is they have no colour.

Different from the photons which are electrically neutral, the gluons have a strong charge, in fact they are coloured (similar to the quarks) and therefore can interact with each other. This gives rise to a more complicated mathematical description than that pertaining to the.electromagnetic theory.

Quarks and leptons interact weakly with each other by exchanging heavy bosons which have a mass approximately eighty times that of the proton.

Two of the bosons have respectively a positive and a negative electric charge ($W^{\pm}$) and another one has no electric charge (Z^0). These are responsible only for the exchanges occurring between the weak forces.

These bosons are, to all intents and purposes, elementary particles likes quarks and leptons; some of them (photons and gluons) are massless, others ($W^{\pm}$, Z^0) have a non-zero mass. Some of these carry an electric charge, other have no electric charge at all. They have one important property in common; their intrinsic angular momentum (or spin) is an integer number in units of Planck's action, i.e. $s = 1\ \hbar$. They are subject to different rules than those which hadrons and leptons are subject to. For example, these last follow Pauli's exclusion principle, while the bosons (photons, gluons, intermediate bosons) do not follow Pauli's exclusion principle. This means, in terms which are not very precise, that two electrons cannot occupy the same quantum state, whereas two photons could very well find themselves in the same quantum state. Pauli's exclusion principle states a fundamental law of Nature, which enables one to explain the great variety of atoms in the Universe. The fact that the photons (and all the other particles with integer spin) do not follows this principle, has among other things, given man the possibility of inventing the laser.

This brief synthesis allows us to state that matter is made up of quarks and leptons (and their antiparticles) and the forces to which

these particles are subject are mediated through the exchange of bosons
which substantially have symmetry properties differing from those of
quarks and leptons.

Although the "Standard Model" unifying the weak and electromagnetic
interactions has been an unqualified success in terms of agreement with
the experiments, it has many unsatisfactory features.

There are some twenty arbitrary parameters in the theory and the unification with the strong interaction is still incomplete. As I have
pointed out earlier, three families of quarks and leptons have been discovered. We don't know as yet what defines their masses or if there are
other families of leptons and quarks, or again if the present picture,
aesthetically an appealing one, is the end of the story. In particular
we don't understand the mechanism which differentiates the masses of the
carriers of the various interactions; further, if the present number of
families of quarks and leptons seems to be a nice small number when compared to the total number of known particles (about 400), the total number of the so-called elementary constituents is still disturbingly large.
In actual fact we have 18 coloured quarks, six leptons, the photon, three
massive vector bosons which carry the electro-weak force of the Standard
Model, eight gluons which mediate the strong interaction and one graviton
which carries the gravitational force (Fig. 4). In total we know of the
existence of at least 37 elementary constituents. We cannot exclude how-

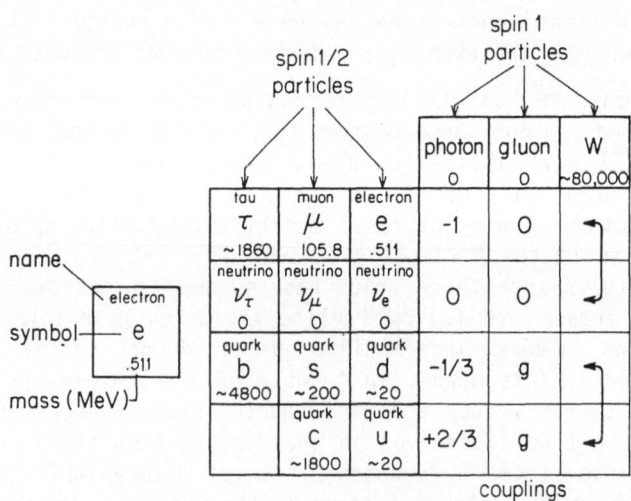

Fig. 4.   The table of the families of the spin 1/2 particles. It is clear
that the repetitions of the spin 1/2 particles will be complete if a particle with the right properties to imply the existence of a new flavour
of quarks (top) is found. Does another repetition at even higher energies
exist? What causes these repetitions remains a mystery. (From R.P.Feynman:
QED. The Strange Theory of Light and Matter, Princeton University Press).

ever that other elementary constituents could exist and therefore the present picture could still not be the final one.

Our present knowledge of the way Nature behaves in the subatomic world is, like all human knowledge, subject to a radical evolution in time. Scientists would be too pretentious if they thought that their present degree of knowledge is definite. What will certainly happen is that the laws which have been discovered up to now and which have been verified by experience, will have to be incorporated in the new theories.

1.3. Outlook in Particle Physics

The three major lines of research in the study of the fundamental theories of Nature today, stem from the evolution of the classical field theories developed by Newton and Maxwell to the grandest theoretical conjectures of our times: unified theory, supersymmetry and problems of families.

I should like to make a few general and certainly incomplete remarks on supersymmetry, that is the symmetry which connects particle of integer and half-integer spin. It is a symmetry which provides a framework for the unification of all known forces between elementary particles, including gravitation. It also unifies the concepts of matter and force into a single framework. It is a space-time symmetry, i.e. an external symmetry. It also leads to the change in spin by half-integer jumps, and therefore changes bosons into fermions and vice-versa. The supersymmetry links matter and the forces, the basic attributes of Nature.

We know that matter is made of fermions and bosons are associated with the forces.

The question is whether the supersymmetry is, or not, an exact symmetry in Nature. If this is the case, then for every boson of a given mass, there must be a fermion of the same mass and vice-versa. To an electron, for example, should correspond a scalar electron (selectron), to a neutrino a scalar neutrino (sneutrino) and so on. This is not the case, because such a degeneracy has not been found in Nature, therefore the supersymmetry is not an exact symmetry. It must be a broken symmetry, i.e. an inexact symmetry. It can be broken in two ways: The Lagrangian contains explicit terms which are not supersymmetric; there is a spontaneous breaking in which the Lagrangian is supersymmetric but the vacuum is not. In both cases, the boson-fermion degeneracy is moved up, but the latter case, similarly to the Higgs boson of the weak interaction symmetry breaking, introduces a new particle, the Goldstone fermion.

The question of major importance for the future experimentation is the scale of supersymmetric breaking. This scale is characterized in terms of the mass splitting between fermions and their boson partners, $\delta^2 = M_B^2 - M_F^2$. Where is this scale to be found? It can be located at the point of the weak scale (in the order of 100 GeV), or it can be located at higher energies.

This is an important question for the future experimentation because if the supersymmetry is broken at a scale of about 100 GeV, many predictions could be verified with the next generation of high-energy accelerators, including LEP. If the scale is situated at higher energies, then the supersymmetry breaking would not necessarily lead to any new low-energy consequence. For more details see reference[6].

PART 2 - THE ACCELERATORS

2.1. Protons versus electrons

The 1 TeV region for the centre-of-mass energy of collision among fundamental constituents is capable of providing the answer to some of the questions mentioned above, certainly leading to new questions.

Since 1930 the energy of the accelerators has grown by an order of magnitude roughly every seven years. Fortunately, as the energy of these machines has increased, the cost per unit energy has decreased almost as fast as the increase in energy. These two facts have determined that while the increase in energy has been dramatic, the cost per new installation has only been of an order of magnitude since 1930 (not corrected for inflation). Another important consideration is that the number of accelerators operating at the forefront in the field has significantly shrunk.

In the first part of my talk we have seen that there are good scientific reasons to maintain the growth in energy. The question we must ask now is whether there are new technologies in sight which would allow a decrease in the cost per unit energy.

The first dilemma an accelerator constructor faces is the use of protons versus electrons. Protons are composite particles: they are made up of two u and one d quarks, plus quark pairs + gluons, i.e., quanta of the colour gauge fields.

The first three u u d are called <u>valence</u> quarks, the remainder is referred to as the <u>sea</u>. Unlike the quarks, the gluons do not interact directly with the electromagnetic and the weak intermediate boson fields. Because the gluons are flavourless, in the sea, for every quark of flavour f, there must be an antiquark of flavour $\bar{f}$. This multitude of constituents (partons) within the proton, share the proton's energy. The proton-proton collision is like having two bags, each containing many constituents, hurtling towards each other. The hard collisions, the ones that lead to the production of phenomena with large mass, are relatively improbable and when they occur, they tend to produce final states particles with large transverse momentum and leave behind a collection of excited constituents of the bags. The centre-of-mass energy in the parton-parton collision is, on the average, much smaller than the centre-of-mass energy of the proton-proton system. Contrary to protons, the electrons are still point-like particles down to dimensions of 10^{-16} cm. Thus electron accelerators of considerably lower energy are equivalent to the corresponding proton accelerators. The energy of the protons has to be reduced by about an order of magnitude relative to electrons in providing the same phenomena into the unknown.

Electron accelerators at ultra-high energies, using know technologies, are becoming expensive and therefore a new strategy is required in order to obtain very high energies (about 1 to 2 TeV) in the centre-of-mass system. One of the major advantages of an electron-positron collider is that all the energy goes into the process we wish to investigate. To be more complete, the comparison between electron-positron and proton-proton colliders must take into account that the total cross-section in proton-proton collisions increases with the energy, while the cross-section for the production of events exhibiting new physical phenomena of interest is expected to decrease as the square of the relevant mass range to be explored. Therefore for a given process to be studied, the signal-to-noise ratio for proton-proton colliders degenerates rapidly with the energy. As a result, the data analysis process in the case of hadron-hadron collider events becomes a very complicated process which has to reject a substantial number of unrequired features of the primary collision. This rejection can be accomplished with confidence in experiments which study well-defined objectives, such as the search for clearly -predicted resonances, but this rejection can be a frightening procedure if one explores regions of energy where the theoretical predictions are not available. In doing so we could reject unanticipated primary discoveries.

Given the choice, most particle physicists would prefer an electron-positron collider of one tenth particle energy to a corresponding proton-proton collider of roughly comparable luminosity.

The major question is to know whether this possible choice exists today. The answer to this is that in the TeV range it does not fall within the present-day technologies.

I am deeply convinced that the study of elementary particles has made an enormous progress in the past thirty years, largerly due to the fact that one has studied the laws of the subatomic world by making use of both proton and electron accelerators. This double possibility of exploring the physical laws must be maintained if possible, even though it is quite obvious that new complementary programmes have to be formulated in the various parts of the world where elementary particle physics is carried out, so as to avoid unnecessary duplications. The very high costs of building and operating accelerators in the energy region of a few TeV and at a luminosity of at least 10^{33} cm^{-2}sec^{-1}, impose full complementarity in all research programmes.

2.2. Proton-proton colliders

The designs of the SCC and the LHC are a step forward in extending the proton-proton collider technologies using superconducting magnets, successfully operating at Fermilab and used in the construction of the electron-proton collider Hera. These proposals do not involve any new technology, but hope to extend the energy range of the decreasing unit cost by economies in the scale and with a better design.

What is required is to reach high luminosities. While the total cross-section for a proton-proton collision is very large, the partial cross-section for the hard collision is very small and depends on the mass of the final state produced. The cross-section for the production of a final state of mass M, plus the excited proton fragments X, has an energy and mass dependence given by:

$$\sigma(M+X) \, \alpha \, \frac{1}{M^2} \quad f \, (\, \frac{M^2}{E^{*2}} \,) \tag{1}$$

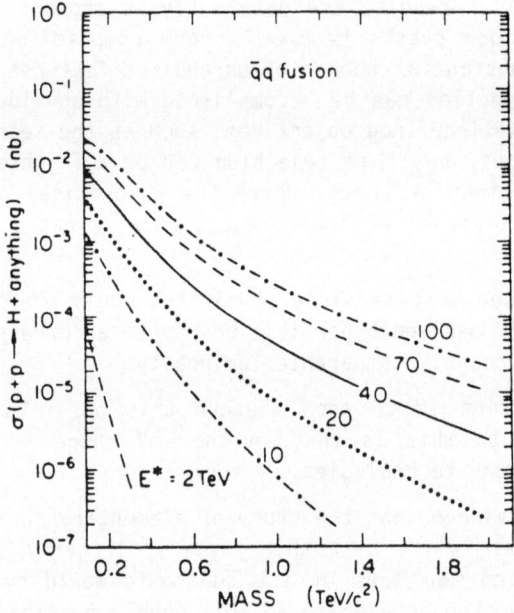

Fig. 5. The total cross-section for Higgs boson production by quark-anti quark fusion in proton-proton collisions as a functions of the Higgs boson mass for various centre-of-mass energies. (From B.R.Richter: Very high energy colliders, SLAC-Pub-3669, 1985).

where E^{*} is the centre-of-mass energy of the proton-proton collision. To illustrate with an example the request for the luminosity of the collider, let me point out that the cross-section for the production of a Higgs boson as a function of Higgs mass for various proton-proton centre-of-mass energies is given in Fig. 5[7]. The cross-section decreases rapidly with increasing mass at fixed centre-of-mass energies, and decreases rapidly with decreasing centre-of-mass energies for a fixed boson mass. A typical luminosity of 10^{33} to 10^{34} $cm^{-2} sec^{-1}$ is required in order to explore the mass regions that are considered to be of interest in the future.

14

If one considers the possibility of building a proton-proton collider to reach energies beyond the SSC so as to explore regions of mass ten times higher than possible with the SSC, it is necessary to increase not only the energy but equally the luminosity.

Equation (1) above shows that in order to increase the limit of discovery by a factor of ten, the centre-of-mass energy of proton-proton collisions must be raised by a factor ten and the luminosity by a factor of one hundred above those of the SSC. These new requirements raise two major problems which need solution. The first deals with the luminosity lifetime and the other with the possibility of building an experimental detector which can stand a very high rate of proton-proton collisions, typically in the order of one hundred proton-proton interactions per beam-beam collision.

The luminosity lifetime is about 20 hours at the SSC; in a Super SSC with an energy ten times higher and about a factor one hundred in higher luminosity, the lifetime goes down by a factor fifteen with respect to the SSC value, i.e., roughly 1.5 hours[7]. This short lifetime will make it difficult to inject the beam into the storage ring. to ramp up to the nominal energy and to collect the data.

At a luminosity of 10^{35} cm^{-2}sec^{-1} it might be hard, and maybe impossible to build detectors capable of withstanding that kind of rate. I am convinced that new ideas are needed in the design of detectors that can stand such a high rate of proton-proton collisions. Research and Development programmes are being considered at CERN and in other laboratories.

In a summary, proton-proton colliders can achieve an effective centre-of-mass energy that is much lower than the proton-proton centre-of-mass energy; the cross-sections are proportional to $M^{-2}f(M^2/E^{*2})$;[7] the SSC has an effective discovery limit[7] of 3 TeV at a luminosity of 10^{33} cm^{-2}sec^{-1} and to extend this limit implies higher energy and higher luminosity; if the luminosity is held fixed, the machine energy must be scaled roughly as the square of the mass limit.

2.3. Electron-positron colliders

W.Schnell will present this subject in greater detail and therefore I shall only outline some considerations.

It is likely that LEP will be the last electron-positron collider to be built along the lines of the well-established storage ring pattern, in fact the cost of a Super LEP will be prohibitive due to the quadratic scaling law of cost versus energy.

The understanding of the physical laws around the energy region of a few TeV using electrons and positrons, will require us to abandon the storage ring technique, unless someone comes up with an invention which will suppress synchrotron radiation. The electron-positron collider of the future must be a linear collider in which two linear machines accelerating electrons and positrons, are aimed at one another to produce the required interaction. One of the great advantages of an electron-positron collider is that there are no "partons" to share the momentum of the primary electron and positron and thus reduce the effective collision energy.

One of the advantages of storage rings which consists in the fact that electrons and positrons collide repeatedly before eventually being lost due to a finite lifetime of the stored beams, is lost in a linear collider. This imposes on the designer of linear colliders a clear-cut condition: the total beam power and the density of the particle beams must become very great in order to obtain adequate reaction rates.

If energyrecovery proves to be unrealizable, then the power consumption inherent in discarding the interacting bunches after collision sets a serious problem.

The physicists has to specify three basic parameters: a) the required beam energy; b) the luminosity; c) the tolerable energy spread produced in the interaction by the beam bremsstrahlung.

It is unavoidable to think that within the current quantum mechanical wisdom, the required luminosities would have to increase with the square of the energy, in order to give reasonable rates of events presenting some interest. A luminosity of at least 10^{33} cm^{-2}sec^{-1} in the 1 TeV region and of 10^{34} cm^{-2}sec^{-1} in the 3 TeV region appear essential if event rates of 10^{2} to 10^{3} per year are anticipated. In addition, requirements on the background due to beam gas interactions or to radiation from upstream and downstream focussing devices must be held within tolerable limits.

The construction of ultra-high energy electron-positron colliders does not involve problems which cannot be solved without violating basic physical principles. In order to produce the high luminosity required for reasonable data rates at high energies, it is likely that the invariant emittance, the bunch length and the β-value will have to be reduced at least by roughly two orders of magnitude with respect to the values of the parameters chosen for the Single Linear Collider (SLC), so as to bring the beam power to what appears to be practical values[8]. New power sources, new final focussing methods, unprecedent alignment techniques, methods of continuous servo-loops, unthought of stability and freedom from noise of basic power supplies must be developed. I have the feeling that all this will not only require new development but new inventive power as well. The future will be very exciting for the people involved in the design of an ultra-high energy linear collider.

2.4. LEP: One of the Colliders under construction

The present-day colliders are the SPS($p\bar{p}$), the Tevatron, Tristan, Petra, Pep. The ones which will come into operation soon is the SLC and in the near future Hera and LEP will also start up.

The Large Electron-Positron collider (LEP), now under construction at CERN, is one of the most powerful and precise scientific instruments upon which further advances in our knowledge of Nature critically depends.

It has become more and more evident, since the advent of the SPEAR Storage ring in 1972, that particle physics, based on electron-positron annihilation, is an exceedingly rich field. The reason for this are:

a) electron-positron collisions provide a very clear way to discover new particles and to study their properties. Past examples are the discoveryof the psi-particle family, of the tau lepton and of charmed hadrons;

b) electron-positron collisions provide a direct method of studying the weak interaction and also the interference between the weak and the electromagnetic interactions. LEP, for example, is an ideal machine to delve deeply into the predictions of the electroweak theory;

c) electron-positron collisions, together with electron-proton collisions (Hera), are a clear way of studying hadron dynamics;

d) electron-positron collisions also offer a direct approach to the study of the very important and rich domain of the Z^0 and the W intermediate bosons and their decay modes.

One of the principal objectives of LEP is the high production rate and the study of the Z^0 vector boson physics. The Z^0 is the neutral carrier of the weak interaction, recently discovered by groups led by C.Rubbia and P.Darriulat at the CERN SPS($p\bar{p}$). The Z^0 is expected to be produced copiously in e^+e^- annihilations. At the luminosity of LEP, the production of about 10'000 Z^0 per day is expected. Measurements of the Z^0 width can be related to the number of neutrino types and thus to the number of heavy leptons and quarks beyond the ones presently known. Production of W^+ and W^- pairs will allow the study of the Z^0, W^+, W^- gauge vectors[9].

LEP will be for many years to come, the largest electron-positron storage ring in the world. In this giant accelerator, matter (e^-) and antimatter (e^+) will be brought into head-on collision and in the resultant mutual annihilation, the total energy will be available for the production of other particles. As is well known, the process through which energy can turn into matter was stated by A.Einstein in the equally famous equation $E=mc^2$, where E is the energy of the particle, m its mass and c the speed of light. Energy and matter are two different aspects of the same entity which can change from one form into the other. While it is easy to transfer matter into energy, the opposite transformation requires special tools, such as LEP.

For colliding beam experiments, the important number is the time-averaged luminosity as given by peak luminosity, luminosity decay time, the frequency and the time required to refill the ring with electrons and positrons. The average luminosity depends on the positron production, on the accumulation rates, the injection energies and the luminosity decay time.

The LEP project includes the construction of two Linacs and an accumulator ring (EPA), the necessary modifications to the existing PS/SPS machine complex to allow the acceleration of electrons and positrons up to an energy of 20 to 22 GeV, and above all the construction of the main ring of LEP[10].

LEP is scheduled to come into operation in the middle of 1989. Four large experiments are in preparation and the sophisticated detecting equipment with the complex associated data-analysis system, will allow the physicists to explore the electroweak theory in detail and to look for the unknown.

LEP provides a tool to perform precise experiments to test the theory, to explore its range of validity and to search for new, fundamental interactions and particles. The experience with the precision tests of QED, fixes the standard by which to judge the validity of the newest field theories.

PART 3 - CONCLUSIONS

What lies ahead

So as to continue our understanding of the laws governing particle physics, we must build new colliders which go beyond the present SPS($p\bar{p}$), Tevatron, Tristan, SLC, LEP, Hera.

Supersymmetries, problems of families, the exploration of the unknown, require not simply high energies but also high luminosities. We cannot exclude in some future discoveries, some new kind of high cross-section physics, but from what we know today we have clear indications that the luminosity has to increase as the square of the centre-of-mass energy. A reasonable luminosity would be 10^{33} to 10^{34} cm^{-2}sec^{-1} at a centre-of-mass energy of about 1 to 3 TeV for an electron-positron linear collider.

The emittance of such a collider must be very small and it will be a real challenge to produce these small emittances and to maintain them during the acceleration cycle. New techniques will have to be studied in order to achieve any future goals. I am fully convinced that the present tradition of building accelerators at a lower cost per unit energy will continue and in-depth studies will have to be carried out to optimize the costs for any future machine where new technologies, as yet not well known, are needed. The requirements of the new machines (particularly the linear colliders) constitute a considerable challenge to machine builders and the accelerator physicists in this audience are facing a severe but most exciting challenge in the years to come.

I have not mentioned in my talk new acceleration methods, such as plasma devices, where a laser beam or laser beams interfering with one another, can produce waves in a plasma, which in turn, will accelerate particles. Such devices appear attractive because of the potentially high gradient they might produce. I am in favour of developing such devices or similar ones, however, I am not too optimistic about their utility, at least in a more or less immediate future (say five to ten years from now), for the following reasons:

a) the overall efficiency of transfer of power source to the beam;
b) the fundamental difficulty in controlling the micro-detail of the plasma to avoid growth in the emittance of the beam.

I still believe that the most predictable performance would involve "conventional" r.f. structures, which must be fed by power sources as yet not developed. These sources are in effect transformers from low voltage, high current devices to the high energy, low current beams to produce the required collisions. The most appealing of these devices is the two-beam system proposed by W.Schnell[11].

Let me conclude by saying that the search for a unified theory may be linked to an old geography problem[5]. Columbus sailed westward to reach India, believing the world had no edge, physicists are searching for a unified theory at shorter and shorter distances, believing the microworld has no edge. Maybe space-time is not continuous and something new will be discovered, something consistent with what we know today.

By making use of the concept of matter and of field in quantum mechanics, physicists have been able to appreciate a simple description of Nature. But this is a simplicity rather difficult to describe and to appreciate because we must formulate physical laws by using a mathematical language which is only learned after many years of study. It is not clear, at least to me, what a simple or unified description of Nature actually means. If by simple we mean easy to describe and to understand, then it is clear that this will remain a dream for ever. If, on the other hand by simple we mean acquiring a highly-specialized education enabling us to be struck by the beauty of the general laws governing natural phenomena, then it is possible that it will become a reality. In any case it is exalting to study the laws of Nature, its symmetrical properties, and it is equally marvellous to be touched by the magnificence of its forms and its manifestations. The appreciation of these forms and manifestations is fortunately within everyone's reach.

ACKNOWLEDGMENTS

I wish to thank E.L. Ratcliff for his assistance in preparing this written report.

REFERENCES

1. T.D.Lee, Symmetries, asymmetries and world of particles (1986).
2. F.Combley and E.Picasso, Some topics in quantum electrodynamics. Metrology and fundamental constants, LXVIII Varenna School (1980), p. 717.
3. M.Gell-Mann, Phys. Rev. 125:1067 (1962); Y.Ne'eman, Nuclear Phys. 26:222 (1961).
4. G.Zweig, CERN Report (unpublished); M.Gell-Mann, Phys. Letters 8:214 (1964).
5. Consult a very clear series of articles in "Particle Physics", in Los Alamos Science N11, Summer/Fall 1984.
6. J.Ellis, Superphysics, 1986 CERN School of Physics, Report CERN 87-02 (1987), p. 276.

7. B.Richter, Very high energy colliders, 1985 Particle Accelerator Conference, Vancouver, B.C. Canada, May 1985 (or SLAC-Pub-3669 (1985)).

8. W.K.H.Panofsky, Limiting technologies for particle beams and high energy physics, AIP Conference Proceedings 153 (1987), Vol. 2, p. 1602.

9. R.D.Peccei, Physics at LEP, 1986 CERN School of Physics, Report CERN 87-02 (1987), p. 276.

10. E.Picasso, The European Electron-Positron Collider (LEP), Nuclear Europe 12:33 (1985).

11. W.Schnell, Considerations for a two-beam R.F. scheme for powering an R.F. Linear collider, CLIC Notes, N7 (1985).
    (For more extensive literature on linear collider subjects, consult: The Report from the Advisory Panel on the Prospects for e^+e^- Linear Colliders in the TeV range, in preparation, April 1987).

ACCELERATOR AMPLITUDE AND PHASE CONTROL

Jörg Kummer

Institut für Angewandte Physik der Universität
D-6000 Frankfurt am Main, Robert Mayer Str. 2-4
Federal Republic of Germany

INTRODUCTION

In the field of physics we here are concerned with, we admire on one side the exciting things done with accelerators (Professor Picasso has shown splendid modern examples), and on the other side the ingenious concepts according to which the accelerators themselves have been built.

One of the latest and most outstanding improvements of the tools, particle physicists are working with, has been the idea and realisation of stochastic cooling. In the first sketch (fig. 1, taken from Simon van der Meer's paper in Physics Reports of 1980) we see, how a signal produced by a bunch of particles passing a PICK UP drives the TRANSVERSE KICKER at that very moment the particles arrive. Assuming correct phase, the particles are in the mean forced closer to the beam axis. Though the term feedback is used and justified since the particles act upon themselves a little bit later via a fast amplifier and control their own motion, we could also call the mechanism feed forward, since the particles and their own correcting signal travel in the same direction.

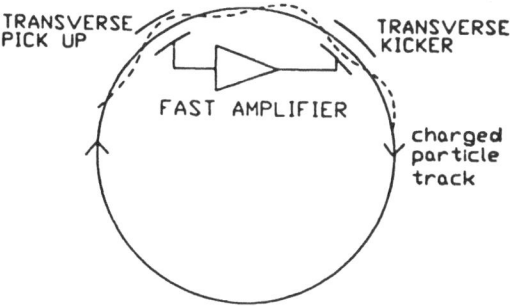

Fig. 1. S. van der Meer's betatron-oscillation cooling scheme

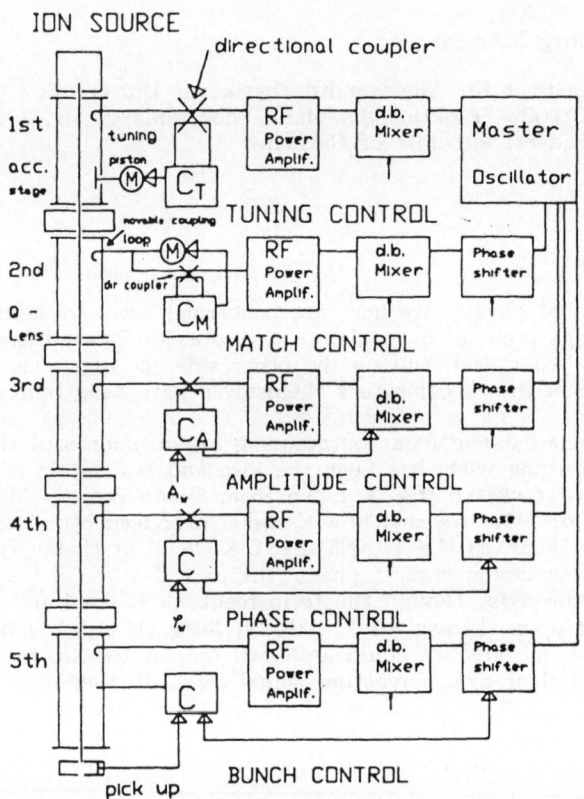

Fig.2. *Examples of physical quantities, that need to get controlled within a multistage linear radio frequency (RF) accelerator.*

For the purpose of controlling certain physical quantities of importance in running an accelerator we will consider feedback systems, in which the direction of the two signals, the one controlling and the other one to be controlled, are opposed to each other. As we shall see, this offers so many advantages, that we may regret that one can't apply (by reason of causality) such "true feedback" to a beam of fast particles.

Let us see (fig. 2), what kind of variables we could control automatically in a linear RF accelerator for example. The first thing to do, would be to tune the cavity to the proper transmitter frequency. Since, e.g. the resonator gets warmer, it will detune slowly. This can be corrected by a piston being moved in and out the cavity by a motor. Likewise the mismatch can be controlled automatically. But whilst these two quantities could also be adjusted (if there are for example no beamloading effects calling for rapid control of tuning and matching) by hand, there will almost always be a need for an automatic control of the RF amplitude and the phase of each accelerator section.

Finally, since very fast sampling circuits are feasible, there is a possibility to measure and compare the duration and shape of particle bunches to preset values and thus to adjust the phases according to what is needed: high time resolution or uniform particle momentum. (In Dr. Raka's review we have heard of further applications of control, e.g. to provide a low impedance for the beam in the resonator slit, in order to avoid self bunching in the storage mode of a circular machine).

FEEDBACK CONTROL

It is the purpose of a controller to bring the actual value of a physical quantity X (like temperature, pressure voltage, amplitude, phase) as close as possible and as fast as possible to a desired value of the quantity W to be controlled and to keep it there. The ability of the controller to approximate the actual value X to the desired value W is indicated by the "control deviation" W-X. Furthermore, the controller should immunize the controlled quantity as well as possible from external disturbances (for example variations of power supply - or, in our case, geometrical changes of a cavity and beam loading).

Fig. 3 shows the scheme of a simple control loop. "Loop" means: a feedback loop: The output of the controlled system, which represents the controlled variable with its actual value X, is subtracted from the desired value of the controlling variable W. The difference of both is fed into the controller, the output of which serves as input into the controlled system (together with the external disturbances, which might be present).

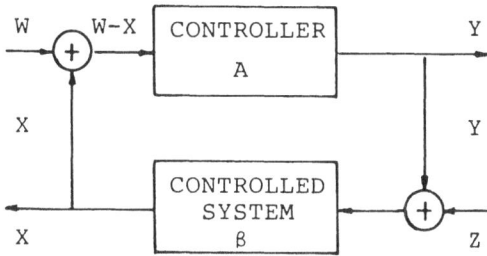

Fig. 3. Feedback control loop

We characterize the two blocks "controller" and "controlled system" by their transfer-functions. These transfer-functions are merely the proportions: output quantity divided by the input quantity. These quantities or signals are complex functions of the signal frequency $\omega = 2\pi f$.

If we call the transfer function of the controller A (in general its amplification), and the transfer function of the controlled system β (which is the feedback factor), we get (if we take into account the signs of X and W):

$$X = \frac{A\beta}{1 + A\beta} W + \frac{A\beta}{1 + A\beta} Z$$

$\uparrow$ $\uparrow$ $\uparrow$

actual   desired   disturbance
value   value

We see:
1.) The actual value of the controlled variable will be closest to the desired value W, if the loop gain $A\beta$ is much larger than 1. $((W - X)/X = 1/A\beta)$
2.) Once $A\beta$ is large, X contains disturbances the less the more $A \gg 1$. $(X \simeq W + Z/A$ for $A\beta \gg 1)$

This suggests to make the amplification as high as possible. But this often is in contradiction to the stability of the control loop, as we shall see. To wit, our control loop has the same feedback scheme that is normally used (without W) to describe a feedback oscillator. The only difference consists in the application of negative feedback in our case.

Often controllers work mechanically or pneumatically or (in biology) chemically. But since physicists are most familiar with electric networks containing resistors, capacitors and coils, we will use these to model any type of control loop adding amplifiers for the controller itself.

PROPORTIONAL CONTROLLER

The most simple controller consists of a proportional amplifier, which does not show any phase shift - at least not at frequencies below a "critical frequency" f_k at which the loop gains drops below 1. All attenuation and phase shift is then caused by the controlled system, which - in general - is given. If we state, the controlled system is of 1st, 2nd, 3rd, or even some higher order, we mean that it consists or can be modeled by one, two, three ... RC (resistor-capacitor) networks, of which I will shortly remind you (fig. 4).

In Bode's diagram (which is double logarithmic) the drop of the output amplitude towards higher frequencies, caused by the capacitor, is given by a straight line, which intersects with the horizontal at the so called "corner frequency" f_c or $\omega_c = 2\pi f_c = 1/\tau$ with $\tau = RC$ being the time constant.

The amplitude drop is affiliated with a phase shift with a tendency towards $- 90^0$ (being $- 45^0$ at ω_c). If we interchange the elements R and C, Bode's diagram shows up the change from "low-pass" to "high-pass" or from "integrator" to "differentiator" (fig. 4).

If we take two RC's behind each other, the amplitude will go down twice as steep, that means by two decades versus one decade in frequency - and the phase will tend to $- 180^0$ at infinite frequency.

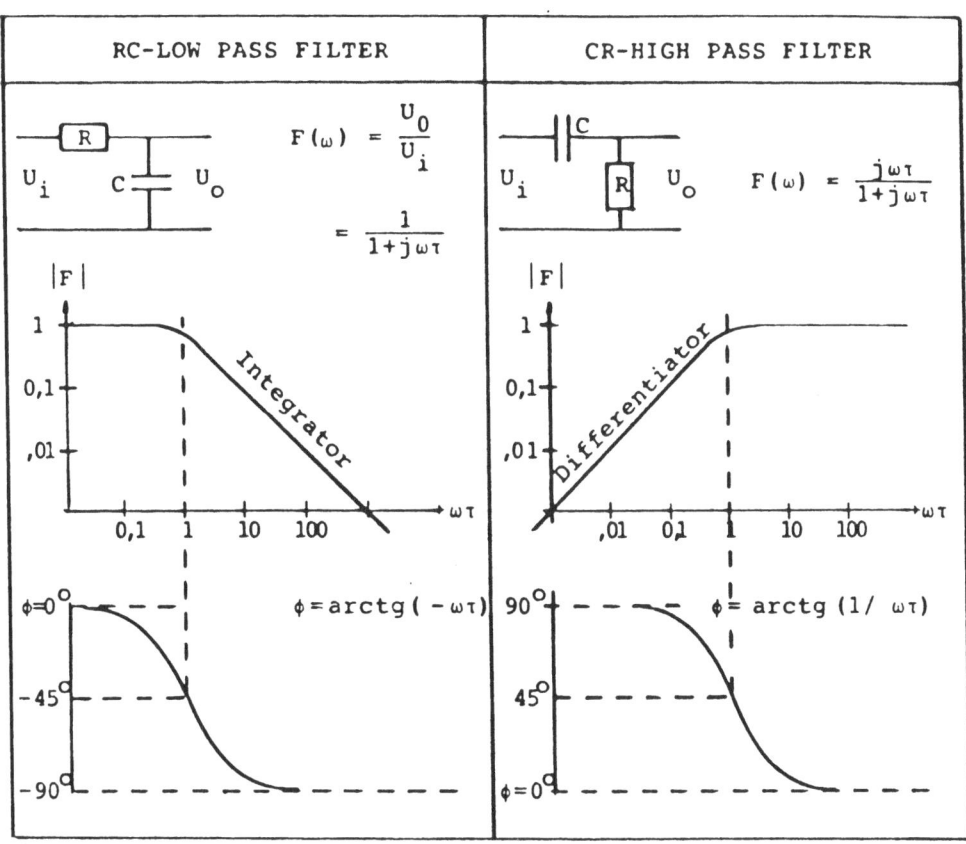

Fig. 4. BODE-plots of the transfer function F of 1st order RC-filters.

With three RCs in series, we shall reach - 180^0 at a finite frequency. If the loss of such an arrangement of three RCs in series is just compensated by the amplification of the controller (i.e. $A\beta \geq 1$) our feedback loop will start to oscillate at that very frequency for which the overall phase shift is zero.

That is how a simple, well known phase shift oscillator works. Therefore: hands off from three RCs in series! Yet, in real life (life of someone, who has to construct a control circuit) three RCs are easily at hand: e.g. two in the controlling system and one in the controller. Fortunately their corner frequencies in general are different.

We will assume a loop with such three corner frequencies separated by a factor of ten concentrated in the controlled system - in order to study the chances a controller has in such a case. As mentioned before, that controller shall be a proportional controller, which is an amplifier with constant gain - independent of frequency. Fig. 5 is the Bode plot of the controlled system alone, of the proportional controller and the combination of both. $f_{1,2,3}$ are the corner frequencies.

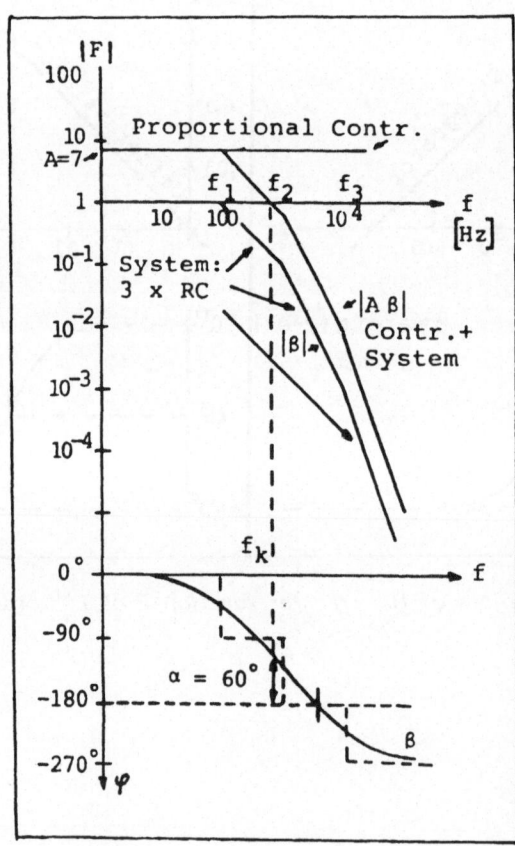

Fig. 5. BODE amplitude and phase plots for a
proportional controller and a system of
3rd order with distributed corner frequencies.
($\alpha = pase\ margin$)

With the critical frequency f_k (where the loop gain is 1) we keep far enough from that frequency, where the loop phase shift is - 180^0, which in our example is at about 3,3 kHz.

Why did I choose A = 7? Let us have a look at the Aβ-dependence of (W-X)/W, the relative control deviation. (Aβ is equal to A at low frequencies, see fig. 6). Why not A = 100?

Table 1. Relative control deviation in % at different values of the loop gain A (Aβ ≈ β at low frequencies)

Aβ	1	2	4	7	10	20	50	100
$\dfrac{W-X}{W}$	50	33	20	12	9	5	2	1

Let us see, what happens when we use step functions instead of sine-signals for the desired values of the quantity to be controlled (fig. 6): the loop behaves like a damped oscillator - at A = 100 even an undamped one. Why this will happen can easily be seen by lifting the curve for Aβ up to A = 100: it will intersect with the frequency axis for A = 1 at about 3,3 kHz, where the phase shift amounts to 180^0.

At A = 7 we observe a slight swing over - the time for reaching the steady state is the shortest one. Here we have a "phase margin" or "phase reserve" of about 60^0 - the "distance" to the critical - 180^0. Nevertheless, the control deviation remains large (12,5 %).

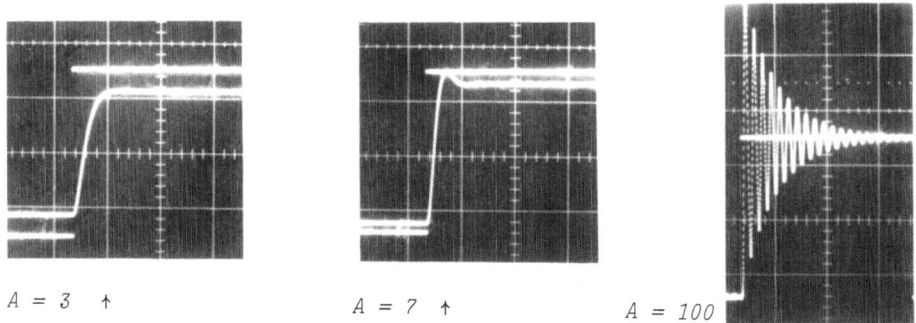

A = 3 ↑   A = 7 ↑   A = 100

Fig. 6. Answer X of the closed loop to a step function signal W

If the loop is of second order, there will be no instability in the sense of free feedback oscillations, but the swinging about the steady state will remain. Only a first order loop allows A to be increased to a rather high value, provided the controller amplifier will then not add attenuation and phase shift himself.

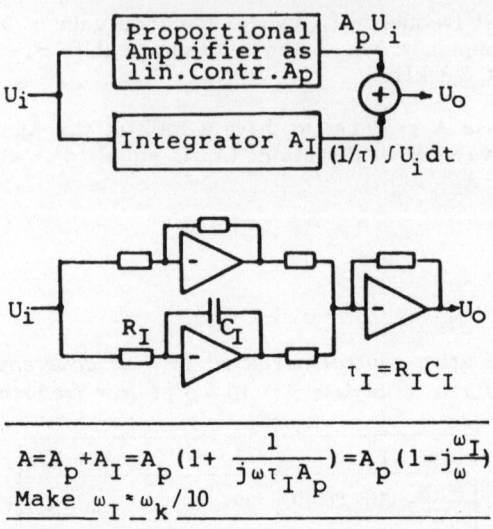

$$A=A_p+A_I=A_p(1+\frac{1}{j\omega\tau_IA_p})=A_p(1-j\frac{\omega_I}{\omega})$$
Make $\omega_I \approx \omega_k/10$

Fig.7. *Proportional-Integral-Controller*

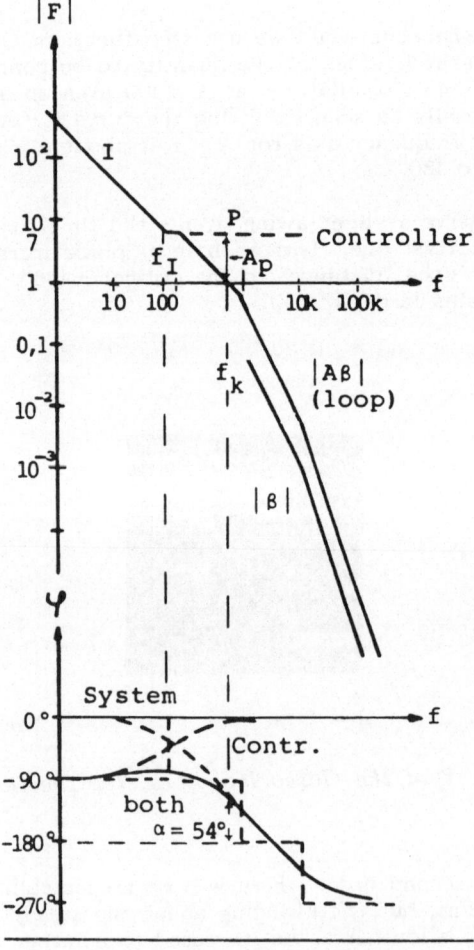

Fig.8. *BODE plot of a feedback loop*
comprising a proportional-
integral controller

PROPORTIONAL-INTEGRAL-CONTROLLER

How can we avoid the shortcomings of the proportional controller? That means: how can we get the control deviation down to zero without oscillations about the steady state value?

The idea is simple: one only has to feed the control deviation into an integrator in order to produce an increasing (or decreasing) voltage, which is used to drive the controlled variable into the wanted value (fig. 7). The problem is analogous to charging a condensor. The corresponding solution

$$(W - X) = (W - X)_{t=0} (1 - e^{-t/\tau})$$

suggests to choose a short integration time constant in order to reach the equilibrium value of the controlled variable as quick as possible.

But again: caution! As we know, an integrator in the loop shifts the phase in the same sense as it is shifted away by our controlled system of three RCs (fig. 8). Therefore, having in mind the stability of the loop the integrator phase shift must have been almost vanished, "before" (on the frequency axis) the phase shift of the system starts to get noticeable. Consequently we have to position the corner frequency f_I for stopping the influence of the integrator low, to – say – one tenth of the critical frequency (which for A = 7 lies at 700 Hz in our example).

Fig. 8 shows the BODE diagram for that case, which is the best compromise between creeping slow and increasing oscillation. What can be seen in fig. 9 is the control deviation W - X for this case as well as for f_I being too low or too high respectively.

Do we have to live with that compromise of low corner frequency and low proportional amplification, which makes our loop relatively slow? If we could find means to elevate the amplification A_p without increasing the oscillations about the steady state value, then the control deviation would be smaller and it would therefore take less integration time to reach the steady state.

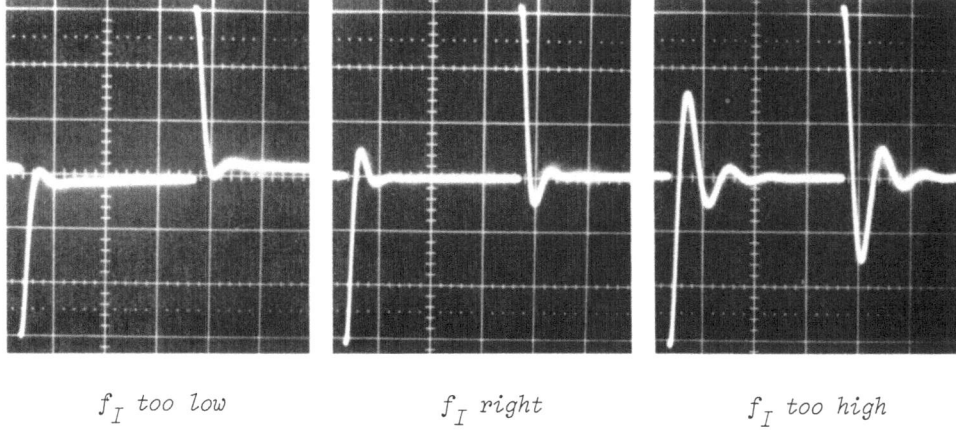

f_I too low $\qquad\qquad$ f_I right $\qquad\qquad$ f_I too high

Fig. 9. Step response of the control deviation W-X for three choices of the corner frequency f_I at which the controller stops integrating.

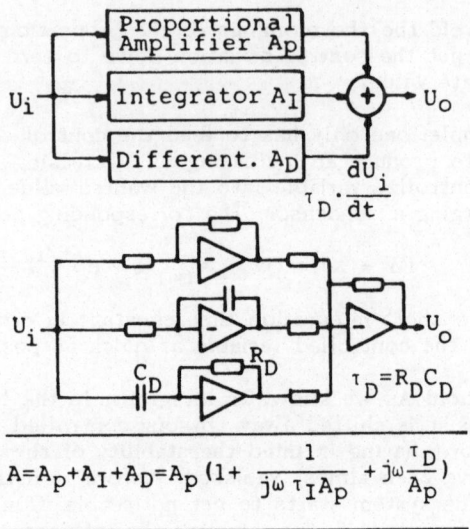

$$A = A_p + A_I + A_D = A_p \left(1 + \frac{1}{j\omega \tau_I A_p} + j\omega \frac{\tau_D}{A_p}\right)$$

Fig.10. Proportional-Integral- Differential-Controller

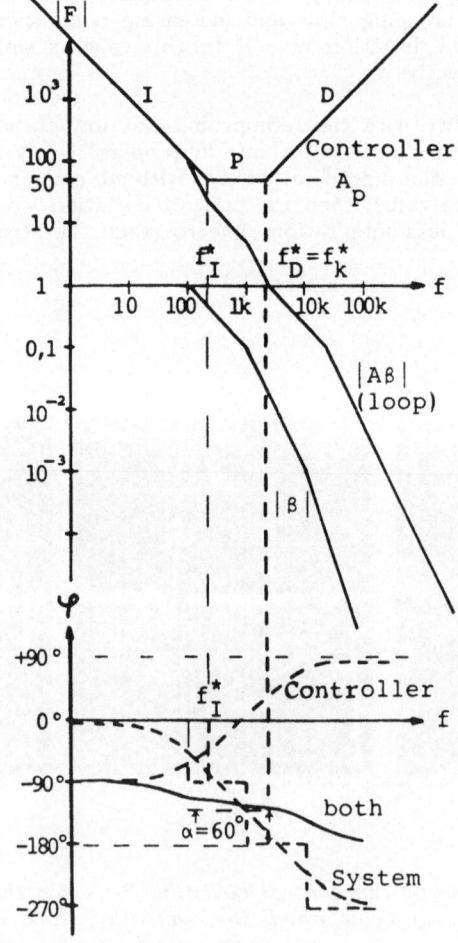

integral-differential controller

PROPORTIONAL-INTEGRAL-DIFFERENTIAL-(PID-)CONTROLLER

By adding a differentiator, parallel also to the proportional controller A_p (fig. 10), we are able to partially compensate the phase lag in the loop (caused by the RCs in the cotrolled system). As is shown in figure 11, this allows us to increase A_p with an accompanying increase of the critical frequency f_k to an f_k^*, which renders the controller quicker. Furthermore we can increase the integrator corner frequency f_I by the same factor without any harm. This also contributes to get into the steady state quickly.

In practice one starts with A_p alone, raising its value until the oscillations, following a step of the desired value, are being only damped weakly. If the phase margin then amounts to, say, merely 15^0, this goes up to safe 60^0 by the addition of a differentiator, the corner frequency of which has been successively lowered to the critical frequency f_k^*. (We remember, that the phase angle of a differentiator amounts to $+ 45^0$ at f_D). Again in practice one looks at the desired value (or the control deviation): the differentiator corner frequency will have its appropriate value, when the signal shows a slight swing over. Then one starts the integrator and increases its corner frequency f_I up to the point, where the controlled variable reaches the steady state value quickly without having added oscillations. Then f_I^* will have a value of about $0,1 \, f_k^*$. This action works without iterations. Fig. 12 demonstrates the improvement with respect to time, when a differentiator is added to the controller.

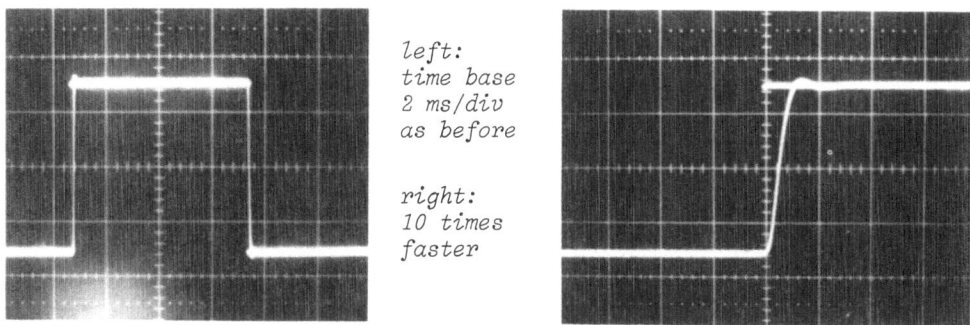

left:
time base
2 ms/div
as before

right:
10 times
faster

Fig. 12. Improvement of the actual value signal X by means of a differentiator.

AMPLITUDE CONTROL

As an example we take the control of the RF amplitude in an accelerator tank (fig. 13). The controller parameters can be found at low RF power as it is compatible with the operating points of the power stages. As a measure of the RF amplitude inside the cavity one might take the forward signal of the directional coupler in the feed from the transmitter to the tank. The controller should act upon the first amplifier stage (controlling its transconductance) or operate ahead of this stage via a double balanced mixer/multiplier.

After having tuned the cavity to the proper frequency and matched for a minimum of reflected power, we close the feedback loop, starting with a low amplification A_p.

Either with a sine signal superimposed on the desired value W of the RF amplitude we can take a Bode plot for the loop gain from measurements of $(W - X)/X = 1/A\beta$, to find out the degree of the loop; or we superpose a small square wave signal and observe the onset of oscillations.

Since the accelerator frequency will be high, the influence of the time constants of the resonant amplifier stages and the cavity itself will be rather weak in the time (or frequency) scale normally considered for automatic control of the RF amplitude; thus only the RCs for filtering out the RF carrier and within the controller amplifier will be of importance.

For a pulsed accelerator we have to store carefully the actual value of the RF amplitude by a sample & hold circuit, driven by the pulser.

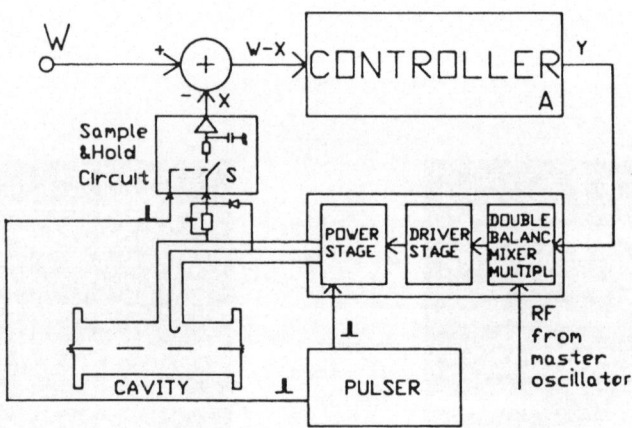

Fig. 13. Control of the RF amplitude within an accelerator cavity.

CONTROLLING THE RF PHASE

We prepare the problem of phase control by considering the control of the frequency of an oscillator (fig. 14). Its tuning should be possible by means of a voltage. (In the name VCO of such voltage controlled oscillator "controlled" is used in the sense of forward and not feedback control).

In fig. 14 the controller is a frequency discriminator, which delivers a voltage signal proportional to the deviation $f_w - f_x$, which is used to correct the frequency of the VCO. In general the frequency of the VCO is already roughly determined by a bias voltage U_{f0}.

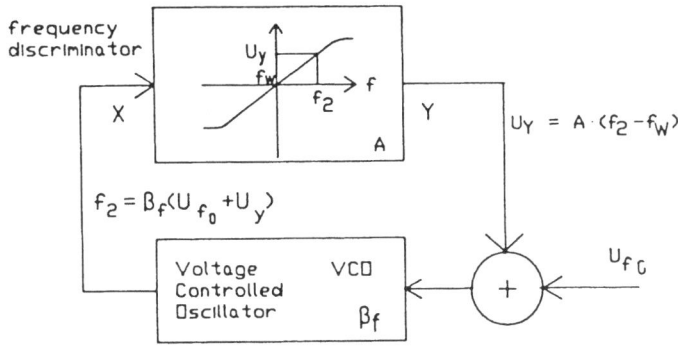

frequency discriminator

$U_Y = A \cdot (f_2 - f_W)$

$f_2 = \beta_f (U_{f_0} + U_y)$

The frequency discriminator might consist of two LC-circuits, slightly detuned up and down resp. against the wanted frequency f_W and of two diodes. The bias voltage U_{f_0} serves to bring the oscillator frequency already close to f_w.

Fig. 14. Control of the frequency of a voltage tunable oscillator.

While in this example the wanted value f_w of the frequency is already built into the controller, it also is possible to deliver a signal with a reference frequency f_w and compare it to the actual VCO frequency f_x by means of a phase detector (in place of the subtractor). Its output voltage U_φ is proportional to the instantaneous phase difference. Using this signal – via the controller A – to monitor the frequency f_x of the VCO this will be done, as we shall see, perfectly without any frequency deviation $f_w - f_x$. Hence we are left to the questions: does a phase deviation remain and: can we prescribe a desired phase?

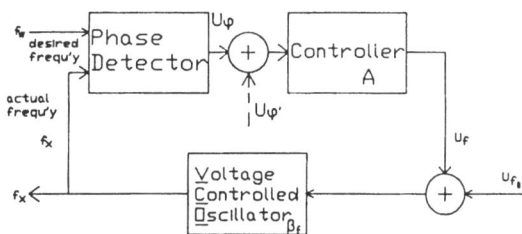

Fig. 15a. Feedback control of frequency by means of a reference frequency. Since the frequency is controlled perfectly, the loop can be used to control the phase (see fig. 15b).

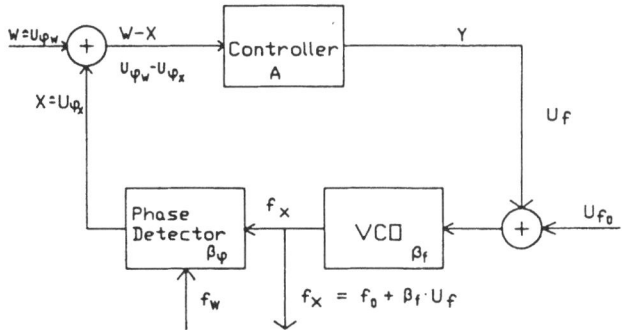

$f_x = f_0 + \beta_f \cdot U_f$

Fig. 15b. For a phase control loop the phase detector is taken down into the controlled system.

Therefore, in fig. 15b our scheme is redrawn by taking the phase detector down into the controlled system to the effect that we can apply a voltage $U_{\varphi w}$ for the desired phase at the place we are familiar with. The control deviation $W - X$ amounts in terms of voltages to

$$U_{\varphi w} - U_{\varphi x} = \frac{U_f}{A} = \frac{f_x - f_0}{A\beta_f}$$

since the voltage U_f controlling the VCO is connected to its output frequency f_x by

$$f_x = f_0 + \beta_f U_f$$

As before, $U_{\varphi x}$ is proportional to the momentary phase difference φ_x of f_w and f_x

$$U_{\varphi x} = \beta_\varphi \cdot \varphi_x$$

Applying the same factor β_φ to $U_{\varphi w}$ (i.e. $U_{\varphi w} = \beta_\varphi \varphi_w$), we get

$$\varphi_w - \varphi_x = \frac{f_x - f_0}{A\beta} \qquad \text{with } \beta = \beta_\varphi \cdot \beta_f$$

$$\uparrow \qquad \uparrow$$
$$\text{(PD)} \quad \text{(VCO)}$$

Setting the desired phase φ_w to zero and assuming f_x to be constant, then φ_x will be constant too: that means: f_w and f_x must be identical, as stated before. Yet, as the formula tells us, there remains a phase deviation. In order to keep it small, $A\beta$ should be large and f_0 should be close to $f_x = f_w$. We might further think of integrating the remaining phase deviation to zero.

$A\beta$ high and use of an integrator?

For the sake of stability, we have to know about the order of the loop: behind the phase detector in general there should be an RC-filter, in order to remove the carrier frequency.

What about the transfer function β_φ of the phase detector itself?

U_φ is a time varying voltage, as long as f_x is not equal to f_w:

$$\frac{U_\varphi(t)}{f_x - f_w} = \frac{U_\varphi(t)}{\varphi(t)} \cdot \frac{\varphi(t)}{f_x - f_w} = \beta_{\varphi_0} \frac{\varphi}{f_x - f_w}$$

Let us now open the loop by removing U_f, replacing it by an a.c. voltage $\hat{U}\cos\omega_m t$. The VCO will answer with a frequency modulated output; that is: the frequency will swing about the carrier frequency by an amount

$$\Delta\omega = \hat{\Delta\omega}\cos\omega_m t$$

We put this into the general expression for a phase difference accumulated in a time t:

$$\varphi(t) = \int_0^t \omega_1 dt - \int_0^t \omega_2 dt = \int_0^t \Delta\omega dt = \frac{\Delta\omega}{\omega_m}\sin\omega t$$

Dividing φ(t) by Δω(t) and taking care of the phase lag of 90^0, we get in complex notation

$$\frac{\varphi}{\Delta\omega} = \frac{1}{j\omega_m}$$

which describes an ideal integrator. This means in the time domain: when Δω jumps, φ will increase (or decrease) continuously.

Thus the transfer function of the complete controlled system is

$$\beta = \frac{U_\varphi}{U_f} = \frac{2\pi}{j\omega_m} \cdot \beta_{\varphi_0} \cdot \beta_f$$

It is in general of higher than first order, since the measurement of the phase takes some time or is accompanied by low pass filtering. This determines the order of β_{φ_0}.

A few remarks before an example of a phase detector and its influence on the time needed for controlling the phase is given. What we have learned is how a phase locked loop works. Phase locked loops (PPL's) are being widely used in the electronics of communication systems as frequency multipliers and generators ("synthesizers") - even in modestly priced modern radio receivers, as you know -, furthermore for frequency- and bit-synchronization, FM- and AM-demodulation.

As phase detectors one can use a simple sample & hold circuit, a multiplier (double balanced mixer) or digital circuits. The latter ones have the advantage of locking the VCO even at large frequency deviations, a problem we are not faced with here.

EXAMPLE:

Let us have a short look at the sample and hold circuit (fig. 16): an electronic switch, activated during a time short compared to the RF-signal period, serves to charge a capacitor. The mean time, after which one has gained phase information, is half the RF-signal period. This brings about an additional phase lag of $\pi f/f_{RF}$ that is $\pi f_k/f_{RF}$ for the critical frequency f_k, where $A\beta = 1$. Thus the phase reserve still amounts to 45^0, if f_k is chosen to be as high as one fourth of the radio frequency. This means, the controller can be made fast.

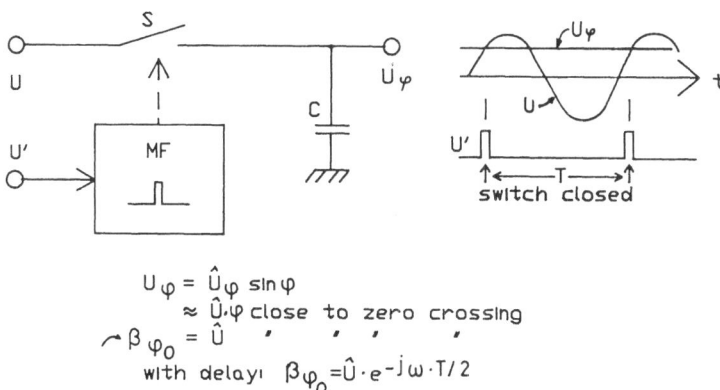

$$U_\varphi = \hat{U}_\varphi \sin\varphi$$
$$\approx \hat{U}\cdot\varphi \text{ close to zero crossing}$$
$$\beta_{\varphi_0} = \hat{U} \quad , \quad , \quad ,$$
$$\text{with delay: } \beta_{\varphi_0} = \hat{U}\cdot e^{-j\omega\cdot T/2}$$

Fig. 16. Sample-and-hold phase detector. (MF = monoflop)

In accelerator technics, where one uses high frequencies, one might in general be content with a relatively slow loop such as is provided by a multiplier phase detector with a (first order) filter behind it. Its corner freqency, which must also be the critical frequency f_k with 45^0 phase margin, should be much smaller than $2f_{RF}$ (2 because of the multiplication of two signals of equal frequency).

In the loop shown (fig. 17), which uses the sample and hold circuit mentioned, the controller is just a voltage divider. The C in it provides the integrator, which brings the phase deviation to zero. (Since the voltage U_φ changes stepwise, a differentiator should not be incorporated into the control loop.)

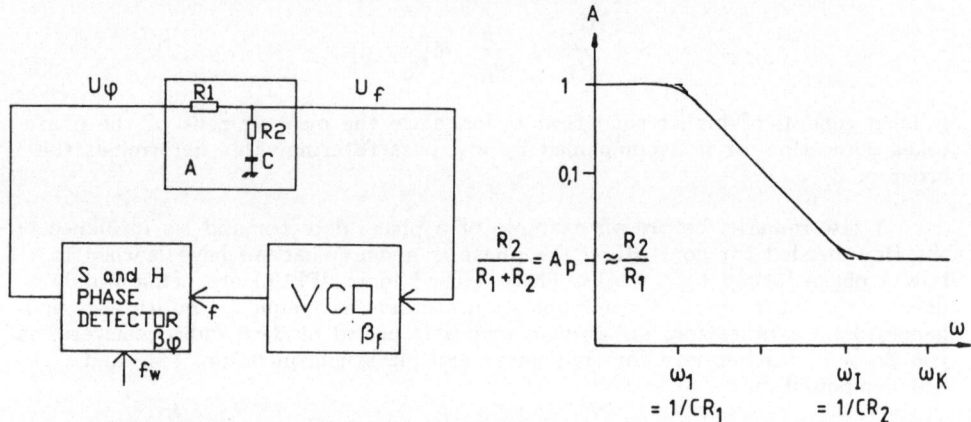

Fig. 17. *Simple Phase Locked Loop (PLL) using a sample & hold phase detector and a voltage divider as PI-controller and its BODE plot.*

At f_k the loop gain per def. equals 1 and $|\beta| = \hat{U}\beta_f/f_m$. Therefore at f_k we get:

$$|A| = f_w/4\beta_f\hat{U}$$

Using reasonable values for the three quantities on the right , $|A|$ will be much smaller than 1. Therefore the controller A can made be a simple voltage divider. Including the integrating capacitor, we have

$$A = \begin{cases} R_2/R_1 \ll 1 \text{ at } \omega_k \ (= \omega_{RF}/4) & \rightarrow \text{ a proportional controller} \\ 1/j\omega CR_1 \text{ for } \omega_1 \ll \omega \ll \omega_I & \rightarrow \text{ an integrating controller} \\ 1 \text{ for } \omega \ll \omega_1 & \rightarrow \text{ a proportional controller} \end{cases}$$

It is helpful that at low frequencies the integrating capability of the controller gets lost: the controller will not run away as long as the loop has not yet locked.

* * *

Instead of using phase locked loops for controlling the RF-phase for each stage of a multistage accelerator (and for setting it, if one wants to change the q/m of the particles), one can also rely upon voltage dependent phase shifters. Since these often do not assure constant RF output amplitudes at different phase angles, an additional amplitude control loop might be necessary.

On the other hand, because of the high frequencies used in accelerators, digital phase shifters require proper frequency down- and up-conversion.

* * *

Literature: Textbooks on Automatic Control and:
U. Tietze and Ch. Schenk, Halbleiter-Schaltungstechnik, Springer-Verlag 1985

RF STRUCTURES FOR LINEAR ACCELERATORS

A. Schempp

Institut für Angewandte Physik, Universität Frankfurt
6000 Frankfurt/M , W. Germany

INTRODUCTION

Accelerators produce beams of various particles in a wide energy range. In linear accelerators (Linacs) particles are accelerated by electromagnetic fields on a linear path. They are used in large numbers as particle sources for low energies, as injectors for circular machines and as accelerator structures in large, synchrotrons and microtrons.

For the acceleration of charged particles (electrons or ions) an electrical field E_z has to be provided along the path of the particle with charge q and mass m. The energy gain than is given by:

$$\Delta T = q \int E_z(z,t)\,dz \qquad 1)$$

In such a general description a linear accelerator is a transformer of power from mains to electromagnetic fields in which particles are accelerated and transported along a linear path.

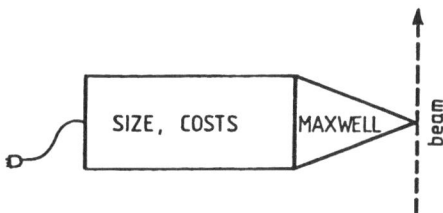

Static accelerating fields, produced in an array of electrodes, are limited by the maximum voltage holding capability of insulating materials and reach values up to 30 MV in large Tandem (Van de Graaf) facilities for heavy ions.

A theoretically unlimited energy gain can be obtained if the accelerating fields E are produced only in the vincinity of the particle resp. the particle bunch as indicated. These E fields then should travel along with the particles having always the same velocity v_p as the particles. Such "pulses" can be produced for a short path length but normally a chain of "pulses" forming an electromagnetic wave travelling with the particles is used.

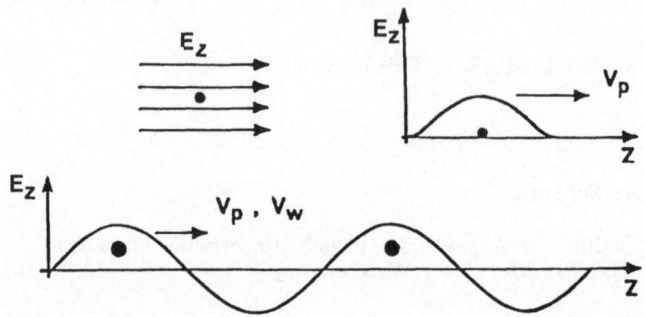

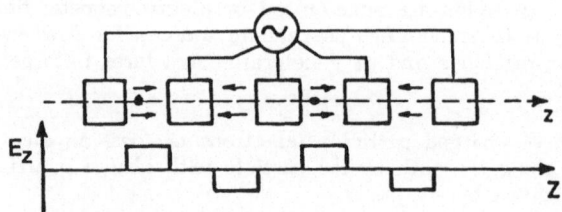

The same effect can be obtained by the fields in an array of electrodes with rf-voltages phased such, that the particles are always accelerated.

The fields in this "Standing Wave" scheme, which has first been proposed by Wideroe [1], can be described as a sum of foreward and backward travelling waves. By "Fourier analysis" the particles are accelerated by the foreward wave component with same velocity as the particle. The phase velocity v_w of the backward wave and all higher (space-) harmonics are so different from the particle velocity v_p that their influence can be neglected.

Linear accelerator structures are electrode or waveguide arrangements which are designed to produce the accelerating synchronous wave for the various applications. They differ in the technical principles applied, in the field values which can be achieved in the beam current they can accelerate and in the efficiency in which electric power is converted to fields and particle energy resp. to particle power.

Examples for rf-linac structures are shown in fig. 1. There is the iris loaded waveguide [2] for relativistic electrons working in pulsed operation and usually in travelling wave mode. The Coupled Cavity structures can be derived from the Iris structure by closing the apertures much more but can also be treated as weakly coupled individual cavities. They are used in high average power electron and proton machines ($v/c \gtrsim 0.5$) and are operated up to cw. Alvarez structures [3], which are used for Protons and Heavy Ions for v/c appr. 5%- 50%, are TM_{010} resonators with drift tubes to shield the particles if the rf - fields are reversed. Wideroe - structures are ion accelerators for $v/c \lesssim 5\%$ which use (resonant) transmission lines to charge adjacent drift tubes with opposite voltages. The cell length (particle path length for one rf period) resp. drift tube length (and the equivalent v_w) will increase according to the particle velocity.

IH – and coupled transmission line resonators [4] use the same synchronism principle: The voltages change their sign when the particles are in the center of the drift tube. Single cell structures and resonators with a small number of cells can be phased independently so that a variable velocity profile along the accelerator can be achieved with const. geometry [5-7]. They are used as Heavy Ion (HI) accelerators as an example. Application of most of these structures at low energies is limited by the insufficient focusing strength of the quadrupoles in drift tubes or between tanks. These low energy structures have a limited focusing strength which limits the application for low energies and very high beam currents. RFQs [8-10] (Fig. 2) are new structures using strong electrical rf-focusing and are applied in low energy (< 10% v/c) high current accelerators e.g as Alvarez injectors. Even though various types of linac structures differ in size, frequency and application, their basic principles are similar. They have to produce a wave travelling along the accelerator axis synchronous with the charged particles.

DESIGN PARAMETERS

Assuming a cylindical volume along the beam path(along z-axis, radius r) free of material, currents and charges Maxwell's equations give field components for the accelerating wave traveling synchronous with a particle (velocity $v_p < c$) :

$$E_z = E_o \, I_o(k_r \cdot r) \, \cos(\omega t - k_z \cdot z) \; , \; k_o = 2\pi/\lambda_o = \omega/v_p \qquad \text{2)}$$

$$E_r = - \, E_o \, \frac{k_z}{k_r} \, I_o(k_r \cdot r) \, \sin(\omega t - k_z \cdot z) \; , \; k_r^2 = k_z^2 - k_o^2 \qquad \text{3)}$$

The axial accelerating field (amplitude E_o) is a function of the radius and the acceleration of particles goes along with defocusing caused by E_r. This is compensated only for particles with $v_p = c$ by the magnetic field H_Θ which is the reason for the simple focusing of electron accelerators:

$$\frac{d}{dt} \frac{m}{q} \, \dot{r} \approx (E_r - v \, B_\Theta) \approx (1 - \frac{v^2}{c^2}) \qquad \text{4)}$$

For $v \ll c$ ($k_r \approx k_z \gg k_o$) radial defocusing by E_r is proportional to the acceleration gradient E_o and to the ratio of operating frequency ω and particle velocity v_p .

$$E_r \sim E_o \cdot r \cdot \frac{\omega}{v}$$

So for ion acceleration the low energy region is especially problematic in respect to focusing. Equ.2,3 for Ez, Er indicate that particle dynamics have a strong influence on parameter choice and set physical boundaries. If physical principles allow acceleration and transport resp. focusing of a particle beam already the next step has to be a look for the cost of a structure, which is after all the real limit.

The cost of a accelerator structure is proportional to the length L, to the radius R_C and the precision P needed for fabrication. The costs of rf power P_{rf} is proportional to the number Ng of generators, the peak power Pp and average power P_a per rf-generator.

$$C \, (DM, \$, \, ...) = (N_c^{\alpha_1} \cdot L_i^{\alpha_2} \cdot R_c^{\alpha_3} \cdot T^{\alpha_4} {}^{\alpha_5} + N_g^{\alpha_7} \cdot P_a^{\alpha_8} \cdot P_p^{\alpha_9} \cdot \omega^{\alpha_{10}}) \cdot N \qquad \text{5)}$$

There are similar cost formulas for each component. The factors and coefficients α_i are depending on the parameters themselves. They might not be even continuous, as can be seen for transmitters which have steps in output power and use several different physical concepts for different output powers and frequencies. A typical problem which has to be solved is the trade off between the number of subunits N and cost of the subunits which interfere with physical effects like field stability, power drop, mode mixing and especially tolerances.

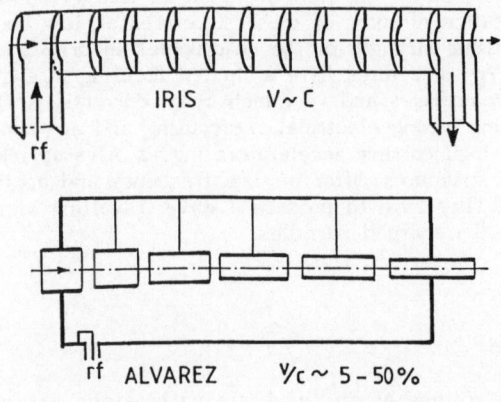

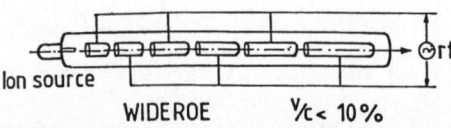

Fig. 1. Scheme of different accelerator structures

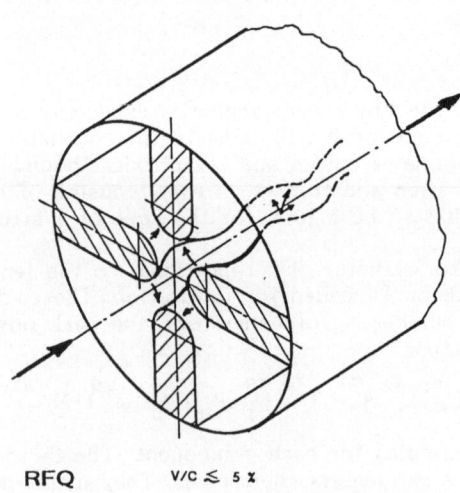

Fig. 2. Scheme of RFQ structure

An important optimisation quantity for room temperature accelerator structures is the shuntimpedance η, which gives the ratio of accelerating field (squared) to the rf power consumption per unit length $P = N/L$:

$$\eta = \frac{E_z^2}{P} \quad , \quad Q = \frac{\omega \cdot W}{N} \quad , \quad \frac{\eta}{Q} = \frac{E^2 \cdot L}{\omega \cdot W}$$

The ratio of shuntimpedance η and Q-value is determined by the geometry of the cavity and describes how good the fields are concentrated near the cavity axis where the particles will be accelerated. These quantities are independent of the field amplitude so they can be easily measured at low power levels without vacuum and cooling.

The pillbox TM_{010} resonator shown schematically in Fig. 3 is a basic accelerator structure. Fields can be calculated explicitly assuming only the TM_{010} mode with axial E-fields. Taking the length $L = \beta \lambda/2$ the shuntimpedance is proportional to the squareroot of the frequency $\eta \sim \sqrt{\omega}$.

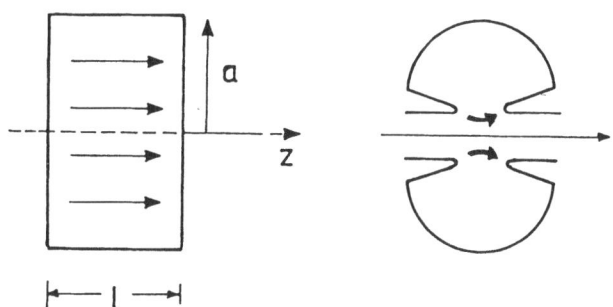

Fig. 3. TM_{010}- Pillbox cavity

For a frequency of 500 MHz a shuntimpedance of $23 M\Omega/m$ (without transittime effects) , a Q value of 75000 and a $\eta/Q = 160 \ \Omega$ are calculated. Optimization leads to another shape of the cavity with a drift tube-like inner part to concentrate the electric fields which has appr. twice the shuntimpedance [11] as shown in Fig. 3a. Similar shapes of the cavities can be seen in all room temperature linacs.

For superconducting cavities the shuntimpedance is not so important. Low maximum field values and a shape which has a small tendency for multifactoring and if possible no higher modes at harmonics of the operating frequency are more important [12] . But the general rule, increasing the ratio of volume to surface is valid also. The impedance depends on the field distribution that means the mode of the resonator. For a TE mode (azimuthal currents) the shuntimpedance is higher for lower frequencies: $\eta \sim \omega^{-3/2}$. Same frequency dependence of the shuntimpedance is obtained for the coaxial resonator and its descendants the Wideroe-, IH- , RFQ- , Spiral and Split-Ring loaded structures, where the shuntimpedance is also dependent on the special drift tube configuration, because how often (N times, $L = N \cdot \beta/2 \cdot v/c$) the same voltage can be used for acceleration depends on $\beta = v/c$. So for a heavy ion IH-resonator a value of up to 1. $G\Omega/m$ has been measured [13] . η can reach really high values if the available power is concentrated in a very short cavity or on a small area. This is the case in most new acceleration schemes where fields up to 1. GeV/m are obtained (sometimes only once) on few centimeters or mostly much less than a mm. Taking the huge drive systems into account the effective shuntimpedance is mostly lower than in conventional systems. Therefore an effective shunt impedance resp. (average E_z)

has to be used for the comparison of different accelerating structures resp. schemes.

Superconducting cavities have "real" shuntimpedances of $T\Omega/m$ with Q improvements of up to 10^6. Even with refrigerator losses taken into account, the improvement factor is enormous. There the Q-value or the average axial accelerating field E_z (always cw- operation) are figures of merit.

The energy gain is simply given by $\quad \Delta T = \sqrt{\eta \ L \ N}$.

Therefore there are three ways to increase the energy gain: The first way is to add length which is expensive and has at last geological limits. Secondly the power N can be increased which can be done for the average power but usually the pulse power is increased by factors of up to 10^4 over the average power for restricted duty cycles. The third way is improving the properties of the structures. The values of are between 20 and 100 Megohm/m depending on application and particle velocity and the progress of improvement is slow. A real step upward is the use of superconductivity with which accelerating fields up to 20MV/m have been measured even without the new miracle materials [12].

MAXIMUM FIELDS

Another general accelerator design criteria is the optimal use of the acelerating fields resp. the voltages U_e on the electrodes. Because of $v/c \ll 1$ for ions and the Besselfunction in equ. 2 the energy gain of the particle will be less than $\Delta T = eU_e$. The integral $\Delta T = q\int E_oI_o(k \cdot r) \ dz$ depends on the k ($k \approx k_r \approx \omega/v_p$) and a, the ratio of aperture a (maximum beam radius a = r_{max}) to applied frequency and particle velocity v_p .

An optimum value is for $\lambda \approx 4$ a . For smaller wavelength ($\beta \ \lambda_o$) the "loss" by the drop of the Besselfunction is significant, for longer wavelengths the average accelerating field is too low resp. the voltage on the electrode which goes quadratic with the rf power needed has to be too high. In addition, if the frequency is higher than the optimum the radial variation of the acceleration field caused by the Besselfunction I_o introduces energy spread and emittance growth. The general optimisation keeps the initial cell length approximately four times the aperture radius which determines the highest possible operating frequency for ion linacs.

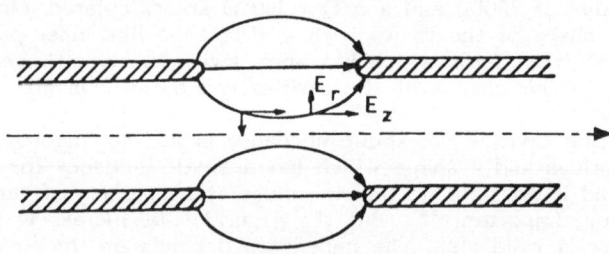

Fig. 3 a.   Linac drift tube geometry

Higher frequencies are only possible if the real beam radius is made smaller, that means the quality of the beam has to be improved. This is the reason that all drift tube geometries look alike. For electrons ($v_p = c$, $k_r = 0$) there is a homogenous field around the axis and the above limitation are not valid. The general optimum ratio beween peak fields on the elektrodes and electrode distance sets a limit for radius of curvature for all electrodes and irises, which asks for a relativly big aperture and also scales all designs.

In general, the diameter, the length and the cost of an accelerator structure can be reduced, if a higher operating frequency is chosen even though the relative precision has to stay the same, which means the absolute precision requirements go up with frequency. This is one reason that future accelerators for electrons are proposed for frequencies as high as 30 GHz, even the structure has an outer diameter of less than 1 cm and Gigawatts of rf power is needed which can be produced for example with another drive accelerator parallel [14] where the second accelerator (not necessarily on the same frequency) generates the rf power. This is a concept which in principle is also applied e.g. in all Klystron driven accelerators, where a low energy high current electron beam (1A, 500 keV) generates microvaves for a low current high energy accelerator (1 μA, 50 GeV).

SPARKING

The length of an accelerator which is sometimes an absolute limit and not only a cost factor can be reduced also by increasing the electrical field strength . The concept of shuntimpedance shows that the power consumption then increases drastically. This has the important consequences that the rf power and the beam are pulsed to save power but also to solve technical problems like cooling.
The cost is proportional to the average power and approximately only to the square root of the peak power. In addition cooling problems are simpler with pulsed operation. Another important parameter limiting the field strength of accelerator structures is the maximum field which can be applied before fields are shorted by a vacuum spark discharging the cavity. The empirical Kilpatrik-Limit [15] :

$$f(MHz) = 1.4 \ 10^4 \cdot E^2 \ e^{-0.085 \ /E} \ (MV/m) \hspace{2cm} 6)$$

For fields higher than given in this formula the probability of sparks increases. Typical values are 12 MV/m for 100MHz and 50MV/m at 3 GHz. The empirical criterion was derivated for cw operation. Experience shows that fields higher by a factor 2 to 3 in cw operation using special surface treatment. Especially for short pulses (< 1 msec) this improvement factor might be as high as 5 or 10 (< .1 μsec) because the time is to short to develop avelanche processes with electrons and ions. The breakdown limit is a complicated function of Voltage, surface properties, gap width, pulse length and frequency. Equ. 6 is a crude approximation which shows schematically that the way for high accelerating gradients is to use higher rf frequencies ($E \sim f^{1/2}$). Gradients up to 1 Gev/m can be achieved in classical sructures before surface melting due to limited cooling capacity (skin depth 3um) sets another limit at frequencies of approximately 30-50 GHz.

High gradient structures operate in travelling wave mode. The load characteristics for the transmitter are better and the ratio of peak to average accelerating fields is higher without the contributions of the backward wave. Cooling problems are simplier too, because approx. half the power is dumped outside the cavity.
For ion accelerators frequencies are in the SW or FM region, power loads are unpractical so the structure is used as resonant load. The matching problems for the transmitters during built up time of the rf pulse have to be controlled. For high average power or cw operation cooling the structure can be the limit too. This limits usually the gradients in existing cw machines to appr. 1 MV/m. This aspect favours lower frequencies where the power densities are lower. Significantly higher gradients can be achieved wih superconducting cavities, where operational fields reach 10MV/m.

SLOW WAVE STRUCTURES

For the acceleration of charged particles an electromagnetic wave must be synchronized with the particle: the phase velocity v_w of the wave must be slowed down to the particle velocity $v_p (v_p < c)$.
Transmission lines and waveguides can't be used directly for acceleration because the electric fields are perpendicular to the particle velocity and $v_w = c$ resp. v_w higher than c . A wave $E_o exp \ i(\omega t - k_z z)$ with $w/k_z = v_p < c$ would

need a static potential in the waveguide which is impossible because the boundary conditions are the same for all velocities.

This can be different with periodic boundary conditions. Using the equivalent circuit shown in Fig. 4, the phase velocities in a transmission line v_a and in a waveguide v_b are:

$$v_a = (L C_s) = c \; ; \; v_b = \frac{\omega}{k} = (L C_s (1 - \frac{\omega^2}{\omega^2}))^{-1/2}; \quad \omega_c = (L \cdot C)^{-1/2}$$

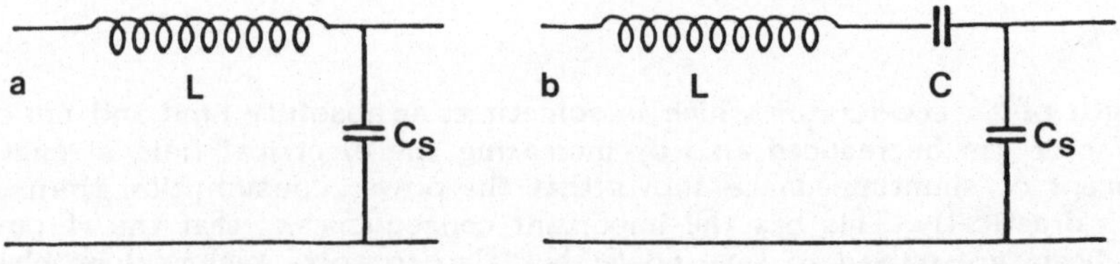

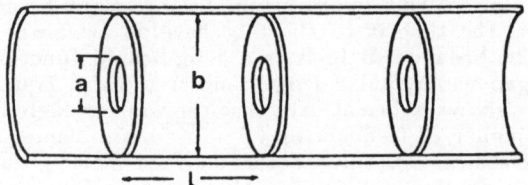

a   L C_s b   L   C C_s

Fig. 4. Equivalent circuits for transmission lines (a) and waveguides (b)

The dispersion relations for these circuits with L, C, C_s taken as distributed quantities (Impedance per length) are :

     a) $k = \omega/c$ b) $k^2 = k_o^2 - k_c^2$

To bring the velocity v_w down, a continuous increase of the capacities C_s would have no effect, the inductivity resp. the frequency ω changes accordingly. An additional periodic capacity C_s with fixed spacing L can reduce the phase velocity because the capacity is increased while the inductivity (appr. the area, resp. the volume) is not changed. This is done by attaching thin irises (inner radius a, outer waveguide radius b) in distance l in a waveguide as shown in fig. 5:

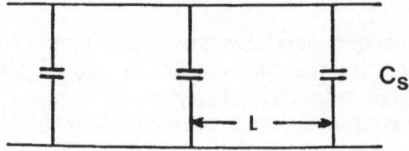

Fig. 5. Iris loaded waveguide

The cut-off frequency is nearly unchanged, because the effect of the C_s will be distributed for long wavelength. For $\omega \gg \omega_c$ the phase velocity can be smaller than in the unloaded waveguide (line). For a ω for which the spacings L equal to $\lambda_o/2$ a standing wave will be excited with $v = L \omega / \pi$. If the shuntcapacity is high enough λ can be smaller than λ_o and $v_w < c$. Smaller aperture (higher cap. load) slows down the wave and the standing wave (π – 180° phase shift betweeen adjacent cells) is reached at smaller $\Delta\omega = \omega - \omega_c$. With very small apertures the "cells" of length l are nearly independent and there will be wave transportation only in a very narrow frequency range. It can be easily seen that the group velocity will be small for small apertures.

     For closed irises the loaded line changes to separate cavities operating at ω_c which have to be idependently exited and by changing the phases between adjacent cavities any equivalent phase velocity v_w can be adjusted.

C_s

$\leftarrow l \rightarrow$

Fig. 6. Periodically loaded waveguide

Using chain matrix formulation for the calculation of the new dispersion relation gives the dispersion relation for a (lossless) waveguide with periodic capacity C_s (Fig. 6): $M = M_1 * M_2 : \{U',I'\} = M * \{U,I\}$

$$M_1 = \left\{ \begin{matrix} \cos kl & -iZ_0\sin kl \\ -1/Z_0 \sin kl & \cos kl \end{matrix} \right\} \qquad M_2 = \left\{ \begin{matrix} 1 & 0 \\ -iwC_s & 1 \end{matrix} \right\}$$

This description of one cell can be used for the periodic chain. If there is no damping: $\det M = 1$. The $\{U,I\}$ vector has harmonic solutions, if the trace of the cell matrix is smaller than 2:

$$\cos \varphi = \cos kl - \omega\, C_s/2\, Z_0 \sin kl ; \quad K = \omega\, C_s/2 \cdot Z_0 \qquad \qquad 7)$$

This is the dispersion relation for the periodically loaded line resp. waveguide. The diagram $\omega(\varphi)$ in Fig. 7 is called the Brillouin diagram.

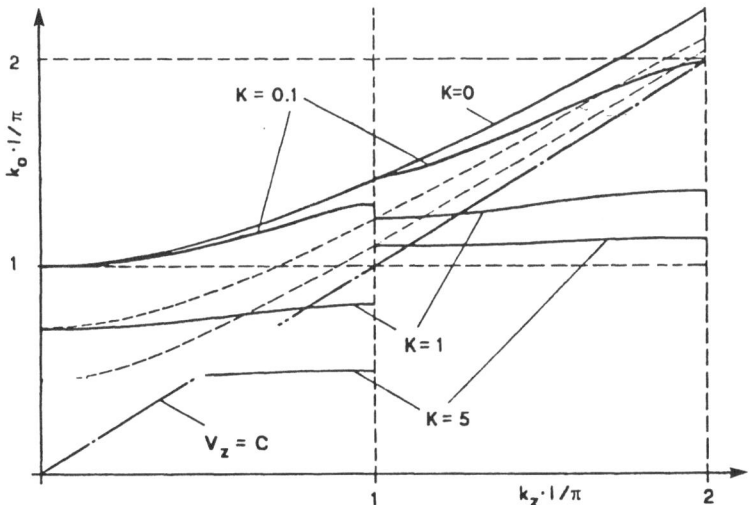

Fig. 7. Brillouin diagram for the periodically loaded waveguide

Curves for the unpertubated waveguide (K=0), for small pertubations (K=0.1) and for small aperture irises (K=1; K=5) are shown. The phase angle φ (φ=kl) is the phase slip of the wave from cell to cell and it will be used to classify the mode. The 0-mode (φ=0, ω= ω_c) has infinite phase velocity as in the unloaded case. π-mode is for kl = $1/\omega_c C_s Z_0$ (kl <<1).
There is no wave until the sin function in equ. 7 has changed its sign. For small C_s the changes are only in the neighbourhood of the π-mode. There will be a small frequency range above ω_π for which there is no wave transmission but damping.
With increasing C_s this "stopband" becomes greater. For smaller apertures of the irises resp high C_s as used in accelerators the frequency band for transmission becomes narrow and the dispersion relation can be written:

$$\omega - \omega_c = \gamma\,(1- \cos \varphi); \quad \gamma = (\omega_c\, Cs\, Z_0\, \pi)^{-1/2} \qquad \qquad 8)$$

So the coupling between the cells is inversely proportional to the C_s and consequently the width of the "conducting" band is taken as a measure for the coupling between the accelerator cells. The Brillouin diagram shows that the fundamental branch of the dispersion relation gives most properties of a

45

periodic structure. But the dispersion relation (equ. 7) for this lowest TM mode has an unlimited number of solutions which are called partial waves (space hormonics) and have the wave numbers resp. phase velocities:

$$k_i = k_o + 2\pi i/l \ ; \ v_i = \ \omega \ / \ (\ k_o + 2\pi i/l)$$

The solution for the fields in the structure is:

$$E \ (z) = \Sigma \ a_i \cdot e^{\ ik_i z} \qquad\qquad 9)$$

The coefficients have to be determined by numeric calculation solving for the specific boundary conditions. The sum over all partial wave than describes the real field also inside the cell. This solution satisfies Floquet's theorem, which gives solutions for periodic boundary conditions for functions:

$$E(z+l) = e^{-\beta l} \ E \ (z) \qquad\qquad 10)$$

The phase velocity v_w is given by $\tan\alpha = v_w/c$. The group velocity v_g is the slope of the dispersion curve for the operation frequency : $v_g = d\omega/dk$. v_g is determined by the aperture a : $v_g \sim (a/b)^4$. Typical values for v_g/c are between 0.1 and 5% for $v_w \approx c$.
The damping length L_o gives the decay of the field amplitude for traveling wave operation: $L_o = v_g \cdot Q/\omega$. This seems to indicate, that a high group velocity is advantageous, because than the accelerating field E_z has a high value along long sections and the number of feed points and transmitters is small.
But the power flow through the structure $E^2 \cdot v_g$ which would determine the size of the transmitters or the ratio of accelerating field to power flow are inversely prop. to the group velocity indicate that a small damping length resp. a small v_g are favourable. Therefore structures with high values for l_o and small v_g are usually standing wave structures. Typical examples are superconducting cavities. A lot more of these questions are discussed extensively in the book of Lapostolle and Septier [16] .
The group velocity resp. the coupling strength between cells is important in another aspect too, the field flatness problem.

FIELD STABILITY

Accelerators operating in π- or o -mode are sensitive against pertubations e.g. fabrication tolerances and temperature inhomogeneities. Using a π-mode chain of N cells a detuning - $\Delta\omega$ results in a phase shift $\Delta\varphi$: $\Delta\varphi = -\sqrt{2} \ \Delta\omega/\gamma$.

The resulting field modulation is proportional to the frequency shift, to the square of the number of cells n and inversely proportional to the coupling strength resp. group velocity :

$$E(n) = \cos (\ \omega t - n\pi)(1 - \Delta\omega/\gamma \cdot n^2) \ ; \ \Delta E/E \sim n^2/\gamma \ \ 11)$$

It corresponds to a concave field distribution in case of a pertubation in the center of the structure and is described also by a contribution of the lower (sine shaped) mode, which is close by for π-mode operation and a large number of coupled cells. So the required field stability sets limits to fabrication tolerances, temperature control and to the number of cells used as subunits. The number of cells per length is proportional to the frequency ($n \sim \omega$), setting another limit to shuntimpedance increase. The field unflatness (Tilt sensitivity against pertubation) can be improved by choosing another operational mode with generally lower shuntimpedance. A good example is the $\pi/2$ mode with linear dependance of φ and $\Delta\omega$. The detuning sensitivity $E \sim \cos (\ \omega t - n\pi/2) \cdot \Delta\omega/\gamma \cdot n$ is smaller compared to π- mode operation and corresponds to an improvement in effective coupling strength from γ to $\sqrt{\gamma}$ ($\gamma \ll 1$).
This property is used for structure "compensation" [17,18], which is important for coupled cavity as well as for Alvarez and even RFQ operation, converting the

operating of o- or π-mode into a π/2 mode [19]. This is done by adding resonators in between the original chain that must have the same frequency but must not have the same shape or need not to be on the beam axis. The same frequency of the coupling resonator fullfills the electrical periodicity requirement.

In π/2 mode the field (standing wave) varies along the accelerator like $E(n) \sim \cos(\omega t)\cos(\pi/2 \cdot n)$ so each second cells is free of (resonant) fields. By choosing proper boundary conditions the additional "coupling cells" can be made field free and do not introduce additional losses. Examples are the side-, the on axis coupled structure, the disk and washer structure shown in fig. 8 and the post coupled Alvarez and the ring-coupled RFQ.

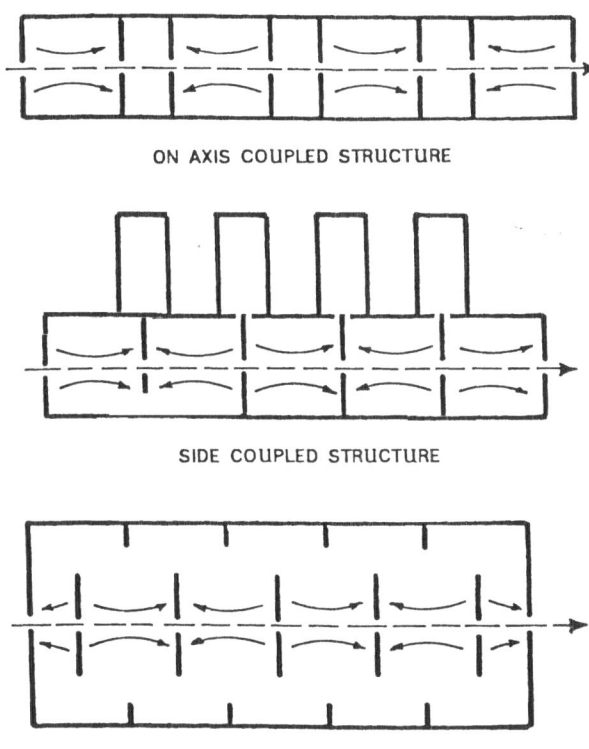

ON AXIS COUPLED STRUCTURE

SIDE COUPLED STRUCTURE

DISK AND WASHER STRUCTURE

Fig. 8. Examples of compensated structures

HIGH CURRENT ACCELERATORS

High current operation adds new constraints to the design. While the optimization for maximum particle energy at limited cost and space leads to relatively short, low duty cycle machines with pulse power to average power ratios of up to 10^5. The average beam current then is very low. Improving this situation the charges per beam pulse will be increased until space charge effects set a new limit.

Another effect which forces the use of lower frequencies is the amount of energy taken by the bunch in relation to the stored energy in the accelerator structure. This ratio goes up with frequency $E^2/W \sim \omega^{1/2}$, so that at higher frequencies the first bunch takes out too much of the stored energy that the following ones get significantly less acceleration. These beam loading and wake field effects have firstly been seen and studied in electron machines especially at SLAC, but their control is essential for the new linear collider machines [14]. These effects are applied in a new accelerator scheme, which uses the fields produced by mirror currents to concentrate fields even more (wake field accelerator [20]). Particles lose energy by this effect, by which also higher order modes will be excited. These modes can be TM as well as deflecting TE or HOM modes. The effects are not as severe as in circular machines where an additional periodic excitation will occur but the bunches following the lead bunch will be affected. This is especially complicated, if the Brillouin diagram is complicated and branches are overlapping. The design of "single mode" cavities, where the excitation of higher modes is minimized is fighting this limitation [21].

Increasing the duty cycle would increase the current proportionally but the particle energy would go down due to the limitations of rf - power sources until the cooling capacity of the accelerator structure sets another limit.

For ion accelerators the problems with power sources are similar but the main limit is set by the current "transport" capability of the accelerator resp. the focusing system. The accelerator in this sense is a pure transport channel and limited by focusing strength. The limiting currents for which the focusing is nearly canceled by the space charge of the bunch (tune depression of 60%) are strongly dependent on the particle velocity [22]:

$$I_r \sim q/m \; \beta^3 \gamma \; N^2 B^2 \omega^{-2} \qquad 12) \; \text{N focusing periodicity, B focusing strength}$$

$$I_l \sim q/m \; \beta \; R \; \varphi_s^3 \qquad 13) \; \varphi_s \text{ stable acc. phase, R beam radius}$$

The "longitudinal" and "radial" currents indicate that the corresponding focusing is the limit for transportation. All contributing parameters are not as separated as Equ. 12, 13 indicate and limiting instabilities are not included as well. As is well known, the frequency has to be lower for low ß particles especially for heavy ions.

Table I. Examples for linac parameters

Type particle	UNILAC Wid./Alvarez	CERN-Inj. Alvarez	LAMPF Alv./Side-CS	SLAC Iris
particle	P - U	P	P, H⁻	e⁺ e⁻
Length (m)	100	30	800	3100
T_f (MeV)	20 MeV/N	50	800	50.000
f (MHz)	27/108	202	201/805	2860
P_{rf} (MW)	11.5	4	66	500
DC (%)	25	.1	6	.1
I_{av} (μA)	10	1	1200	10

The limits imposed by quadrupole gradients and cell length (the product is prop. to the focusing phase advance) sets the bottleneck for Alvarez injectors at their first cells. Higher starting energy for the Alvarez linac and high currents at the same time was not possible until RFQs started to replace static injectors. The RFQ uses strong electric rf focusing which is velocity independent, so that the current limit is only prop. to β instead of β^3.

RFQs can efficiently ($\eta \sim (v/c)^{-2}$) accelerate up to $\sim$ 2 MeV (MeV/N) so the high current Alvarez acceleration (higher ß at the entrance) will be improved. RFQs are also successfully used as heavy ion injectors and now even high current heavy ion injectors (appr. 20 mA of U^{2+} ions) are being built. An example for a system with extremely high ion currents is the injector linac "tree" for HIBALL[23] an heavy ion fusion study. Fig. 9 shows that, starting with eight RFQs at 13.5 MHz (20 mA U^{2+} each), the beams will be funneled and the frequency doubled multipally to use the full current transport capability of all stages. One prototype of such an injector is being built at GSI and will be used as high current injector (Maxilac [24]) for the heavy ion synchrotron SIS.

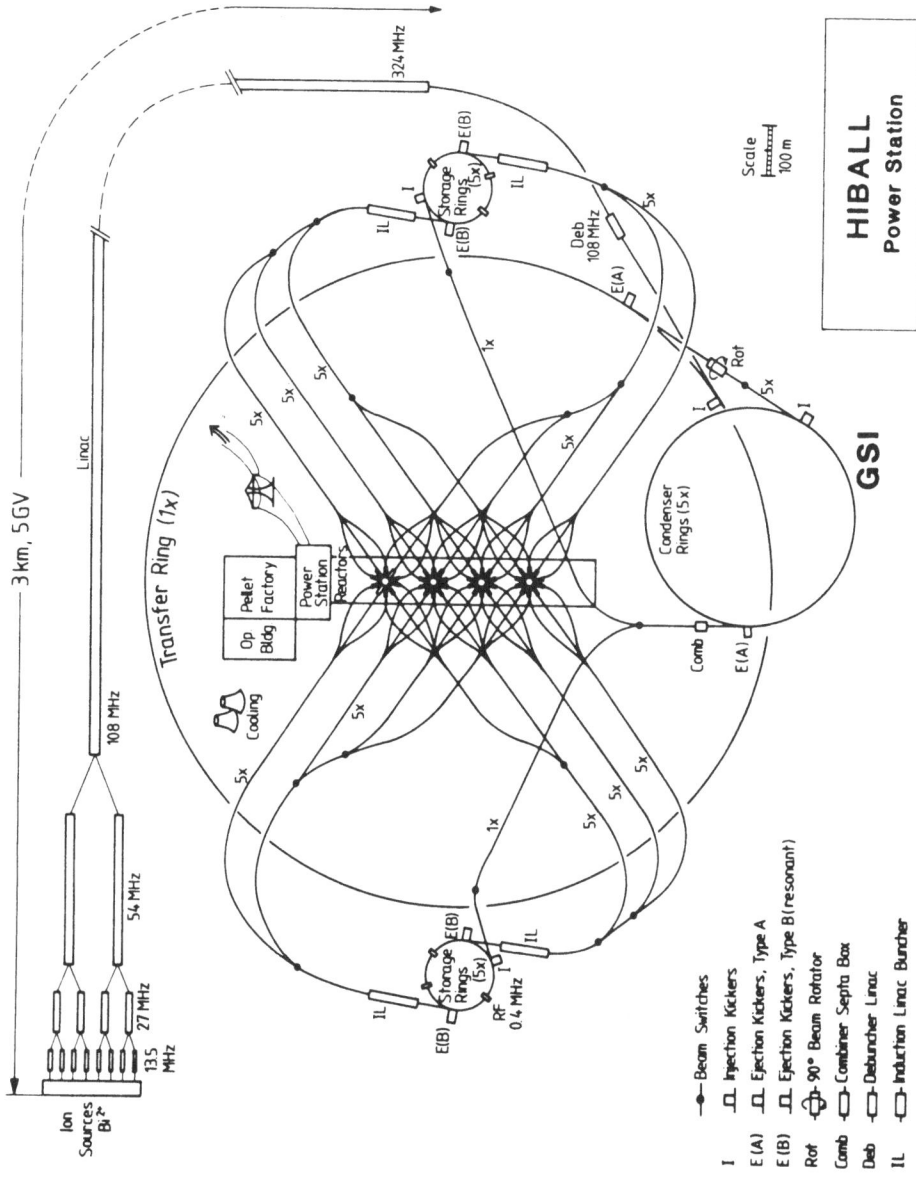

Fig. 9. HIBALL heavy ion fusion power plant design[23]

Table I shows data for typical linacs. The SLAC e- linac which is now extended to the first linear collider for 50 GeV e^+ e^-, the CERN Injector Linac-II as high current proton injector for a synchrotron, LAMPF as high current proton linac with a average beam power of appr. 1 MW and the UNILAC of the GSI as most powerful heavy ion linac.

REFERENCES

1 R. Wideröe, Archiv für Elektrotechnik, 21, (1928) 387
2 E.L. Chu, W.W. Hansen, Journ. Appl. Phys. 18 (1947) 996
3 L. Alvarez et al, Rev. Sci. Instr., 26 (1955) 111
4 M. Bres et al., IEEE Trans. Nucl. Sci., NS-16, 3 (1969) 372
5 C. Schmelzer, Linac 68, BNL 50120 (1968) 335
6 E. Jaeschke, Linac 84, GSI 84-11 (1984) 24
7 A. Schempp et al., Linac 79, Brookhaven Nat. Lab., BNL-51134 (1980) 159
8 I. M. Kapchinskiy, V. A. Teplyakov, Prib. Tekh. Eksp., No. 4 (1970) 17, 19
9 H. Klein, IEEE Trans. Nucl. Sci., NS-30, No. 4 (1983) 3313
10 A. Schempp, Eloisatron-Workshop, Erice, Sicily 1986
11 J. J. Manca, E.A. Knapp, Los Alamos Nat. Lab., LA-7323 (1978)
12 H. Piel, this workshop
13 T. Weis et al., IEEE Trans. Nucl. Sci., NS-30, No. 4 (1983) 3548
14 W. Schnell, CERN-LEP-RF/86-06
15 W. D. Kilpatrick, Univ. of California, Berkeley, UCRL-2321 (1953)
16 Lapostolle, Septier, "Linear Accelerators", North Holland Publ.Co., 1969
17 E. Knapp et al, RSI 39, (1969) 979
18 G. Dome , in [16]
19 A. Schempp, Linac 84, GSI 84-11 (1984) 338
20 G. Voss, Linac 84, GSI 84-11 (1984) 250
21 T. Weiland, IEEE NS-32,No.5, (1985) 2738
22 T. Wangler, LANL 8388, 1980
23 HIBALL, KfK 3202, Karlsruhe (1981)
24 R. W. Müller et al, Linac 84, GSI 84-11 (1984) 77

RF DRIVEN LINEAR COLLIDERS

W. Schnell

LEP Division
CERN
1211 Geneva 23
Switzerland

INTRODUCTION

The main problem encountered in our attempts to produce conceptual designs for e⁺e⁻ colliders in the TeV range is the very large luminosity required, which should increase with the square of energy and should exceed 10^{33} cm^{-2}s^{-1} - at least - for 2 TeV in the centre of mass. Making the main linacs superconducting would be ideal, since the high efficiency of transferring power to the beam would permit very high beam power. Unfortunately, the limited accelerating gradients attainable with superconducting cavities - and also the high cryogenic losses associated with really high gradients if these were attainable - would lead to a total length in the 100 km range if not beyond, considered excessive by most. Some of the advantages of RF superconductivity may, however, be maintained by using a superconducting drive linac in a two-beam scheme, as discussed below.

With normal-conducting radio-frequency structures accelerating gradients of several hundred megavolts per metre are possible in principle. In practice, maximum attainable gradients are given by considerations of efficiency and limitations of peak power rather than by electrical breakdown. Another fundamental problem is presented by self-deflection and self-deceleration due to the electromagnetic wakefields left behind by the particles.

Travelling wave structures offer the important advantage of presenting a matched load to a short pulse of RF power at a single feed point per section. Thus we are led once again to base ourselves on the well-known principles of the travelling-wave electron linac. The parameters and the technology required will, however, have to be extrapolated far beyond the present state.

1 **Normal conducting linacs, definitions**

It is now well established that copper structures, operated at the high microwave frequencies which are mandatory anyhow for minimizing stored energy, can support accelerating gradients of many hundred MV/m. Therefore, practicable accelerating gradients are determined by available peak power and consideration of average power efficiency rather than break-down. Most of the discussion presented in this and the subsequent sections will, therefore, be concerned with peak and average power. The following simple model will be used for this discussion.

The linac is made of travelling wave sections of section length L, obtained by coupling resonators to each other so as to obtain the phase velocity c at the desired operating frequency $f = \omega/2\pi$. Each section is energized by a square power pulse of duration τ and peak power $\hat{P}_L$. This makes a wave front propagate in the section with group velocity v. The parameters are chosen so as to make τ the group delay or "fill time" of the structure which means that

$$\tau v = L \quad . \tag{1}$$

The power is shut off when the wave front has reached the end of the section. At this moment the bunch to be accelerated (or the last one of a train of bunches) is made to pass. For simplicity $v/c \ll 1$ will be assumed (although it may easily reach 10% in practice) so that the beam traverses a section in a time which is short compared with τ and the dilution of the wave front due to dispersion will be neglected. Constant group velocity (i.e. a "constant impedance structure") is assumed unless stated otherwise. Basic structure constants are the quality factor Q (the e-folding decay time of stored energy measured in RF radians) and the "R over Q per unit length" defined as

$$r' = \frac{R'}{Q} = \frac{E^2(z)}{\omega W'(z)} \tag{2}$$

where E(z) is the accelerating field and W'(z) the stored energy per unit length at location z within a section. (Instead of r' the "fundamental frequency loss factor" $k_0 = \omega r'/4$ is often used.) The shunt impedance per unit length R' is defined by

$$R' = \frac{E^2(z)}{P'_D(z)} \tag{3}$$

where $P'_D(z)$ is the power dissipation per unit length.

Finally a dimensionless attenuation constant α for **stored energy** may be conveniently defined by

$$\alpha = \frac{\omega L}{Qv} = \frac{\omega}{Q}\tau \quad . \tag{4}$$

If a given structure geometry is scaled to different wavelengths,

$$Q \propto \omega^{-\frac{1}{2}}, \quad R' \propto \omega^2, \quad r' \propto \omega \text{ and } \tau \propto \omega^{-\frac{3}{2}} \quad .$$

The travelling wave linac assumed throughout this paper has the great advantage of offering a matched load at a single feed point per section. At very short wavelengths (well below 1 cm say) individually fed resonators might be necessary. Such arrangements which rely on a distributed power source and coupling device and entail power reflection during the fill time are outside the scope of this discussion.

2 Normal conducting linacs, basic parameters

Clearly the peak input power $\hat{P}_L$ per section must exceed $E_\Sigma^2 L/R'$ for an average accelerating gradient E_Σ. As this power is very high, a normal conducting electron linac is always pulsed with a very small duty cycle. On the other hand, as the Q-factor is much too small to conserve any appreciable fraction of stored energy over a realistic repetition period one is led to assume that the average RF input is roughly given by

$$\frac{\langle P_b \rangle}{\eta} \tag{5}$$

where $\langle P_b \rangle$ is the average beam power and η the fraction of stored energy extracted by the beam pulse. The beam power is given by

$$\langle P_b \rangle = bNf_r eU \tag{6}$$

and determines luminosity, b being the number of bunches per pulse, N the bunch population, f_r the repetition frequency and eU the final energy. The fraction of energy extracted

$$\eta = \frac{bNe\omega r'}{E_0} \tag{7}$$

[from equation (2)] cannot be large, certainly not above 10%, if the beam's energy spread - about $\eta/2$ - should remain acceptable or at least correctible. Note that $\eta \propto \omega^2/E_0$ (because of $r' \propto \omega$) and that equation (7) contains only basic design parameters which have to be chosen just right in order to achieve the desired compromise between efficiency and energy spread. The requirement of making η approach 10% for good efficiency leads to the choice of very high frequencies, much above the customary 3 GHz.

In reality dissipation **during** the fill time is by no means negligible. If this is included the correct expression for the peak input power $\hat{P}_L$ per section of length L is found as

$$\frac{\hat{P}_L}{L} = \frac{E_0^2}{g^2 \alpha R'} \tag{8}$$

where

$$g = \frac{1 - e^{-\alpha/2}}{\alpha/2} \tag{9}$$

On the other hand, by combining equations (6) to (9) the average RF power is found to be given by

$$\langle P_{RF} \rangle = \frac{\langle P_b \rangle}{g^2 \eta} \tag{10}$$

instead of expression (5). Thus, the RF to beam efficiency is ηg^2. Clearly a compromise has to be made between economy in peak power and average efficiency by a suitable choice of α (and hence $\tau = \alpha Q/\omega$). The classical choice is for the minimum of peak power occurring at $\alpha = 2.5$ and $[g^{-2}\alpha^{-1}]_{min} = 1.23$. But this implies $g^{-2} = 3.1$ and, hence, an intolerable wastage of average power in the context of a large collider. Since average power is of basic importance here a smaller value of α must be chosen, in spite of the concomitant increase of peak power. For all numerical examples in this paper $\alpha = 0.5$ (hence $g^{-2} = 1.28$) will be chosen, implying that the peak power per section length is 2.56 times E_0^2/R', which is twice the minimum. The RF to beam efficiency is, thus, 0.78 η.

For the small values of attenuation constant proposed here the use of a constant gradient (graded group velocity) changes very little. In equation (8) the factor $g^2\alpha$ in the denominator is replaced by α_0, the attenuation constant at the input. In equation (10) the factor g^2 in the denominator is replaced by $\alpha_0/\ln(1-\alpha_0)^{-1}$. But making $P_b/\langle P_{RF}\rangle = 0.78\eta$ as before (by choosing $\alpha_0 = 0.4$) makes $P_L R'/E_0^2 L = 2.50$, only 2.4% below the value of 2.56 found above for a constant impedance structure.

In Table 1 two examples of basic parameters are given. In both cases the frequency is 29 GHz, the beam power 5 MW, the top energy (per linac) 1 TeV and the bunch population $N = 5.4 \times 10^9$ giving about 10^{33} cm^{-2}s^{-1} luminosity with $\sigma_r^* = 65$ nm beam radius and the rather modest pinch enhancement of 3.5. As the energy extraction per beam pulse is taken as 8% the average RF input power per linac equals 80 MW in both cases, i.e. the RF to beam efficiency is 6.25%. In case A the combination of gradient, frequency and bunch population leads to the desired energy extraction with a single bunch per pulse – the safest choice possible. In case B the gradient has been doubled to 160 MV/m, requiring a fourfold increase of peak power per unit length $\hat{P}/L$. Moreover, since an increase of frequency does not seem advisable because of the wake field problem discussed under 8, two bunches per beam pulse have to be used in order to obtain the quoted efficiency.

TABLE 1

Main linac parameters for two accelerating gradients.
Parameters for one linac.

Case	A	B	
Final energy eU	1	1	TeV
Frequency f	29	29	GHz
Average accelerating gradient E_0	80	160	MV/m
Total active length L_{tot}	12.5	6.25	km
Shunt impedance per unit length R'	170	170	MΩ/m
Quality factor Q	4150	4150	
R'/Q = r'	41	41	kΩ/m
Attenuation constant for power α	0.5	0.5	
Fill time τ	11.4	11.4	ns
Peak power per unit length $\hat{P}_L/L$	96	384	MW/m
Bunch population N	5.35	5.35	$\times 10^9$
Energy extraction per pulse η	0.08	0.08	
Number of bunches per pulse	1	2	
Repetition rate f_{rev}	5.8	5.8	kHz
Average RF power $\langle P_{RF}\rangle$	80	80	MW
Beam power $\langle P_b\rangle$	5	5	MW
Beam radius at collision σ_r^*	65	65	nm
Disruption D	0.91	0.91	
Pinch enhancement H	3.5	3.5	
Beam-beam radiation loss δ	0.19	0.19	
Bunch length σ_z	0.3	0.3	mm
Luminosity	1.1	1.1	$\times 10^{33}$cm^{-2}s^{-1}
Fractional average critical energy T	0.28	0.28	
Normalized emittance $\varepsilon_n(\beta^* = 3$ mm)	2.8	2.8	$\times 10^{-6}$ rad m

A substantially lower value of frequency - at or below 10 GHz, say - would greatly alleviate the wake field problem and facilitate the construction of the accelerating structure at acceptable tolerances.   Also, the more conventional solution of multiple d.c. to RF power converters would be favoured (while the auxiliary beam system actually proposed below would be rendered difficult).   However, as η is proportional to ω^2/E_0 the common effort of many successive bunches (e.g. 8 at 10 GHz) would be required already at 80 MV/m to achieve 8% extraction.   This is not a good solution since it is not clear whether transverse wake fields and the final focus permit multiple bunches.   And if such multibunching were to become possible it would be very desirable to increase the total energy extraction per pulse considerably beyond 8% by means of the bunch-to-bunch compensation of energy errors described in section 3.   One is led to conclude that the requirements of acceptable RF to beam efficiency and an accelerating gradient at least in the neighbourhood of 100 MV/m lead to the necessity of short wavelength - about 1 cm - and that the possibility of using several bunches per pulse is desirable.

3   High efficiency operation

The 6.3% RF to beam efficiency postulated in Table 1 is already above present-day routine.   It is dependent on the achievement of high energy-extraction and short fill time.   A further increase of efficiency might be obtained by one of the two following methods.

Firstly, the choice of $\alpha = 0.5$ implies that in the absence of beam loading 60% of the input RF energy reappears at the output of the acceleration section.   With $\eta \sim 10\%$ about 50% will be left so that a factor two in efficiency could be gained if it were possible to recover this leftover energy and store it in a suitable way.   The two-stage RF scheme described in section 7 holds the promise of doing just this without much extra expense.   A more remote possibility might be a rectifying load.   Conversion from microwave power to d.c. was in fact achieved [1] a decade ago, albeit with a continuous wave at the level of tens of kilowatts, with about 80% efficiency.

Secondly, a more substantial increase of efficiency may be obtained by employing a train of bunches distributed over a certain fraction of the fill time τ.   The charge of each bunch is limited by the maximum value of η and concomitant energy spread, but the bunch interval is adjusted so that the fresh influx of RF energy restores the average accelerating field from bunch to bunch [2,3].   In order to operate a given linac in this way one may pass the first bunch at time $\chi\tau$ when the propagating wavefront has filled the fraction χ of the accelerating structure.   Clearly, this reduces the final energy by the same fraction (or necessitates lengthening of the entire installation by χ^{-1}) and $\chi = 0.8$ (say) may be an acceptable choice.   It turns out that $\tau_b = \eta\tau/2$ is the right bunch interval for first order compensation of the bunch to bunch energy difference.   The last bunch is passed at time τ when the wave front has reached the end and this makes the number of bunches b equal to $1 + 2(1-\chi)/\eta$.   For $\chi = 0.8$ and $\eta = 0.08$, b = 6.

Table 2 shows the two linacs of Table 1 extended by 25% in length and operated in compensated multibunch mode.   Case B illustrates the flexibility of this scheme in accommodating higher accelerating gradients without increase of operating frequency or decrease of RF to beam efficiency which is as high as 30% in both cases listed.

Two classes of fundamental problem must, however, be solved before compensated multibunching can be attempted.   Firstly, a final focus scheme

TABLE 2
Table 1 modified for compensated multibunch operation
Parameters for one linac

Case	A	B	
Final energy eU	1	1	TeV
Frequency f	29	29	GHz
Accelerating field E_0	80	160	MV/m
Filling factor for first bunch χ	0.8	0.8	
Average accelerating gradient χE_0	64	128	MV/m
Total active length L_{tot}	15.6	7.8	km
Peak power per unit length $\hat{P}_L/L$	96	386	MW/m
Bunch population N	5.35×10^9	5.35×10^9	
Energy extraction η per bunch	0.08	0.04	
Energy spread within bunch	5%	2.5%	
Number of bunches per pulse b	6	11	
Repetition rate f_{rev}	5.8	3.2	kHz
Average RF power $\langle P_{RF} \rangle$	100	100	MW
Beam power $\langle P_b \rangle$	30	30	MW
Structure fill time τ	11.4	11.4	ns
Bunch interval τ_b (not adjusted for integer $\tau_b f$)	0.456	0.228	ns
RF cycles between bunches $\tau_b f$ approx.	13	7	
Beam pulse duration $(b-1)\tau_b$	2.28	2.28	ns
Luminosity	0.6×10^{34}	0.6×10^{34}	$cm^{-2} s^{-1}$

must be found that can cope with multiple bunch crossings starting at a few centimetres' distance from the main collision point.   Secondly, higher-order longitudinal wake fields must be minimized and their time dependence tuned in such a way as to make the bunch-to-bunch variation of effective accelerating field tolerable.   Modulations of input power, bunch population or bunch interval τ_b might be elements of freedom.   Contrary to the energy spread within the bunch the residual bunch-to-bunch energy variation is unlikely to be a monotonic function of time and, hence, difficult to correct.

4   Accelerating structures

Of the three structure constants defined above R' determines peak power, r' the energy extraction and Q the fill time.   Achieving the largest possible values is, therefore, important in all three cases.   The values listed in Tables 1 and 2 are obtained from scaling SLAC-type disc-loaded structures and are likely to represent the very best available in practice.   Disc-loaded structures for about 1 cm wavelength have, indeed, been manufactured and tested[4].   Unfortunately, if such a structure is directly scaled from the customary dimensions at 3 GHz the low group velo-city - of the order of v/c = 0.01 - leads to the inconveniently short sec-tion length of about 3 cm for the short fill time required here.   larger values of group velocity can be obtained, albeit at the cost of a moderate deterioration of R' and r', by increasing the coupling aperture and this would also be very effective in reducing the wake fields.

Structures with even higher group velocity can certainly be built. The Jungle Gym is the most extreme example though one that does not seem to lend itself well to the very small transverse dimensions (5 mm diameter) required here.   Several variants of related (bar loaded) structures are being studied at CERN[4] with a view to linear collider application but preliminary results indicate that good shunt impedance cannot be obtained for the large beam apertures which are required because of the transverse wake fields discussed in section 8.

At the average power levels discussed here rather intense water cooling will be required. It is likely that the entire accelerating structure will have to be surrounded by tightly fitting, permanent magnet quadrupoles in order to cope with transverse wake fields. The manufacturing tolerances of a 1 cm disc-loaded structure are in the micrometre range[5] and similar tolerances are likely to be required for transverse alignment of structures and quadrupoles.

5 RF power converters

With linac parameters such as those shown in Tables 1 and 2 a very serious problem is the generation of peak RF power. The conventional solution is to power each linac section - or small group of sections - by an individual d.c. to RF power converter, each one containing its own pulsed high-voltage input, cathode, electron-gun, RF structure and collector.

The main limitation is in the electron gun where d.c. power must be converted to kinetic energy of electrons whose initial velocity is very low. Clearly the peak output power cannot exceed the product of input voltage, cathode current density and cathode area. The voltage is limited by breakdown, 0.5 MV now being considered an ultimate limit in practice. The limit for current density is given by space charge. In steady state and for a narrow gap the current density is inescapably determined by "Child's Law" as

$$2.3 \times 10^{-6} \; V^{\frac{3}{2}} d^{-2}$$

where V is the voltage and d the gap width. Different geometry and the shortness of the pulse modify this but a basic limitation remains. The cathode dimension is related to the wavelength λ which means that the output power of a given design tends to scale with λ^2. Impressive quantitative progress is, however, being made using beam compression after the cathode, overmoded RF cavities to interact with the beam or, possibly, distributed RF structures.

The classical pulsed klystron has reached output powers[6] above 100 MW at 3 GHz but the λ^2 scaling does not make this look promising for an order of magnitude higher frequency. Laser driven optical cathodes will permit generation of the RF beam structure as well as the pulse envelope by suitable modulation of the laser, leading, in principle, to a very compact design not requiring an external modulator. Such lasertrons are being developed at SLAC, KEK and LAL Orsay. The expectation that optical cathodes can be applied to distributed RF structures has led to the concept of the Micro Lasertron[7]. For frequencies much above 3 GHz the most advanced solution at the present time is the Gyroklystron. The longitudinal bunching of the normal klystron is replaced by gyration of the electrons at the cyclotron frequency in a longitudinal magnetic field and consequential bunching in transverse phase space. The gyrating electrons interact with TE-mode cavities of large aperture.[8].

In the likely event that a given power source produces insufficient peak power but more than the necessary pulse length, pulse compression may be applied. The most promising scheme is Binary Pulse Compression[9]. Power sources are used in pairs and their outputs are combined in a hybrid junction possessing two inputs and two outputs. A phase reversal in the low power drive of one of the power sources halfway through the pulse switches the power from one output to the other so that power pulses of half the original duration appear successively at the two outputs. Delaying one of these pulses with respect to the other and recombining them in another hybrid junction creates the desired output of double power and half pulse length.

In spite of all these developments it does not appear easy to power a 1+1 TeV RF collider with individual d.c. to RF converters. Column A of Table 1 indicates a total peak power of 2.4 TW at 80 MV/m gradient, necessitating many thousands if not tens of thousands of power converters. And this number would **increase** in inverse proportion to the accelerator length if this were shortened by using a higher gradient. The auxiliary beam schemes described in the next two sections are inspired by these considerations.

6   Two-beam accelerators

Instead of the multitude of pulsed d.c. generators, cathodes and electron guns a continuous drive beam running along the main linac (or at least a good fraction of it) may be employed. The drive beam supplies energy to the main linac at regular intervals via transfer structures. The drive beam energy is restored, at the same or different intervals, by accelerating structures forming a "drive linac". Free electron lasers (FEL) and direct RF decelerating sections have been proposed as transfer structures, induction units and superconducting RF accelerating cavities as drive linacs. These possibilities will be discussed in this and the following section.

The first auxiliary beam scheme was proposed by Sessler[10,11] under the name of Two Beam Accelerator (TBA). This is also the scheme that has, so far, seen the most extensive experimental development. In the TBA the drive linac is formed by induction units, driven by spark gaps or, preferably, being saturable induction devices[12]. The transfer devices are tapered FEL wigglers delivering RF power at 1 cm wavelength to a continuous rectangular waveguide from which it is coupled to the main linac (consisting of disc loaded sections running at very high gradients) at regular intervals. Among the impressive results achieved so far are the attainment of well over one GW RF power at 35 GHz from a FEL section, the fabrication of 35 GHz disc loaded accelerating sections and corresponding feeder waveguides and the actual attainment of 180 MV/m accelerating gradient[12].

A basic feature of the FEL transfer structure is the limitation of drive beam energy to values below 100 MeV due to the fact that the generated wavelength is proportional to γ^{-2} times the wiggler wavelength. As the drive beam is only moderately relativistic the phase slip with respect to the main beam and the possibility of phase modulation causes problems. A variant of the TBA, the "Relativistic Klystron"[13] replaces the FEL units by decelerating cavities directly extracting energy by the action of longitudinal fields on a bunched drive beam. It would appear that maintaining the induction units as drive linac still limits the drive beam energy and the total drive linac voltage gain for economic reasons.

7   Two-stage RF schemes

The drive linac of an auxiliary beam system may be formed by RF cavities. Since efficiency is of paramount importance here and energy extraction is again limited by energy spread, the RF drive linac has to be super-conducting. This, however, is quite acceptable provided the drive linac's gradient and operating frequency is made sufficiently low. Since the superconducting drive linac is being run in CW regime d.c. to RF power conversion can be accomplished with large (1 to 2 MW per unit) high-efficiency klystrons of proven design. The beam repetition rate is determined by pre-injector considerations only and can be very high (above 5 kHz), as is indeed required for high luminosity. The possibility of energy recovery from the main linac is an additional advantage.

Energy transfer to the main linac may be via FEL units[14] or RF decelerating sections[15]. The former solution requires bunching of the drive beam at the (low) drive frequency only but limits the drive beam energy to below 100 MeV just as in the TBA case. The latter scheme requires the drive beam to be very tightly bunched at the main linac frequency. It opens, however, the possibility of GeV drive-beam energies, thus eliminating all phasing problems. Proper phasing of several tens of thousands of main linac sections is automatically assured by the highly relativistic drive beam, all adjustments of drive-power, phase and timing being carried out at the drive beam pre-injector.

In either case the basic scheme is that shown in Fig. 1. The mains input power is converted to RF power at UHF frequency by means of large CW klystrons and distributed via low-cost sheet metal waveguides at atmospheric pressure. The klystrons deliver power to a series of superconducting cavities very similar to those developed for circular e^+e^- colliders at CERN[16] and elsewhere. Drive beam pulses of a duration equal to the main linac fill time τ have their energy periodically restored by passing through this superconducting drive linac.

Energy conservation along the drive beam demands that the "transformer ratio", i.e. the ratio of the accelerating gradient, E_0, in the main linac to that, E_1, in the drive linac, be given by

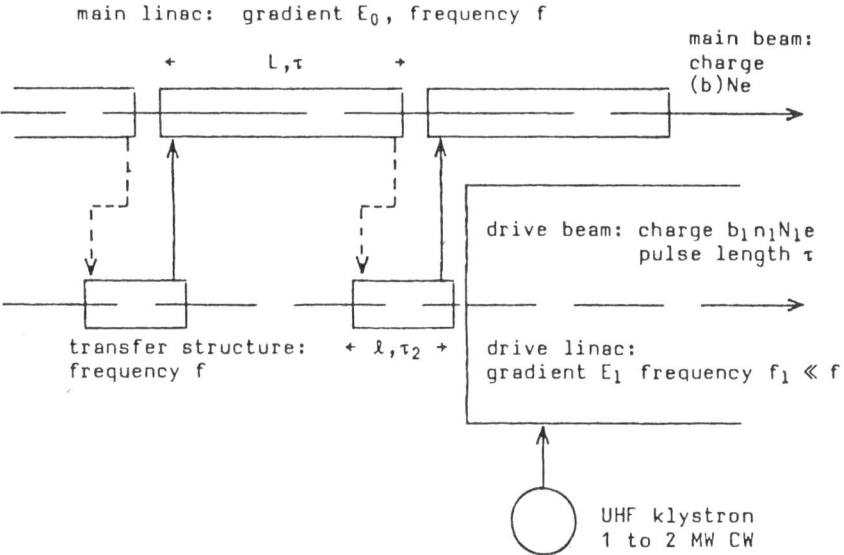

Fig. 1. Two-stage linear accelerator composed of a superconducting CW drive linac at UHF frequency and a microwave main linac.

$$\frac{E_0}{E_1}^2 = \eta_1 \eta_2 mg^2 \frac{\omega r'}{\omega_1 r_1'} \propto \frac{\omega^2}{\omega_1^2} \tag{11}$$

where η_1 is the energy extraction in the drive linac, η_2 the transfer efficiency, m the fraction of main linac length occupied by drive sections (an economic choice), $\omega_1/2\pi = f_1$ the drive frequency and r_1' the drive linac R over Q per unit length. Introducing the superconducting cavities' quality factor Q_1 one finds

$$\langle P_1 \rangle = \frac{E_0 U}{\omega r' g^2} \frac{\omega_1}{Q_1 \eta_1 \eta_2} \tag{12}$$

for the drive linac dissipation determining the input to the cryogenic system via the overall cryogenic efficiency η_{cr}. Note the absence of m or E_1 in equation (12) and the importance of keeping f_1 low. An interesting choice may be 350 MHz, the frequency of the superconducting cavities for the second stage of LEP. In fact, these cavities could be used at their present state of development without any change. Table 3 gives a few drive linac parameters.

TABLE 3

Case		A	B	C	
Main linac energy	eU	1	1	1	TeV
Main linac frequency	f	29	29	29	GHz
Main linac accelerating gradient	E_0	80	160	445	MV/m
Main linac active length	L_{tot}	12.5	6.25	2.24	km
Drive linac voltage gain	U_1	15	12	33.6	GV
Drive linac frequency	f_1	350	350	350	MHz
Drive linac R over Q parameter	r_1'	270	270	270	Ω/m
Drive linac accelerating gradient	E_1	6	15	15	MV/m
Drive linac active length	mL_{tot}	2.5	0.8	2.24	km
Drive linac quality factor	Q_1	5×10^9	5×10^9	5×10^9	
Cryogenic input power ($\eta_{cr} = 0.2\%$)	$\langle P_1 \rangle/\eta_{cr}$	33	67	186	MW

The first column is for the main linac of case A of Table 1. The corresponding drive linac parameters, $E_1 = 6$ MV/m and $Q_1 = 5\times10^9$ at 350 MHz are present-day performances[16]. The second and third columns show the impressive improvement which can, in principle, be expected from an increase of E_1 to 15 MV/m, a development that is, in fact, likely to occur within a few years' time. Clearly case C is an extreme example but one that illustrates the ultimate potential of this method of peak power generation. Multiple bunches are definitely required in the main linac in this case. Since the drive linac and main linac have the same length two drive beams, each equipped with half the drive linac sections, must be made to run along either side of the main linac powering it alternately from the left and from the right.

If the energy transfer to the main linac is by direct deceleration in an RF transfer structure the drive beam energy can be raised to a few GeV (3 to 5 GeV, say) by a modest addition to the superconducting drive linac. This would assure rigid drive bunches and the absence of any phase slip.

The required drive charge, unfortunately, is rather large and it has to be bunched. To be synchronous with the drive wave there must be $n_1 = \tau f_1$ bunch trains ($n_1 = 4$ for the parameters chosen). To be synchronous with the main linac each of the n_1 trains must consist of b bunchlets at f^{-1} interval. And in order to match the build-up of voltage in the transfer structure the b_1 bunchlets must coincide with the rising slope of the sinusoidal drive wave as shown in Fig. 2. The charge per drive bunch is given by

$$eN_1 = \frac{4\pi^2}{1-\cos\phi} \frac{E_0 U}{\alpha g^2 \omega R' \eta_2 U_1} \qquad (13)$$

where $\phi/2\pi = b_1\omega_1/\omega$ is the fraction of the drive wave over which the b_1 drive bunches are distributed.

For the parameters of the first column of Table 3 and $\phi = \pi/2$, $b_1 = 10$ and each bunchlet has to contain a population $N_1 = 4\times10^{11}$ (within about 1 mm bunch length). Phase modulation of the bunch train[17] permits matching of the energies received from the drive linac and delivered to the transfer structure over as much as half a drive period. This brings about a reduction of the population to 6×10^{10} but necessitates a larger number ($b_1 = 40$) of shorter bunchlets. In either case the generation and acceleration to relativistic energies of the very dense and multiple drive bunches appears to be the main difficulty with this scheme, a difficulty, however, that is confined to the injector.

The transfer structures can take the form of short travelling wave sections, each one coupled to the input of a corresponding main section via a short run of waveguide. The group delay of a transfer section τ_2 should approximately equal f_1^{-1} for continuous power flow. The required R' over Q of the transfer structure is given by

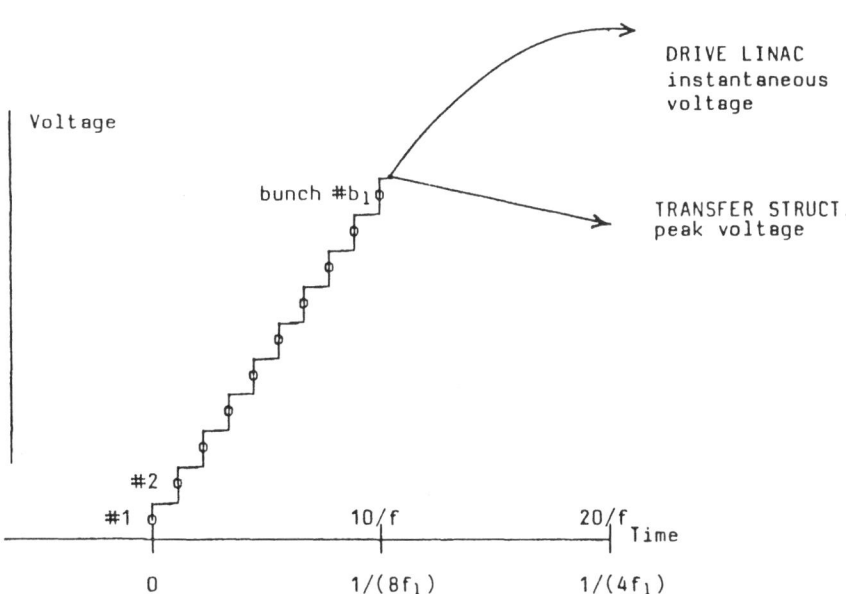

Fig. 2. Matching of transfer voltage in a two-stage RF scheme with RF transfer structure.

$$r'_2 = \frac{vr'}{v_2}\left(2g\eta_2 n_1 \frac{U_1}{U} \sin \frac{\phi}{2}\right)^2 \qquad (14)$$

where v_2 is the group velocity of the transfer structure. The required impedance, r', of the transfer structure turns out to be very low, less than 1% of the main linac r' for the parameters chosen. This is fortunate since it will permit the design of a structure with large enough aperture to cope with the longitudinal and transverse wakefields due to the intense drive beam. Figure 3 shows the scaled-up model of a transfer structure which is being studied at CERN[18].

Connecting the output port of each main accelerating section to an input of the following transfer section permits recovery of the leftover stored energy after the beam passage and, hence, a gain by a factor of two in power economy as pointed out in section 5. To do this, a recovery pulse has to follow the drive pulse. It has to be phased for acceleration in the transfer structure, deceleration in the drive linac and a reversal of the matching process.

8 Wake fields

Each bunch induces longitudinal and transverse-deflecting wake fields as it passes through the accelerating structure. The wakes left behind by leading particles act on the trailing part of the same bunch. Longitudinal wakes lead to energy loss and energy spread. Dipole wakes may amplify the amplitude of accidental transverse oscillations (due to misalignment of accelerating structures or quadrupoles) so as to cause severe emittance blow-up or even beam loss.

For axially symmetric structures fairly accurate computations of longitudinal and dipole wakes are available[2,19]. They are based on the superposition of many modes of cavity resonance completed by extrapolation to very short wavelength with the help of an optical resonator model. The

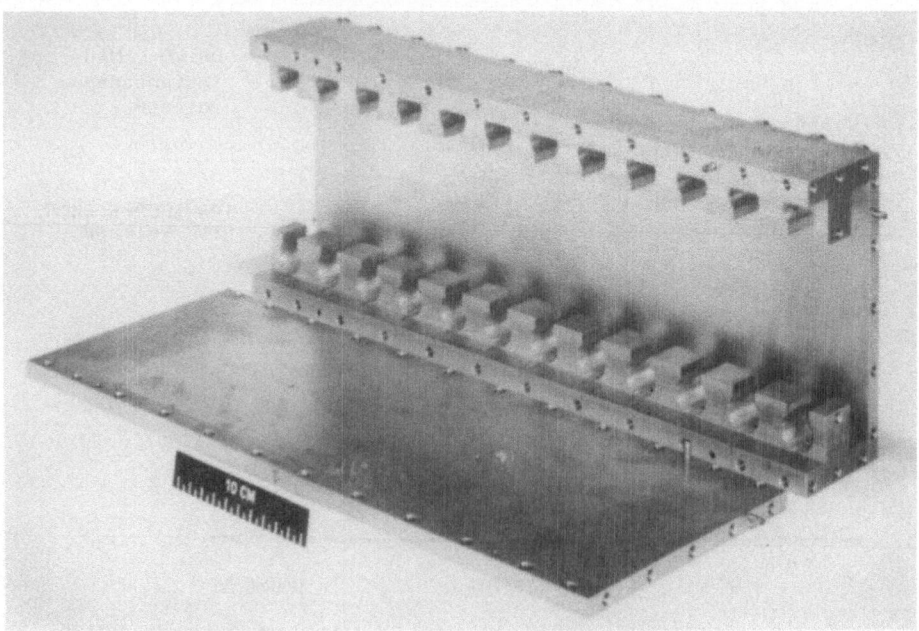

Fig. 3.   Scaled-up model of transfer structure

longitudinal point charge wake (in V/As of energy loss per cell and unit of charge) starts with a very high value and comes down steeply with increasing distance (given in ps) from the inducing charge. The dipole point charge wake (in V/Asm of transverse energy per cell and unit dipole moment of excitation) rises from zero to a maximum a few ps later. Folding these curves with a Gaussian charge distribution and integrating one obtains the longitudinal and transverse wake potentials, $W_\parallel(s)$ and $W_\perp(s)$ at position s within the bunch.

In the absence of any spread in transverse oscillation frequency within the bunch the action of transverse wakes is downright fatal as any accidental oscillation at the entry of a long linac is amplified by an enormous factor (typically 10^7).

The situation can, however, be rescued by the introduction of an energy spread within the bunch (or by tolerating the energy spread due to longitudinal wakes) so as to create a spread of transverse oscillation frequencies via the energy dependence of focussing[20,21]. If the combination of energy spread and focussing strength is below a certain critical value the wake-induced blow-up of an initial oscillation is only reduced - helpful but not sufficient in practice. If, however, energy spread and focussing strength are strong enough to fulfill a criterion[20-22] given by

$$eN\dot{W}_\perp \sigma_z = 2\xi \, dU/\lambda_\beta^2 \qquad (15)$$

the wake fields provide damping at the centre of the bunch and any blow-up is avoided. In equation (15) $\dot{W}_\perp$ is the slope of the wake field within the bunch (it is approximately constant within $\pm\sigma_z$), λ_β is the wavelength of transverse oscillations divided by 2π and d is the fractional energy deviation (assumed linear over the bunch) at σ_z. The particle energy is eU and $-\xi$ is the chromaticity of the focussing system, near unity in all practical cases. Note that it is U/λ^2 that matters. This means that the criterion is simultaneously fulfilled all along the linac if λ is made to grow with $\sqrt{U}$ by maintaining constant quadrupole strength. For $\dot{W}_\perp = 4.8 \times 10^{21}$ VA^{-1}s^{-1}m^{-3} (transverse energy gain per dipole moment, per running metre of linac, per distance s behind the bunch head) N = 5.35×10^9, ξ = 1.27, σ_z = 0.3 mm and d = 1.3% one finds U/λ_β^2 = 19 GVm^{-2} corresponding to a focussing wavelength $2\pi\lambda$ = 3.2 m at 5 GeV. This means very strong focussing but is probably just feasible by means of small-aperture permanent-magnet quadrupoles surrounding the accelerator structure. The assumed value of $\dot{W}_\perp$ corresponds to a scaled SLAC structure with 2.3 mm holes. In reality somewhat larger holes may be necessary for acceptable group velocity (cf. section 4) making the situation more favourable at the expense of some loss in shunt impedance. Extensive computations[19] carried out at CERN and computations[23] carried out for TBA parameters essentially confirm equation (13).

The very strong focussing required to stabilize the wake fields leads to extremely tight alignment tolerances for the quadrupoles and the large momentum spread is deleterious for the final focus system. Two ways of alleviating these problems are being studied at present.

The first method[24] consists of combining a beam where emittance is much larger horizontally (say) than vertically with an RF structure whose apertures have the form of vertical slits. In the horizontal direction the wake fields necessitate very strong focussing but the large emittance makes the tolerances for quadrupole alignment and jitter manageable. In the vertical direction the emittance is small but wake fields are weak and the focussing can be weak also (i.e. the amplitude function is larger vertically than horizontally).

The second proposed method[25] also relies on RF structures with slits as apertures but for a different purpose. The asymmetric apertures give the electromagnetic fields a strong transverse quadrupole component. By orienting the slits alternately vertically and horizontally an RF FODO structure can be created, alone or in conjunction with conventional quadrupoles. Among the advantages of such schemes is the possibility of creating a large spread in transverse oscillation frequencies – as required for stabilizing transverse wake fields – without much energy spread and without excessively small values of the transverse wavelength itself. Pure RF focussing also permits the particle energy to be changed over a large range simply by varying peak power, the quadrupole strength adapting itself automatically. With permanent magnet quadrupoles the requirement of covering a wide range of final energy poses a problem.

References

1 W.C. Brown NASA CR-135194 (August 1977).
2 P. Wilson, SLAC-PUB-3674 (May 1985).
3 P. Wilson, SLAC-PUB-3985 (May 1986).
4 J.-P. Boiteux et al., CLIC Note 36.
5 D.B. Hopkins, R.W. Kuenning, IEEE NS-32-5, p. 3476 (October 1985).
6 T.G. Lee et al. SLAC-PUB-3619 (April 1985).
7 R. Palmer, Seminar on New Techniques for Future Accelerators, Erice (1986).
8 V.L. Granatstein et al., IEEE NS-32/5, p. 2957 (October 1985).
   V.L. Granatstein, A. Mondelli, Physics of High Energy Accelerators (SLAC Summer School 1985).
9 Z.D. Farkas, SLAC-PUB-3694 (May 1985).
10 A.M. Sessler, Laser Acceleration of Particles AIP Conf. Proc. 91, 163 (1982).
11 A.M. Sessler, D.B. Hopkins, LBL-21613.
12 D.L. Birx et al., IEEE NS-32-5, p. 2743 (October 1985).
13 R. Marks, LBL-20918/UC-34A (September 1985).
14 U. Amaldi, C. Pellegrini, CERN CLIC Note 16 (June 1986).
15 W. Schnell, CERN-LEP-RF/86-06 (February 1986). Also: CLIC Note 13.
16 Ph. Bernard et al., CERN/EF/RF/85-6 (November 1985).
17 W. Schnell, CERN-LEP-RF/86-14 (June 1986). Also: CLIC Note 21.
18 T. Garvey et al., CERN-LEP-RF/86-35 (November 1986). Also: CLIC Note 28.
19 H. Henke, 1987 Particle Accelerator Conference, March 1987, Washington D.C.
20 V.E. Balakin, A.V. Novokhatsky, V.P. Smirnov, 12th Internat. Conf. on High Energy Accelerators (Fermilab 1983), p. 119.
21 K.L.F. Bane, SLAC-PUB-3670 (1985).
22 H. Henke, W. Schnell, CERN-LEP-RF/86-18 (September 1986). Also: CLIC Note 22.
23 F. Selph, A. Sessler, LBL-20 083 (August 1985).
24 R. Palmer, Private communication.
25 W. Schnell, CERN-LEP-RF/87-24: also CLIC Note 34 (March 1987).

FREQUENCY SCALING AND GYROKLYSTRON SOURCES FOR

LINEAR COLLIDERS WITH SLAC-TYPE LINAC STRUCTURES*

M. Reiser[1], W. Lawson[2], A. Mondelli[3] and D. Chernin[3]

[1]Electrical Engineering Department and
[2]Laboratory for Plasma and Fusion Energy Studies
 University of Maryland, College Park, MD 20742 (USA)

[3]Science Applications International Corporation
 1710 Goodridge Drive, McLean, VA 22102 (USA)

ABSTRACT

This review of e^+e^- linear colliders is for the most part tutorial, but it will also present some new results that are summarized in Section 1 and discussed in more detail in Sections 4 and 5. The paper is restricted to colliders with SLAC-type linac structures. The scaling with frequency in the range from 8 to 20 GHz is studied for a 500 GeV on 500 GeV accelerator system of fixed length (2 × 3 km) and luminosity $L = 10^{33}$ cm^{-2}s^{-1}, as presently being considered at Stanford. Two scenarios, one which is limited by beamstrahlung where the number of particles per bunch, N, is approximately constant and one in which N is proportional to wavelength λ, are investigated. While the peak rf power requirement for each feed decreases with frequency identically in both cases, average ac power decreases with ω in the first case but increases in the second. Gyroklystron design parameters are presented that would meet the peak power requirements in X-band.

*Work supported by US Dept. of Energy Contract No. DE-AC05-84ER0216.
 Lecturers: Sections 1-4 M. Reiser; Section 5 W. Lawson.

1. INTRODUCTION AND OVERVIEW

Future high-energy accelerators are becoming too large and expensive to build unless new cost-saving technologies and acceleration techniques can be developed. Many panels, special meetings, workshops and summer schools have been concerned with this issue during recent years. The Superconducting Supercollider (SSC) that was just approved by the United States Government represents the largest scientific project in the world and will provide the research tool for $\bar{p}p$ particle physics in the last decade of this century and beyond.

With regard to e^+e^- colliders, the situation is fundamentally different. Due to the limits imposed by synchrotron radiation, storage rings become too expensive beyond about 100 GeV. The only viable option is to build linear colliders which do not have the excessive energy losses due to synchrotron radiation suffered by circular machines. The 50 GeV linear collider project (SLC) at SLAC is a pioneering effort designed to explore the path towards a future e^+e^- collider at much higher energy (~ 1 TeV). A considerable amount of research work at SLAC and other places has been carried out over the past years defining the physics and technology requirements. A major goal of this ongoing research is to find ways of reducing the size and cost of such a future supercollider.

Reducing the length of the accelerator implies that one needs to achieve higher accelerating gradients than in existing machines. The linear accelerator at Stanford, for instance, is approximately 3 km long. It operates at S-band frequency of 2856 MHz, and the improved klystrons powering the rf structures produce an average gradient of $E_a \approx 17$ MV/m with 60 MW peak power to achieve the 50 GeV of energy. Using the same technology for a 1 TeV supercollider would imply a total length of $2 \times 60 = 120$ km for the two opposing linacs. Since capital costs are roughly proportional to the length of the accelerator, such a design would undoubtedly be too expensive.

The gradients that can be achieved in SLAC-type linac structures are limited by voltage breakdown (in the linac as well as in the klystrons) and by the peak power capability of the klystrons. Both are functions of frequency which, unfortunately, go in opposite directions: the electric breakdown limit increases with frequency while the peak power of the klystrons decreases with frequency. The breakdown field E_b scales roughly as $E_b \sim \omega^{7/8}$, i.e., it rises almost linearly with the rf (circular) frequency $\omega = 2\pi\nu$. For a SLAC structure at $\nu = 2856$ MHz, the theoretical limit (known as the "Kilpatrick limit") is about 50 MV/m. Under ideal conditions a fivefold greater peak surface field strength has been achieved in experiments at SLAC. However, in practice, one wants to remain safely below the breakdown threshold, which implies operating at frequencies higher than SLAC. Recent studies have been exploring the frequency range between about 8 and 30 GHz. A research effort is underway at SLAC[1] to determine whether klystrons can be developed which meet the higher peak power requirements at least at the lower end of this frequency range. The most promising candidate for satisfying the microwave power needs of SLAC-type linac structes in X-band is the 10-GHz gyroklystron being developed at the University of Maryland. We will discuss this work in Section 5 of this paper. Among other options being pursued are the free electron laser (FEL) at the Lawrence Livermore Laboratory[2] and the superconducting drive linac at CERN[3] both of which have a design frequency of 30 GHz. (The CERN project is discussed by W. Schnell at this meeting.)

The most recent review of rf-driven linear colliders was presented by P. B. Wilson at the 1987 Particle Accelerator Conference.[1] Figure 1.1 is a schematic diagram of such a linear collider. It shows only one of the two opposing linacs and it is similar to a viewgraph shown by P. B. Wilson in his talk. The text in the figure highlights some of the most important considerations and parameters. Thus, the electron gun with thermionic cathode must produce on the order of 10^{11} electrons in a 2 ns pulse with a normalized emittance of $\epsilon_N \pi \sim 10^{-5}$ π m-rad. This pulse is then bunched by a factor of 100 or more to achieve a final bunch length in the main linac which is on the order of $\sigma_z \sim 0.01 \lambda$ where λ is the linac wavelength.

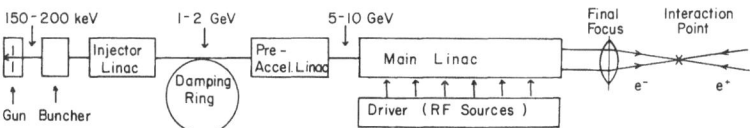

Gun: 150 - 200 kV, ~ 5-10 A, ~ 2 ns → N = 10^{11} electrons, $\epsilon_N \sim 10^{-5}$ m-rad

Buncher: Produces small bunch size ($\sigma_z \sim 0.01 \lambda$ in main linac) for small ΔE/E

Injector: Used for e$^+$ production in the e$^+$ acceleration for system, may operate at lower frequency than main linac

Damping Ring: Reduces emittance to $\epsilon_N \sim 10^{-6}$ m-rad by radiation cooling

Pre-accelerator: May operate at lower frequency than main linac to reduce wakefield effects

Main Linac: High gradient, high frequency, high peak power to reduce length, constraints imposed by wakefield effects and electric power

Final Focus: Must focus beam to extremely small diameter < μm, requires small ΔE/E

Interaction: Point Beam pinching enhances luminosity, beamstrahlung produces energy loss (classical vs. quantum regime)

FIG. 1.1. Schematic of e$^+$e$^-$ linear collider (two opposing linacs—only one is shown).

Research is in progress with photocathodes capable of producing several hundred amperes/cm^2 of electron current. Such photocathodes could replace the electron guns with thermionic cathodes and eliminate or reduce the bunching system which causes considerable emittance growth. In the opposing linac, the electron gun and buncher are the same; however, the electron bunch is used to produce the positrons (e$^+$) in the injector linac. Damping rings at 1-2 GeV energy would be required in both linacs to reduce the emittance of the e$^-$ and e$^+$ bunches to about 10^{-6} π m-rad by radiation cooling. To minimize detrimental wake field effects, the injector linac as well as the pre-accelerator linac following the damping ring could operate at a low frequency, e.g., that of SLAC. However, the bunch length σ_z would have to be shorter than the SLAC case to fit the 0.01 λ requirement (for low energy spread) of the main linac which would operate at the desired high frequency. As mentioned already, research is in progress at various places to develop rf sources with high peak power (hundreds of megawatts) necessary to drive the main rf linac.

The most important parameter in a linear collider is the luminosity at the interaction point. The beams must be focused down to dimensions of a micrometer or less, requiring very low emittance and energy spread. In addition, two effects illustrated in Fig. 1.2 strongly influence the interaction point physics. Prior to crossover the particles in both beams experience a very small defocusing force since the repulsive electric force due to the space charge is almost cancelled by the attractive magnetic force due to the current. However, when the two beams overlap the space charge is neutralized and the force abruptly changes from defocusing to focusing with the strength increasing by a factor of $2\gamma^2$. This results in pinching of the beams which enhances the luminosity significantly. A measure for the pinch effect is the so-called disruption parameter $D = \sigma_z/F$, where σ_z is the RMS half width of the beam (assumed to have a Gaussian profile) and F the focal length of the thin lens equivalent to the pinch effect. The luminosity is then defined by

$$ L = \frac{N^2 f_r H_D}{4\pi\sigma_y^2} \tag{1.1} $$

for a round Gaussian beam with RMS transverse half width of σ_y. N is the number of particles in the bunch, f_r the number of bunches per second, and H_D a function of the disruption parameter which approaches the value $H_D \approx 6$ when $D > 2$. Numerical studies have shown that the beam becomes unstable when $D > 10$.

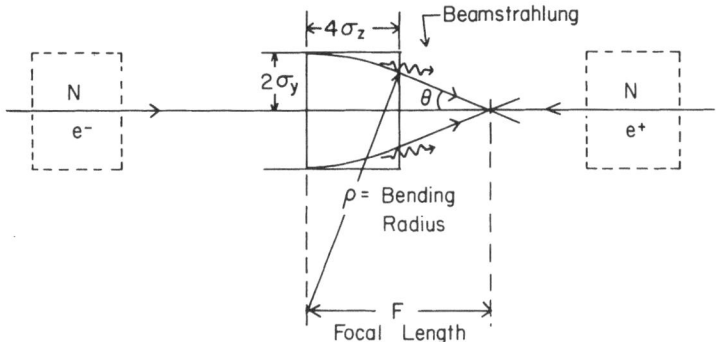

Defocusing force before interaction: $F_{r,o} \sim \dfrac{e^2 N}{\sigma_y \sigma_z}(1-\beta^2)$

Focusing force during interaction: $F_r \sim -\dfrac{e^2 N}{\sigma_y \sigma_z} 2\beta^2 = -2\gamma^2 F_{r,o}$

Disruption parameter: $D = r_e \dfrac{N\sigma_z}{\gamma \sigma_y^2} = \dfrac{\sigma_z}{F}$

Magnetic field at interaction point: $B \sim \dfrac{e\,N}{\sigma_y \sigma_z}$

Bending Radius: $\rho = \dfrac{\gamma mc}{eB} \sim \dfrac{\gamma \sigma_y \sigma_z}{e^2 N}$

Beamstrahlung parameter: $\delta_{BS} \sim \dfrac{N^2 \gamma}{\sigma_y^2 \sigma_z}$ (classical)

$\delta_{BS} \sim \left(\dfrac{N^2 \sigma_z}{\gamma \sigma_y^2}\right)^{1/3}$ (quantum regime)

FIG. 1.2. Pinch effect and beamstrahlung at interaction point.

The bending of the particle trajectories due to the strong self-magnetic field at the interaction point brings back a detrimental effect which linear colliders were supposed to avoid: synchrotron radiation! In this context, where the magnetic field is produced by the beams themselves, it is appropriately called "beamstrahlung." Its effect is measured by the beamstrahlung parameter δ_{BS} which is approximately three times the RMS energy spread introduced in the beam by the beamstrahlung. This energy spread should not be larger than about 10% so that $\delta_{BS} = 0.3$ is generally taken as an upper limit for this effect. Most studies to date have shown that beamstrahlung is the most serious constraint for the beam physics and luminosity in the interaction point of a linear collider. Fortunately, at very high energies above 1 TeV, mother nature provides some relief from this constraint. In this region the peak photon energy of the synchrotron radiation becomes comparable to the electron energy, and the classical model has to be replaced by a quantum description of the effect. The results is that $\delta_{BS} \sim N^{2/3}/\gamma^{1/3}$, versus $N^2\gamma$ for the classical case, i.e., δ_{BS} is decreasing with energy rather than increasing as in the classical regime. A very important requirement from the high-energy physics point of view is that the luminosity should increase as γ^2, which requires extremely small beam dimensions σ_y and high repetition rates as the energy is increased. The disruption and beamstrahlung effects can be reduced by making the beams flat (rather than round) in the interaction region. The pinch enhancement factor H_D as well as the beamstrahlung parameter δ_{BS} then become functions of the aspect ratio $R = \sigma_x/\sigma_y$ of the flat beam, as will be discussed in the next section.

Apart from the interaction point physics, there are other constraints for linear colliders, the two most important ones being average power requirements and wake field effects in the linear accelerator. The average power per beam

$$\overline{P}_b = eNf_r \gamma m_o c^2 = \eta \overline{P}_{AC} \qquad (1.2)$$

cannot exceed a few megawatts since the overall efficiency, $\eta = \overline{P}_b/\overline{P}_{AC}$ (where $\overline{P}_{AC}$ is the average total power), is very small (typically in the range of a few percent). Thus, the cost of electricity to run the collider poses a limit to luminosity that may be more stringent than any beam physics effects. This point was emphasized by Lawson[4] in his recent review of this subject.

The wake field effects are illustrated in Fig. 1.3. When a bunch of particles travels through a disk-loaded linac structure, the wake field from the front of the bunch acts on the particles in the rear end. Basically, the wake fields constitute waveguide modes that are excited by, and co-moving with, the bunches. The force $\vec{F} = e\vec{E} + e\vec{v} \times \vec{B}$ acting on the particles due to the fields ($\vec{E}$ and $\vec{B}$) of these waves may lead to serious beam deterioration. The most important force components are those of the lowest-order modes, i.e., the longitudinal wake field W_z (m = 0 mode) which leads to energy loss of the particles and the transverse dipole wake field W_t (m = 1) which leads to transverse displacements and emittance growth. Most likely, the transverse quadrupole field (m = 2) will also affect the beam adversely, if the beam intensity is large enough or the structure is small enough. Obviously, effects of these forces are strongest at the low-energy part of the linac, since the change of slope of a particle trajectory with $\beta \approx 1$ is proportional to the force and inversely proportional to the particle energy. Wake field effects will be discussed in Section 3.

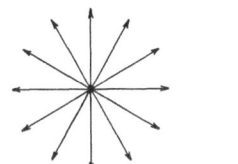
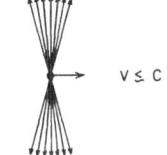

Electric field of point
charge at rest in
free space

Electric field of point
charge moving at relativistic
velocity in free space

$v \leq c$

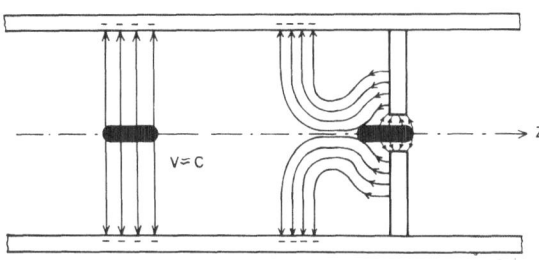

$v = c$

Bunch of positively charged particles moving with
v=c inside a linear accelerator structure

FIG. 1.3. Electric fields of a point charge in free space and a bunch of
particles inside a linac structure (wake field).

In view of the necessity to go to higher gradients and frequencies (to
decrease the length of the linacs) the question of how beam physics and
linac parameters scale with frequency is of utmost importance. Most of the
studies to date have focused on the interaction point physics and linac
design parameters. Scaling of SLAC-type structures to higher frequencies,
for instance, reduces the stored energy ($\sim \omega^{-2}$), the dissipation in the
walls ($\sim \omega^{-1/2}$), the filling time ($\sim \omega^{-3/2}$), and the peak rf power per
meter ($\omega^{-1/2}$). Assuming that the number of particles per bunch and the
luminosity remain unchanged, one then finds that the average rf power
decreases with frequency as ω^{-1}, i.e., overall efficiency η increases.
This assumption is valid when one operates in the transition region between
classical and quantum beamstrahlung and keeps δ_{BS} fixed. Thus, everything
favors going to higher frequency, with 30 GHz ($\lambda = 1$ cm) being considered
an upper limit set by manufacturing considerations.

Unfortunately, the implicit assumption of constant particle number per
bunch $N = N_o = $ const., in this scaling scenario must be questioned from the

beam physics point of view. For instance, to keep the energy spread fixed, the bunch length in the main linac must be decreased linearly with the wavelength, i.e., $\sigma_z \approx 0.01 \lambda \sim \omega^{-1}$. There will undoubtedly be some limit in phase-space density of the bunches due to fundamental constraints imposed by the source and by space-charge effects in the buncher and in the low-energy section of the accelerator system (including the damping ring). The number of particles per bunch due to this density limit will scale as

$$N = N_1 \sim \lambda \sim \omega^{-1} . \tag{1.3}$$

Potentially more serious may be the wake field effects. As will be discussed in Section 3, the longitudinal wake fields scale as ω^2 when all dimensions of the linac structures are reduced by the same factor proportional to λ. There will be a limit in the number of particles per bunch, say at a certain wavelength where control of this effect becomes impractical, and the particle number N must then be reduced with decreasing wavelength as

$$N = N_2 \sim \lambda^2 \sim \omega^{-2} , \tag{1.4}$$

when rf frequency ω is increased. The dipole wake field scales as ω^3, and is considered a more serious problem than the longitudinal wake field. If we assume that control techniques, such as Landau damping via phase programming, reach a limit at some frequency, then N must be scaled as

$$N = N_3 \sim \lambda^3 \sim \omega^{-3} , \tag{1.5}$$

To illustrate these threshold effects, the three curves [Eqs. (1.3)-(1.5)], together with the beamstrahlung limit ($N = N_o$ = const.), are plotted qualitatively in Fig. 1.4 with arbitrary scales reflecting our judgement that the dipole effects represent the practical limit at high frequencies. In actual fact, we do not know the relative magnitudes and frequency thresholds of these effects until a specific scenario is studied in more detail. But the message of this little exercise is clear, namely that the beam physics considerations just discussed will reduce the number of particles that can be contained in a bunch as the frequency is increased. The consequences of this scaling scenario are that the repetition rate f_r and average beam power $\overline{P}_b$ will increase with frequency if the luminosity is to remain unchanged. Take the linear scaling [Eq. 1.3)], for

72

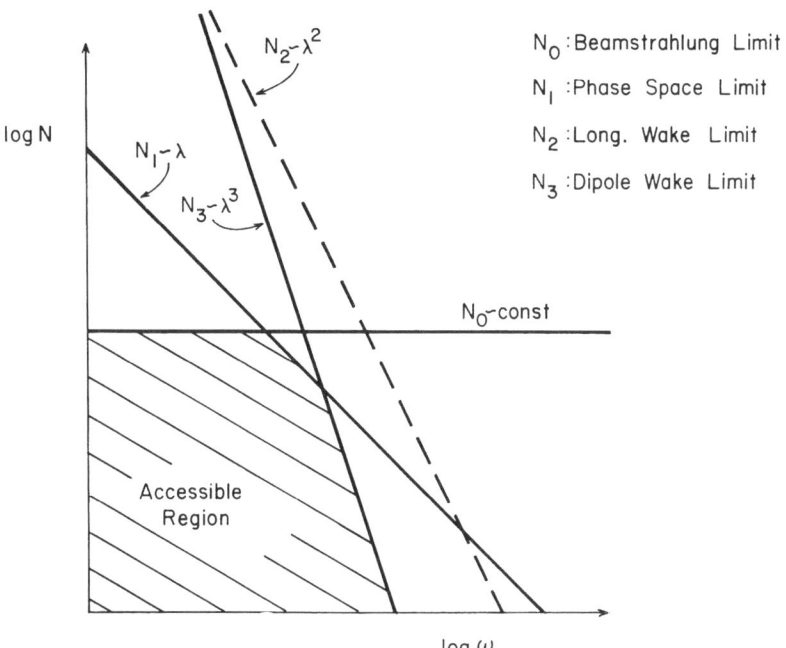

FIG. 1.4. Illustration of several beam physics effects limiting the number of particles per bunch as rf frequency is increased.

example. From Eq. (1.1) follows that

$$L \sim N^2 f_r \sim const.$$ (1.6)

if the effect of H_D is neglected. But $N \sim \omega^{-1}$, hence

$$f_r \sim \omega^2,$$ (1.7)

and average beam power increases as

$$\overline{P}_b \sim N f_r \sim \omega .$$ (1.8)

In Section 4, we report the results of an analysis with our computer model of the linear collider scenario presently being discussed at SLAC. This is a 500 GeV on 500 GeV accelerator system with a total length of 2 × 3 km (that would fit on available land at Stanford) and with a luminosity of $L = 10^{33}$ cm^{-2}. The average gradient required is thus 170 MV/m. In our scaling studies we varied the frequency between 8 and 20 GHz and treated two scenarios. For the first case, the parameters are constrained by beamstrahlung ($\delta_{BS} = 0.3$), which is the usual approach and which turns out to leave N approximately constant. The second case uses the linear decrease of N with frequency ω. The two results are drastically different. In the first case, the average ac power decreases as expected. In the second case, on the other hand, the average beam power increases even more strongly with frequency than Eq. (1.8) indicates, since the lower number of particles diminishes the pinch enhancement effect represented by the factor H_D, which we neglected above. Furthermore, the average ac power increases with frequency, in contrast to the first case. In both cases, though, the peak power per feed decreases with frequency in identical fashion. Attempts to scale N as λ^2 led to completely unrealistic power levels. It is obvious from these results that a considerable research effort is called for to determine the fundamental beam physics limits and to invent ways of circumventing or minimizing the constraints imposed by them. Since average ac power decreases with frequency along the beamstrahlung line ($N_o \sim$ const in Fig. 1.4), the ideal operating point would be the highest frequency compatible with the other limits. Thus, if the phase-space density limit $N_1 \sim \lambda$ were the first constraint encountered, as depicted in Fig. 1.4, one would choose the frequency corresponding to the intersection of the $N_o \sim$ const and $N_1 \sim \lambda$ lines for the rf linac. The gyroklystron appears to be a good candidate to meet the peak power requirements of such a high-frequency linac. In Section 5, a gyroklystron design will be discussed which, in conjunction with pulse compression, would yield the desired peak power level at 10 GHz. It is shown that peak power of the gyroklystron scales linearly with wavelength λ (as compared to the λ^2 scaling of klystrons) so that a higher frequency would be even more favorable.

References

1. P. G. Wilson, "Design of RF-Driven Linear Colliders," 1987 Particle Accelerator Conference, to be published as Conference Record 87 CH 2837-9.
2. A. M. Sessler and D. B. Hopkins, "The Two-Beam Accelerator," 1986 Linac Conference, SLAC-303, p. 385.
3. W. Schnell, "A Two-Stage RF Linear Collider Using a Superconducting Drive Linac," CERN-LEP-RF/86-06 and CLIC Note 13, 13.2.1986.
4. J. D. Lawson, "Collider Constraints in the Choices for Wavelength and Gradient Scaling," 1986 Linac Conference, SLAC-303, P. 599.

2. BASIC SCALING LAWS FOR THE INTERACTION POINT AND THE LINEAR ACCELERATOR

The scaling of a linear collider separates into two almost disjoint studies; viz. (1) the scaling at the final focus where the colliding beams intersect, and (2) the scaling laws for the beams in the linear accelerators. The interaction point scaling is determined by the requirements of the users, and is independent of the means used to accelerate the beams. Given the specification of the beams at the final focus, an accelerator model can be used to determine the requirements for generating the beams using a particular accelerator type.

2.1 Beam Interaction-Point Equations

The beams at the final focus may be specified in terms of eleven quantities. These are the particle energy (γmc^2), the number of particles per bunch (N), the number of bunches per rf pulse (b), the repetition rate (f), the bunch size ($\sigma_x, \sigma_y, \sigma_z$), the total luminosity (L_T), the disruption parameter (D), the beamstrahlung energy loss (δ_{BS}), and the average beam power ($\overline{P_b}$). Here, the two opposing bunches are assumed to be equal in size and in number of particles per bunch, and they are assumed to collide head-on.

For an e^+e^- collider, the space charge of the opposing bunches will cancel when the bunches overlap, but the currents in the opposing bunches will add constructively, thereby causing the bunches to focus (or "pinch") more tightly than they would otherwise. The pinch effect will lead to an unstable disruption of the bunch if the focal length of the pinch (treated as a thin lens) is small compared with the bunch length, σ_z. The disruption parameter, D, is just the ratio of σ_z to the focal length. For $D \ll 1$, the pinch effect is negligible. For $D \sim 1$, the pinch effect leads to an enhancement of luminosity by a factor $H_D(D,R)$, as shown in Fig. 2.1, where $R = \sigma_x/\sigma_y$. For $D > 10$, simulations[1] indicate that the bunch will become unstable, thereby reducing the luminosity as the bunches disrupt. The scaling studies described below are all carried out in the stable regime, where the pinch enhancement factor, H_D, is greater than unity.

The disruption parameter may be expressed as

$$D = (\frac{r_e N \sigma_z}{\gamma \sigma_y^2})(\frac{2}{1 + R}) \ , \tag{2.1}$$

where

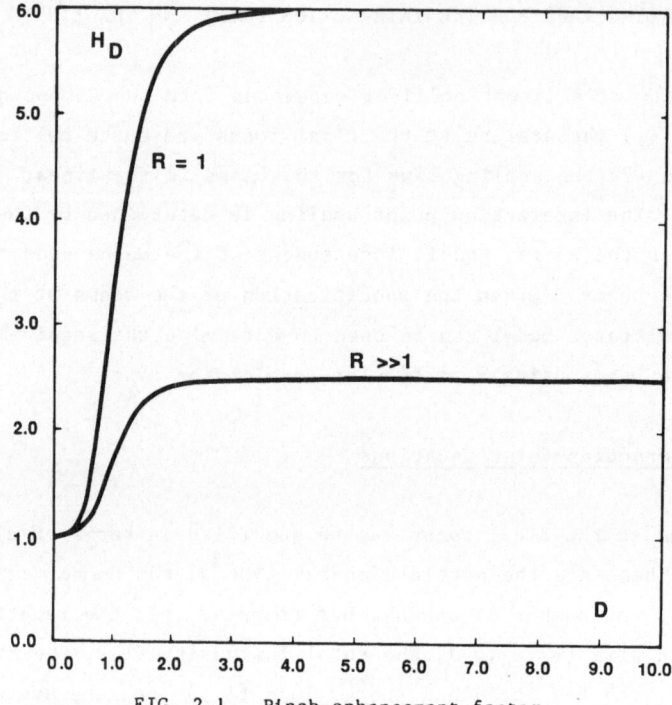

FIG. 2.1. Pinch enhancement factor.

$$r_e = \frac{e^2}{4\pi\varepsilon_o mc^2} = 2.82 \times 10^{-15} \text{ m}$$

is the classical electron radius, and R is the beam aspect ratio (defined
to be R > 1). The luminosity is enhanced by the pinch, and may be
expressed as

$$L_T = \frac{bN^2 f H_D(D,R)}{4\pi\sigma_y^2 R} .$$ (2.2)

The required luminosity depends on the center-of-mass energy of the
collision ($E_{cm} = 2\gamma mc^2$). Since the cross-sections for collisions of
interest to high-energy physicists decrease as E_{cm}^{-2}, the luminosity (which
is the counting rate per unit cross-section) must increase as γ^2 to
preserve the event rate as the energy increases.

During the pinch at the final focus, the particles will radiate by
synchrotron emission. This radiation, or "beamstrahlung," leads to both an
average loss of energy per particle and an energy spread. The
beamstrahlung loss parameter, δ_{BS}, represents the average loss of energy
per particle, normalized to the incident particle energy. The RMS energy

spread of the colliding particles, given by $\delta_{BS}/2\sqrt{2}$, must be constrained to < 10% for the experiments to have reasonable energy resolution. It is, therefore, reasonable to assume δ_{BS} < 30% for collider scaling studies. Using the classical synchrotron radiation formulation,[2] the beamstrahlung spectrum will peak at a critical energy which is proportional to γ^3. As γ increases, therefore, the energy of emitted photons will eventually overtake the electron kinetic energy, leading to an unphysical result. To avoid this situation, it is necessary to use the quantum formulation for the synchrotron emission.[3] The quantum theory treats the synchrotron radiation as a train of photons and limits the emitted spectrum to the particle energy. In the quantum regime, therefore, the amount of beamstrahlung is reduced compared with the classical results. This effect becomes particularly important at energies of several TeV.

The resulting expression for the beamstrahlung energy loss,[4] including the quantum correction[5] and the correction for flat beams,[6] is given by

$$\delta_{BS} = (\frac{2r_e^3}{3})(\frac{N^2\gamma}{\sigma_y^2\sigma_z})[\frac{F(R)}{R}] \, H_D H_\delta ,$$ (2.3)

where $F(R)$, shown in Fig. 2.2, describes the dependence of δ_{BS} on the beam aspect ratio. $F(R)$ takes the value 0.328 at $R = 1$, and asymptotically approaches the curve $F(R) \sim 1.3/R$ for $R \gg 1$.

$H_\delta(T)$, shown in Fig. 2.3, is the quantum correction to the beamstrahlung. H_δ depends on the quantum parameter, T, which is defined as

$$T = \frac{\gamma B}{B_c} ,$$ (2.4)

where B is the effective magnetic field which bends the particles and

$$B_c = \frac{m^2c^2}{e\hbar} = 4.4 \times 10^9 T .$$

Using the average magnetic field on the bunch as computed by Wilson[7] and the generalization for flat beams derived by Amaldi,[8] the average value of T is given by

$$\bar{T} = \frac{eNZ_o H_D\gamma}{16\pi\sigma_x\sigma_z B_c} [3RF(R)]^{1/2} ,$$ (2.5)

where $Z_o = 377 \, \Omega$ is the free-space impedance.

For small values of T, $H_\delta \approx 1$, which is the "classical" beamstrahlung regime. When $T \gg 1$, H_δ approaches the curve, $H_\delta \approx 0.556 \, T^{-4/3}$, which is the "deep-quantum" regime. In the classical regime, the beamstrahlung

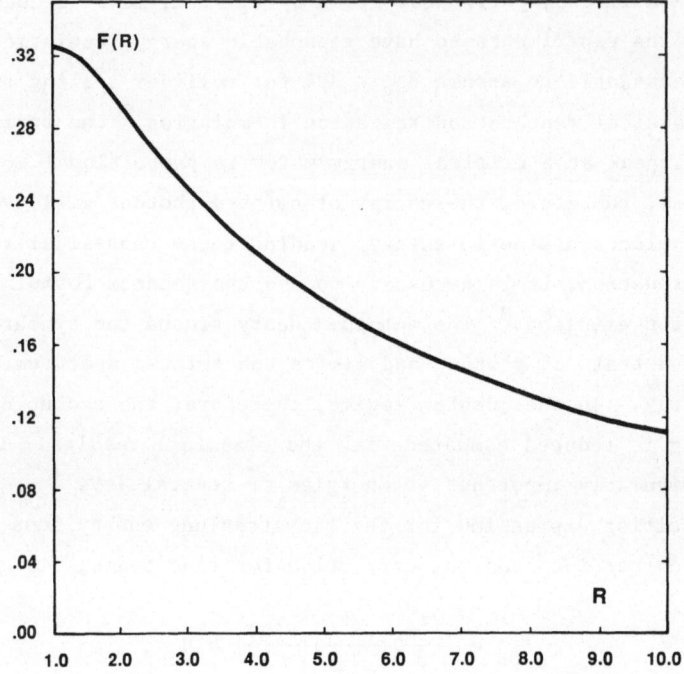

FIG. 2.2. Beamstrahlung aspect-ratio function.

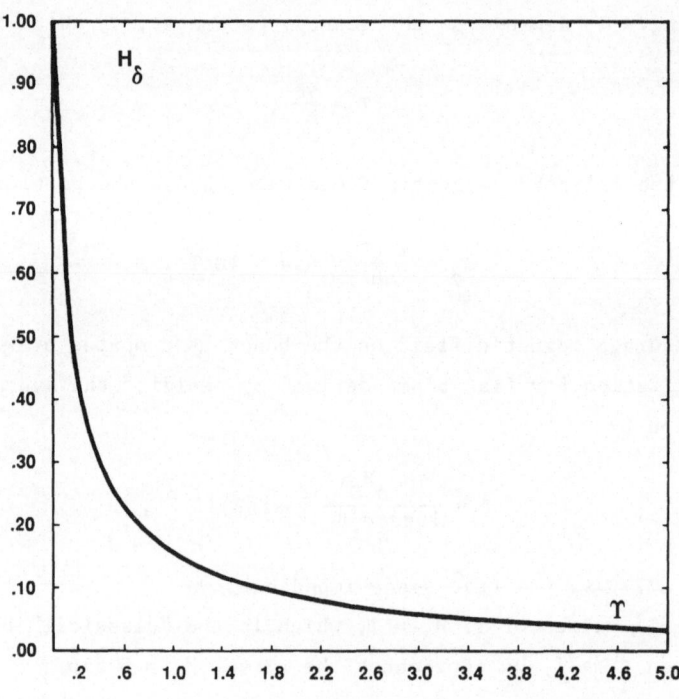

FIG. 2.3. Quantum reduction function for beamstrahlung.

scales as $\delta_{BS} \sim \gamma/\sigma_z$, favoring long bunches. The deep-quantum scaling, on the other hand, is given by $\delta_{BS} \sim (\sigma_z/\gamma)^{1/3}$, which favors short bunches. For short bunches, Eq. (2.5) shows that T will become large, thereby pushing the collider toward the quantum limit. In practice, collider scenarios at energies between 0.5 and 5.0 TeV (0.1 < T < 100) lie neither in the classical nor the deep-quantum limit. They are all in the transition region, where Wilson[9] has pointed out that $TH_\delta(T) \simeq$ const. Using this approximation, the beamstrahlung is independent of σ_z.

The equations for L_T, D, and δ_{BS}, together with the expression for the average beam power,

$$\bar{P}_b = \gamma mc^2 bNf = (\frac{e^2 \sigma_z L_T}{\varepsilon_o DH_D})(\frac{2R}{1 + R}) , \qquad (2.6)$$

constitute four relationships between the eleven quantities which specify the interaction point. Seven of these quantities may therefore be specified. There are 330 possible combinations of eleven quantities with any seven specified. If γ, R, b, and L_T are always assumed to be specified, then the model requires the additional specification of three additional quantities from the set $\{\sigma_z, \sigma_y, \delta_{BS}, \bar{P}_b, D, N, f\}$. There are 33 combinations of these parameters, any of which can constitute the basis of a scaling model. [The total number of combinations is $_7C_3 = 35$. The sets $(\bar{P}_b, N, f)$ and $(\bar{P}_b, D, \sigma_z)$, however, cannot be satisfied in general, as seen in Eq. (2.6).] Two interaction-point scaling scenarios are explored below; one is based on the specification of $(\delta_{BS}, \sigma_y, \sigma_z)$, and the other on (N, σ_y, σ_z).

2.2 Linear Accelerator Parameters and Electrical Power Requirements

The accelerator model for an rf linac consists of a model for power flow in the linac, and a description of the rf cold-test properties of the linac, including wake fields. Fig. 2.4 illustrates the various channels for energy flow in an rf linac. This model has been described in detail elsewhere,[10] and will be only briefly summarized below. The ac power, $\bar{P}_{AC}$, is converted to quasi-dc pulses by the modulator, which typically operates at an efficiency $\eta_M \simeq 85\%$. The dc-to-rf power conversion occurs in the microwave sources, at an efficiency $\eta_K \simeq 50\%$, which is typical both of klystrons at S-band and gyroklystrons at X-band. The transmission of rf power from the sources to the accelerating structure involves several sources of power loss, including mode conversion from the internal operating mode of the source to a

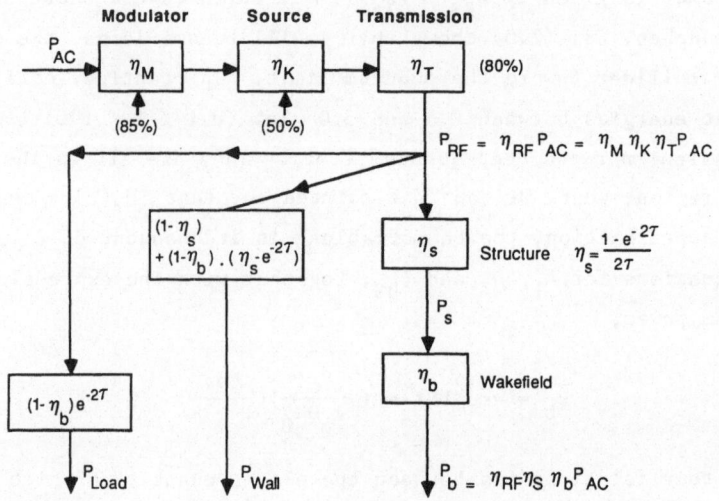

FIG. 2.4. Efficiencies and power flow in an rf linac.

propagating waveguide mode to the accelerating mode in the structure, propagation losses in the waveguide, and losses due to inefficiencies associated with pulse compression,[11] which may be used to obtain higher peak rf power for a shorter pulse duration. These various processes have been lumped into a single transmission efficiency, $\eta_T \simeq 80\%$. The average rf power to be supplied per linac is, therefore,

$$\bar{P}_{RF} = \eta_{RF}\bar{P}_{AC} = \eta_M\eta_K\eta_T\bar{P}_{AC} \ , \tag{2.7}$$

where $\eta_{RF} \simeq 34\%$ is assumed to be independent of frequency.

The accelerating structure may be characterized by the parameter, τ, where $\tau = \omega t_F/2Q$, with ω being the rf (circular) frequency, t_F the fill time, and Q the rf quality factor for the accelerator cavities. Only a fraction, η_S, of the incident rf power ends up as stored rf energy in the structure. The rest is dissipated in the walls during the fill time. The structure efficiency, η_S, is given by

$$\eta_S = \frac{1 - e^{-2\tau}}{2\tau} \tag{2.8}$$

for a constant-gradient structure. For $\tau = 0.57$, $\eta_S = 60\%$, and the fraction of the ac power which is stored as rf energy in the accelerator before the arrival of the beam pulse is $\eta_{RF}\eta_S = 20\%$.

The beam pulse consumes a fraction η_{bT} of the stored rf energy, with the remainder of the energy dissipated in the walls and in the rf loads after the beam pulse has passed through the structure. The incident rf energy, therefore, ultimately goes to the beam, the walls, and the load, as illustrated in Fig. 2.4. The accelerating structure acts as an intermediate energy store and a power conditioning element. The overall efficiency of the accelerator for conversion of ac power to beam power is $\eta = \eta_{RF} \eta_S \eta_{bT} \approx \eta_{bT}/5$, which is typically < 5%.

The scaling model requires either direct specification or subsidiary models for each of the component efficiencies: η_M, η_K, η_T, η_S, and η_{bT}. The first three of these are specified directly at the present time, while the η_S is determined from τ, and η_{bT} is computed as described below. Given $\overline{P}_b$ at the final focus, the energy flow diagram may be followed in reverse to infer the power requirements for each accelerator component.

The cold-test properties of fundamental importance in a traveling-wave rf accelerator (such as the SLAC linac) are the normal-mode frequencies, the Q's, and the shunt impedances (r) for each synchronous traveling-wave mode supported by the structure. From these quantities the loss parameters, k_n, may be expressed as

$$k_n = \left(\frac{\omega_n}{4}\right)\left(\frac{r_n}{Q_n}\right) = \frac{E_{on}^2}{4 w_{sn}} , \qquad (2.9)$$

where E_{on} is the rf amplitude, w_{sn} is the energy stored per unit length, r_n is the shunt impedance per unit length, and Q_n is the quality factor for the nth mode. The k's are related to the wake fields,[12] as will be seen below.

For modular scaling of all accelerator components with rf wavelength, the Q varies as $Q \sim \omega^{-1/2}$, while the shunt impedance per unit length varies as $r \sim \omega^{1/2}$. The k's therefore will vary as $k \sim \omega^2$. If the structure parameter, τ, is held fixed with rf frequency, the fill time will vary as $t_F = (2Q/\omega)\tau \sim \omega^{-3/2}$. At high frequencies, therefore, the fill time becomes very short, thereby allowing pulse compression techniques to be more effectively utilized for enhancing the peak rf power delivered to the accelerator.

The wake fields are discussed fully in the next section, where a simulation model for a single bunch in an rf linac will be described. The so-called delta-function wake field, i.e., that due to a single point charge traversing the accelerator, may be written as a superposition of the synchronous modes. The longitudinal 2m-pole δ-function wake field, $W_{mz}(\theta)$, and the transverse 2m-pole δ-function wake field, $W_{m1}(\theta)$, may be written as

$$W_{mz}(\theta) = \Sigma_n \left(\frac{2k_n}{a^{2m}}\right) \cos\left(\frac{\omega_n \theta}{\omega_1}\right) \qquad (2.10a)$$

$$W_{m\perp}(\theta) = \Sigma_n \left(\frac{2mk_n c}{\omega_n a^{2m}}\right) \sin\left(\frac{\omega_n \theta}{\omega_1}\right) , \qquad (2.10b)$$

where θ is the phase angle on the fundamental rf mode (with frequency ω_1) between the leading point charge and the trailing field point, and a is the largest clear aperture (such as the iris aperture in a disk-loaded waveguide). The cold-test properties of the accelerating structure are therefore sufficient to generate the wake fields.[12]

The δ-function wake fields, due to a point charge, may be treated as Green's functions in calculating the total integrated wake field due to a distribution of charge. Assume that $f_b(\theta)$ represents the single-bunch particle distribution in phase on the fundamental rf wave, normalized to unity,

$$\int_{-\infty}^{\infty} f_b(\theta)d\theta = 1 . \qquad (2.11)$$

For a Gaussian distribution, f_b may be expressed as

$$f_b(\theta) = (2\pi\theta_z^2)^{-1/2} \exp\left[-\frac{(\theta - \theta_o)^2}{2\theta_z^2}\right] , \qquad (2.12)$$

where $\theta_z = 2\pi\sigma_z/\lambda$ is the standard deviation, and the beam is centered at phase θ_o relative to the crest of the rf wave. The integrated monopole wake field may then be expressed as

$$W_L(\theta) = \int_{\theta}^{\infty} f_b(\theta') W_{oz}(\theta' - \theta)d\theta' = \int_0^{\infty} f_b(\theta + \theta') W_{oz}(\theta')d\theta' . \qquad (2.13)$$

The total loss factor due to the longitudinal wake field is given by

$$k_L = \int_{-\infty}^{\infty} f_b(\theta) W_L(\theta)d\theta . \qquad (2.14)$$

A particle at phase θ, relative to the crest of the rf wave, will experience a gradient given by

$$E(\theta) = E_o \cos\theta - NeW_L(\theta) , \qquad (2.15)$$

and the average accelerating gradient, E_a, for the entire bunch is

$$E_a = \int_{-\infty}^{\infty} E(\theta) f_b(\theta) d\theta = E_o [\exp(-\frac{\theta_z^2}{2}) \cos\theta_o - \frac{Ne k_L}{E_o}] , \qquad (2.16)$$

where the second expression holds for a Gaussian distribution.

The ratio of the energy gained per unit length to the rf energy stored per unit length of accelerating structure is the beam efficiency,

$$\eta_b = \frac{NeE_a}{w_s} = \frac{4Ne k_1}{E_o} [\exp(-\frac{\theta_z^2}{2}) \cos\theta_o - \frac{Ne k_L}{E_o}] , \qquad (2.17)$$

where Eq. (2.9) has been used to replace w_s by $E_o^2/4k_1$. The external rf sources (ideally) drive only the fundamental traveling-wave mode; hence, w_s is written in terms of k_1. The beam loses energy to the wake fields, which are a superposition of all the synchronous traveling-wave modes. The expressions for E_a and η_b, therefore, contain a term having the total loss factor, k_L, in place of k_1.

The efficiency in Eq. (2.17) is the single-bunch beam efficiency. There is also an energy sag from bunch to bunch. In the study described below, the rf pulse length is longer than the fill time by an amount which allows the rf sources to compensate for this energy sag between bunches. The extra rf energy that must be injected between bunches increases the rf energy stored per unit length to $w_{sT} = w_s[1 + (b - 1)\eta_b]$. The total beam efficiency for b bunches is then given by

$$\eta_{bT} = \frac{b\eta_b}{1 + (b - 1)\eta_b} . \qquad (2.18)$$

This quantity determines the overall energy requirements of the accelerator, as described above.

Particles at different phases of the rf wave will experience different accelerating gradients, which leads to an energy spread within the bunch. The longitudinal wake field causes particles near the head of the bunch to experience a higher gradient than those near the tail. Positioning the bunch ahead of the crest of the rf wave, however, can partially compensate for the energy spread due to wake fields at the cost of reduced average gradient. The single-bunch energy spread, σ_E, may be expressed as a relative energy spread, $\delta = \sigma_E/E$, as

$$\delta^2 = E_a^{-2} \int_{-\infty}^{\infty} [E^2(\theta) - E_a^2] f_b(\theta) d\theta = (\frac{E_o}{E_a})^2 \{0.5[1 + \exp(-2\theta_z^2) \cos(2\theta_o)]$$

$$+ (\frac{Ne}{E_o})^2 I_T - (\frac{2Ne}{E_o})[I_c \cos\theta_o - I_s \sin\theta_o]\} - 1 , \qquad (2.19)$$

where

$$I_T = \int_{-\infty}^{\infty} W_L^2(\theta) \, f_b(\theta) d\theta \; , \qquad (2.20a)$$

$$I_c = \int_{-\infty}^{\infty} \cos\theta \, f_b(\theta) \, W_L(\theta) d\theta \; , \qquad (2.20b)$$

$$I_s = \int_{-\infty}^{\infty} \sin\theta \, f_b(\theta) \, W_L(\theta) d\theta \; . \qquad (2.20c)$$

The four quantities $(E_o, \eta_b, \delta, \theta_o)$ are related through Eqs. (2.17) and (2.19). Specifying any pair of them allows the other pair to be determined. Alternatively, one may specify that δ lie at its minimum value, and any one of the remaining three quantities will then determine the entire set.

2.3 Description of the Computer Model

A computer code has been constructed to analyze the interaction point and accelerator models described above, and to plot the results of the models as the rf frequency is varied.

Two interaction point scenarios are implemented in the code. Both assume that L_T, R, b, σ_y, σ_z and γ are specified, and both compute f, D, and $\bar{P}_b$. One scenario completes the seven-parameter specification with δ_{BS} and computes N, while the other specifies N and computes δ_{BS}.

The scenario with N specified is straightforward. Given N, D is determined from Eq. (2.1), and used to determine H_D. The equation for luminosity, Eq. (2.2), then determines f. Equations (2.3)-(2.6) are then completely specified and determine the quantities, δ_{BS} and $\bar{P}_b$.

The scenario where δ_{BS} is specified requires an iterative solution. From the ratio, L_T/δ_{BS}, the repetition frequency may be expressed as

$$f = (\frac{8\pi}{3}) r_e^3 (\frac{L_T}{\delta_{BS}}) [\frac{\gamma F(R)}{b\sigma_z}] H_\delta \; . \qquad (2.21)$$

With an initial guess for H_δ, e.g., $H_\delta^{(0)} = 1$, Eq. (2.21) gives a value for f, which may then be used to combine Eqs. (2.1) and (2.2), eliminating N, to obtain,

$$D^2 H_D = [\frac{16\pi R}{(1 + R)^2}] (\frac{r_e \sigma_z}{\gamma \sigma_y})^2 (\frac{L_T}{bf}) \; , \qquad (2.22)$$

which may be solved for D and H_D. N then follows from Eq. (2.1). Knowing

N and H_D, $\bar{T}$ may be obtained from Eq. (2.5) and used to evaluate H_δ. A new iteration may then begin. The iterative process continues until all quantities have converged to a specified relative error, which is measured as the relative change in that quantity from one iteration to the next. After convergence, the average beam power is computed from Eq. (2.6).

At the conclusion of the interaction-point model, all eleven beam quantities are known. The energy flow in the accelerator is then computed in reverse, using the energy flow model and efficiencies described in the previous section, to arrive at the requirements in rf power and ac power to achieve the average beam power at the final focus.

The specified structure parameters in the model are τ and Q, from which t_F is computed. The group velocity, v_g, is specified and determines the length of an rf feed, $v_g t_F$. Given ω, Q, and the loss parameter, k_1, for the fundamental mode, the shunt impedance r is obtained. The longitudinal wake field, with the parameters specified above, then determines any two of the set $(E_o, \eta_b, \theta_o, \delta)$ if the other two are specified.

The peak rf power required per feed is determined from the cold-test properties of the structure. The energy stored per unit length of structure is $w_s = QE_o^2/\omega r$. The rf energy stored per feed is, therefore, $w_s v_g t_F$. The rf energy that must be delivered to the structure in order to store this energy is $w_s v_g t_F/\eta_s$. Since this energy is delivered in a fill time, the peak rf power per feed is

$$\hat{P}_{feed} = \frac{w_s v_g}{\eta_s} .\qquad (2.23)$$

If the scaling study is carried out so that $E_o \lambda = const.$, thereby using higher frequency operation to reduce the length of the accelerator, the peak rf power per feed will be constant. If the rf amplitude is independent of frequency, on the other hand, Eq. (2.23) shows that the peak rf power per feed will decrease as λ^2. Since the power handling capability of most rf tubes decreases with frequency, the second scaling scenario is preferred.

References

1. R. Hollebeek, NIM 184, 333 (1981).

2. J. D. Jackson, Classical Electrodynamics (Wiley, New York, 1975), Chap. 14.

3. A. A. Solokov, N. P. Klepikov, and I. M. Ternov, JETP 23, 632 (1952);
    A. A. Solokov and I. M. Ternov, Synchrotorn Radiation (Pergamon, New
    York, 1968), p. 99.

4. H. Wiedemann, SLAC-PUB-2849 (1981).

5. T. Himel and J. Siegrist, SLAC-PUB-3572 (1985).

6. M. Bassetti and M. Gygi-Hanney, LEP Note 221 (1980).

7. P. B. Wilson, SLAC-PUB-3674 (1985).

8. U. Amaldi, CERN-EP/85-102 (1985).

9. P. B. Wilson, Proc. 1987 Part. Accel. Conf. (to be published).

10. V. L. Granatstein and A. Mondelli in Physics of High-Energy Particle
    Accelerators (AIP Conf. Proc. No. 153, 1987), pp. 1508-1528.

11. Z. D. Farkas, SLAC-PUB-3694 (1985).

12. P. B. Wilson in Physics of High Energy Particle Accelerators (AIP
    Conf. Proc. No. 87, 1981), pp. 508-528.

3. WAKE FIELD EFFECTS IN THE LINAC

The quality of a bunch (its emittance and energy spread) produced by a
linac strongly affects the attainable luminosity at the interaction
point. This bunch quality, in turn, is affected by virtually every detail
of the linac design including the injector and damping ring, the linac
structure, the focusing system, the rf accelerating wave, and the overall
system alignment. For large bunch intensities, beam quality is also
affected by wake fields, which are fields generated in the linac structure
by the passage of the bunch which act back on the bunch. The dynamical
consequences of wake fields can place severe constraints on the linac
design, especially on injection jitter and magnet misalignment tolerances.

For a bunch in a linac moving very nearly at v = c the direct electro-
and magneto-static interaction among its constituent charges is negligible
compared to the interaction of the bunch with co-moving waves excited by
the bunch in the periodic structure of the accelerator. The "wake fields"
act in two ways; they extract energy from and induce energy spread in the
bunch (longitudinal effects) and they deflect and deform the bunch
(transverse effects).

For cylindrically symmetric structures, a general formulation of the
problem is possible.[1] We consider the fields produced by a single point

charge of charge Q, located at $r = r_p$, $\theta = \theta_p$ and $z = ct$ traveling through an infinitely repeating structure and define the fields at a distance ξ behind it at r, θ [MKS units]

$$f_r = E_r - cB_\theta \; , \tag{3.1a}$$

$$f_\theta = E_\theta + cB_r \; , \tag{3.1b}$$

$$f_z = E_z \; . \tag{3.1c}$$

These fields are assumed to have been averaged over a structure period. Bane, et al.[1] show that the quantities

$$W_z(r,\theta,\xi) \equiv -\frac{1}{Q} f_z \ell \; , \tag{3.2a}$$

$$\vec{W}_\perp(r,\theta,\xi) \equiv \frac{1}{Q} \vec{f}_\perp \ell \; , \tag{3.2b}$$

where ℓ is the length of a structure period, take the form

$$W_z(r,\theta,\xi) = \sum_{m=0}^{\infty} (\frac{r_p}{a})^m (\frac{r}{a})^m \cos m(\theta - \theta_p) \, a\psi_m'(\xi) \; , \tag{3.3a}$$

$$\vec{W}_\perp(r,\theta,\xi) = \sum_{m=1}^{\infty} m(\frac{r_p}{a})^m (\frac{r}{a})^{m-1}$$

$$\times \, [\hat{r} \cos m(\theta - \theta_p) - \hat{\theta} \sin m(\theta - \theta)] \, \psi_m(\xi) \; , \tag{3.3b}$$

where a is the maximum clear aperture of the structure. The $\psi_m(\xi)$ are the delta-function wake potentials (units: volts/coulomb/period). They vanish by causality for $\xi < 0$. These functions must be calculated numerically by launching a particle into a periodic structure and solving Maxwell's equations, then averaging the fields over a structure period; alternately they may be calculated by a normal mode expansion. The advantage of the multipole expansions in Eqs. (3.3a) and (3.3b) is that they automatically give the transverse dependence of the wake fields. Only the longitudinal dependence [$\psi_m(\xi)$] needs to be calculated numerically.

For particles close to the axis one typically keeps only the leading terms in Eqs. (3.3a) and (3.3b):

$$W_z(r,\theta,\xi) \simeq a\psi_0'(\xi) \tag{3.4a}$$

$$\vec{W}_{\perp}(r,\theta,\xi) \approx \frac{r_p}{a} \hat{r}_p \, \psi_1(\xi) \, . \tag{3.4b}$$

Note that the $m = 0$ (monopole) longitudinal wake is independent of both the position of the leading particle and that of the trailing test particle. The $m = 1$ (dipole) transverse wake is proportional to the displacement of the leading particle and independent of the displacement of the trailing particle. Under some circumstances, the $m = 2$ (quadrupole) term in the transverse force may be important,[2] but we will neglect it here except to say that it does become increasingly important to consider as linac structures are scaled to higher frequency operation (reduced in transverse dimension). The functions $a\psi_o^{\tilde{}}$ and ψ_1 are given for the SLAC structure in Fig. 3.1. (The abscissa is ξ/c.)

For any charge distribution, it is straightforward to sum up for each particle those wake forces due to all particles ahead of it. When this is done for a bunch, one obtains equations of motion governing the evolution of the bunch. Specifically, keeping only the lowest order wake fields, one finds that the bunch centroid, considered as a function of distance ξ back from the head of the bunch satisfies

$$x^{\prime\prime}(z;\xi) + \frac{\gamma^{\tilde{}}}{\gamma} x^{\prime}(z;\xi) + k_{\beta}^2(z;\xi) \, x(z;\xi)$$

$$= - \frac{e}{m\gamma c^2} \int_0^{\xi} d\xi^{\tilde{}} \, \lambda(\xi^{\tilde{}}) \, x(z;\xi^{\tilde{}}) \, W_t(\xi - \xi^{\tilde{}}) \, , \tag{3.5a}$$

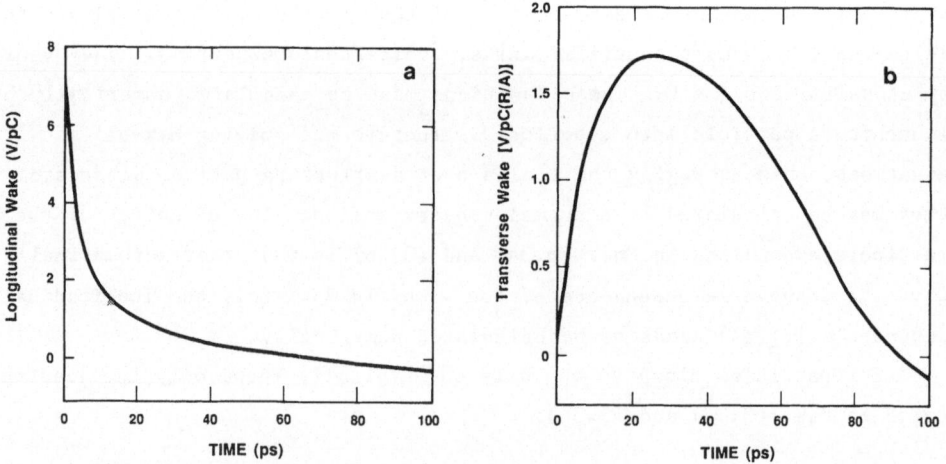

FIG. 3.1. Longitudinal (a) and transverse dipole (b) δ-function, wake functions for the SLAC linac structure.

$$\gamma'(z;\xi) = -\frac{e}{mc^2} \left\{ E_o \cos[\phi_o(z) - \frac{2\pi\xi}{\lambda_{rf}}] - \int_0^\xi d\xi' \; \lambda(\xi') \; W_\ell(\xi \; \xi') \right\}, \tag{3.5b}$$

where z measures distance along the accelerator, the prime denotes $\partial/\partial z$, k_β^2 is the local betatron wavenumber squared. (It depends on ξ since it is energy dependent.) $\lambda(\xi)$ is the bunch line charge density, E_o is the peak accelerating field, λ_{rf} is the rf wavelength, and we have defined $W_t(\xi) = \psi_1(\xi)/\ell a$ and $W_\ell(\xi) = a\psi_o'(\xi)/\ell$. An equation identical to Eq. (3.5a) holds for the y-coordinate.

Similarly one may, by taking a moment of the single particle equations of motion, find evolution equations for the variances of the transverse position (x) and slope (x') about their mean values:

$$\sigma'_{xx} = 2\sigma_{xx'}, \tag{3.6a}$$

$$\sigma'_{x'x'} = -2\frac{\gamma'}{\gamma}\sigma_{x'x'} - 2k_\beta^2\sigma_{xx'}, \tag{3.6b}$$

$$\sigma'_{xx'} = -\frac{\gamma'}{\gamma}\sigma_{xx'} - k_\beta^2\sigma_{xx} + \sigma_{x'x'}, \tag{3.6c}$$

where $\sigma_{xx} = \langle x^2 \rangle - \langle x \rangle^2$, x is the transverse position of a single particle, and the average is taken over particles at fixed ξ. Note that the σ's are unaffected by the dipole transverse wake; the dipole wake affects, and is affected by, only the centroid motion. If it were included, the quadrupole wake function would appear in Eqs. (3.6b) and (3.6c), modifying k_β^2.

Equations (3.6a)-(3.6c) admit a constant of motion, the normalized emittance ϵ_N which we define as

$$\epsilon_N^2(\xi) = 16\gamma^2(\sigma_{xx}\sigma_{x'x'} - \sigma_{xx'}^2). \tag{3.7}$$

If we define a beam radius (in the x-direction) as

$$r_x = 2\sigma_{xx}^{1/2}, \tag{3.8}$$

then Eqs. (3.6)-(3.8) may be shown to combine to give the usual envelope equation:

$$r_x'' + \frac{\gamma'}{\gamma}r_x' + k_\beta^2 r_x - \frac{\epsilon_N^2}{\gamma^2 r_x^3} = 0. \tag{3.9}$$

One may also define a total normalized emittance of the bunch (not just at a position ξ) as

$$\varepsilon_N^B = 16\overline{\gamma}^{-2}(\sum_{xx} \sum_{x'x'} - \sum_{xx'}^2) \qquad (3.10)$$

where $\sum_{xx} = \overline{\sigma}_{xx} + \overline{x^2} - \overline{x}^2$, etc. and an overbar indicates an average over ξ. In general, ε_N^B will __not__ be conserved, that is it will depend on z. There are two contributions to its growth; one is due to misalignment of the different "slices" of the bunch (the $\overline{x^2} - \overline{x}^2$ contribution) and the other is due to distortion of the shape of the slices (the $\overline{\sigma_{xx}}$ contribution). By integrating Eqs. (3.5) and (3.6), one may follow the evolution of the normalized emittance and energy spread (the bunch quality) along the linac as the bunch is acted upon by the accelerating field, the external focusing field, and the wake fields. Further refinements, including effects of magnet and cavity misalignments, are easily included in the formulation but will not be discussed here. An example of results of an integration of Eqs. (3.5) and (3.6) will be given in Section 4.1.2.

Some compensation of the monopole longitudinal wake's effect on the bunch energy spread is possible by positioning the bunch slightly ahead of the crest of the rf accelerating wave. In this way the energy loss of the trailing particles is partially compensated for by the slightly higher accelerating gradient they experience. The exact phase location which minimizes the resulting energy spread depends on $W_\ell(\xi)$, the bunch density distribution $\lambda(\xi)$, and the accelerating gradient.

Some degree of compensation of the dipole transverse wake effect is also possible. Since the betatron oscillations in one section of the bunch are essentially driven by the oscillating wake fields of those sections ahead of it, some reduction in the amplitude of oscillation may be realized by inducing a spread in betatron wavelengths along the bunch. This is most simply accomplished by inducing an energy spread in the bunch, using the focusing lattice chromaticity to advantage. The energy distribution should be __increasing__ from the tail to the head of the bunch so that the tail is slightly __more__ strongly focused than the head; this extra focusing tends to compensate for the wake field "kick" received by the tail. A two particle model[3] suggests the value of energy spread required for this cancellation in an actual bunch will be

$$\Delta E \simeq \frac{eQW_t(\xi_o)}{4k_\beta^2}, \qquad (3.11)$$

where $Q = eN$ is the charge in the bunch and ξ_o is the bunch length. The resulting reduction in the disruption of the bunch, and in the emittance growth, is known as Landau damping. Its use will probably be essential to minimize sensitivity to injection jitter in a large linear collider.

Since the rf vacuum electric field breakdown threshold increases with increasing rf frequency, frequencies higher than S-band are under active investigation for use in a large collider. With increasing frequency, the corresponding accelerator structures become smaller (transversely) and consequently wake field effects generally become more severe. One may show that if any cylindrically symmetric structure is "scaled" to operation at higher frequency, i.e. all dimensions (cell length, aperture size, aperture thickness, etc.) are reduced by a factor of the frequency ratio, ν, then the delta function wake fields obey the scaling laws

$$W_\ell(\xi) = \nu^2 W_{\ell o}(\nu\xi) \tag{3.12a}$$

$$W_t(\xi) = \nu^3 W_{to}(\nu\xi) \tag{3.12b}$$

where $W_{\ell o}$ and W_{to} are the wake fields at the "old" design frequency. For scaled versions of the SLAC structure, for example, one may simply relabel the axes of Fig. 3.1 according to Eqs. (3.12a) and (3.12b). [We may mention here that the quadrupole wake function (units: volts/coul/m^2) scales as ν^5.]

Since Eq. (3.12) is so unfavorable for high frequency operation, consideration has been given to opening the aperture slightly of a scaled SLAC structure. This has the effect of reducing the wake fields, though at some cost in shunt impendance. The changes in the delta function wake fields for small changes from the SLAC value in the aperture a have been cited by Wilson.[4] Specifically, if we define $\alpha = a/a_{SLAC}$ then

$$W_\ell(0) = \alpha^{-1.68} W_{\ell s}(0) , \tag{3.13a}$$

$$\xi_o \simeq \alpha\xi_{os} , \tag{3.13b}$$

$$W_t^-(0) \simeq \alpha^{-3.48} W_{ts}^-(0) , \tag{3.13c}$$

$$\xi_m \simeq \alpha\xi_{ms} , \tag{3.13d}$$

$$W_t(\xi_m) \simeq \alpha^{-2.25} W_{ts}(\xi_{ms}) , \tag{3.13e}$$

where ξ_o = location of half height of $W_\ell(\xi)$, ξ_m = location of first maximum for $W_t(\xi)$, and a subscript s denotes the value for the SLAC structure (Fig. 3.1). The relations in Eqs. (3.13a)-(3.13.e) are expected to hold when α is not too far from 1.0. When α is large or when considering entirely new structures, one must resort to numerical techniques[5] to obtain the wakes.

Once the wake functions are known, one needs to study the effect on beam emittance and energy spread, that is, one must solve Eqs. (3.5) and (3.6). This must be done numerically in general, but certain simplified cases have been solved analytically. Other than the two-particle model, perhaps the most interesting of these solutions is due to Chao, Richter, and Yao[6] who have studied the transverse wake effect on a uniform bunch with no energy spread. They have found that the displacement of the bunch centroid grows along a linac in a way that depends on the dimensionless parameter

$$\eta \equiv (\frac{eQZ}{k_\beta}) \frac{\ln(E/E_i)}{(E - E_i)} W_t(\ell)(\frac{\xi}{\ell})^2 , \qquad (3.14)$$

where Q is the charge of the bunch, ℓ is its length, E is the energy at z, E_i is the energy at z = 0, and k_β is the betatron wave number, taken to be fixed during the acceleration. The parameter η gives a useful measure of the importance of transverse wake effects. It illustrates the competition between wake effects, acceleration, and focusing. For example, if one were to consider scaling a specific structure to higher frequency operation, Eq. (3.14) indicates how bunch intensity would need to be decreased and/or accelerating gradient and/or focusing strength would need to be increased in order to maintain the same level of transverse wake effects.

For the case of a continuum bunch with energy spread, no analytic solutions to Eqs. (3.5) and (3.6) exist and numerical solutions are necessary. These are carried out by dividing the bunch into slices longitudinally and tracking each slice in z, according to Eqs. (3.5) and (3.6). Some results of such calculations are given in the next section.

References

1. K. Bane, T. Weiland, and P. Wilson, in AIP Conference Proceedings No. 127, pp. 875-928 (1985).

2. A. Chao and R. Cooper, Part. Accel. 13, 1 (1983).

3. V. Balakin, et al., Proc. 12th Int. Conf. on High Energy Accelerators, Fermilab (1983). K.L.F. Bane, IEEE Trans. Nucl Sci. NS-32, 2389 (1985).

4. P. B. Wilson, in AIP Conference Proceedings No. 87, pp. 524-525 (1981).

5. T. Weiland, DESY M-82-015 (1982). K. Bane and P. B. Wilson, Proc. 11th Int. Conf. on High Energy Accelerators, Basel (1981). See also Ref. 1 and references cited there.

6. A. Chao, B. Richter, and C-Y. Yao, Nucl. Instrum. Methods 178, 1 (1980).

4. COMPUTER RESULTS FOR A FIXED-GRADIENT 500 GeV on 500 GeV COLLIDER AT LINAC FREQUENCIES OF 8 TO 20 GHz

Computer calculations of collider scaling and simulations of wake field effects have been carried out for a 500 GeV on 500 GeV collider operating at a fixed rf amplitude of 170 MV/m. Each linac will therefore be approximately 3 km in length. Two separate scenarios have been examined. They both correspond to colliders with total luminosity of 10^{37} $m^{-2}s^{-1}$. The accelerator is assumed to be a modified scaled SLAC structure, having $a/\lambda = 0.146$ (compared with the SLAC structure $a/\lambda = 0.111$). The first scenario corresponds to a "beamstrahlung-limited" case, where $\delta_{BS} = 30\%$ is assumed. The second case is a "density-and-beamstrahlung-limited" scenario, where $\delta_{BS} < 30\%$ and $N = N_o(\lambda/\lambda_o)$ is specified to scale linearly with rf wavelength. In both cases a normalized emittance of $\epsilon_N < 3 \times 10^{-6}$ m-rad is implied. The input parameters for these two cases are summarized in Table 4.1.

4.1 Results for the Collider Constrained by Beamstrahlung

4.1.1 Results of the Scaling Model

In this scenario, the interaction-point parameters depend on the rf frequency only because the bunch length, σ_z, has been specified as $\sigma_z/\lambda = 0.01$. In the transition beamstrahlung regime, using Wilson's prescription $TH_\delta(T) = $ const., the beamstrahlung loss is independent of σ_z. Since the luminosity is also independent of σ_z, N will be independent of σ_z (and hence independent of λ) in this approximation. Figure 4.1 shows N vs. ν (the rf frequency). N varies by less than 20% while ν varies by a factor 2.5. The curves labeled "A" in these figures correspond to R = 1 (round beam), while those labeled "B" represent the results at R = 10.

TABLE 4.1. Specified parameters for scaling scenarios.

	Beamstrahlung–Limited	Density–Limited
ID	W16	W17
γ	10^6	10^6
$L_T\ [m^{-2}s^{-1}]$	10^{37}	10^{37}
R	1, 10	1, 10
$R\sigma_y^2\ [m^2]$	2.5×10^{-15}	2.5×10^{-15}
σ_z/λ	0.01	0.01
δ_{BS}	0.30	0.30
$N/\lambda\ [m^{-1}]$	--	1.90×10^{11}
b	5	5
τ	0.57	0.57
$E_o\ [MV/m]$	170.	170.
$Q/\lambda^{1/2}\ [m^{-1/2}]$	43,470.	43,470.
a/λ	0.146	0.146
$v_g\ [m/s]$	9×10^6	9×10^6

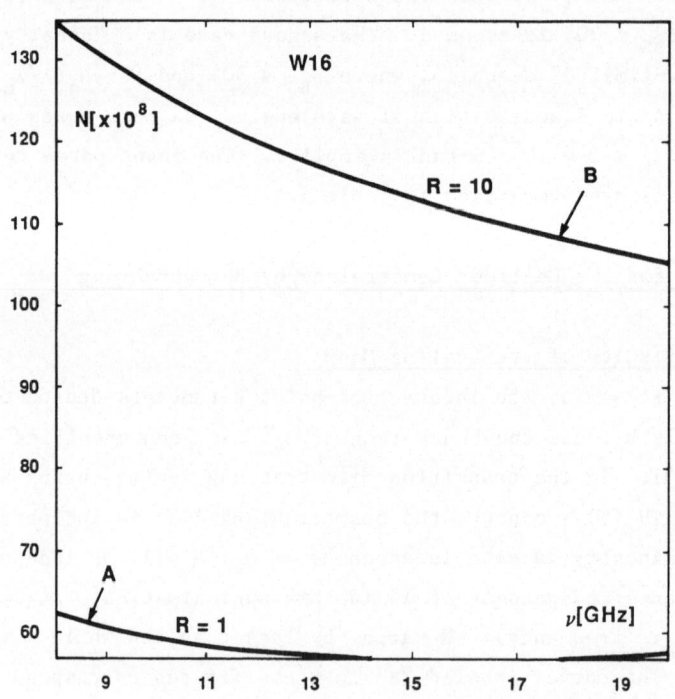

FIG. 4.1. N vs. ν for beamstrahlung–limited scaling scenario.

Here $R\sigma_y^2$ is assumed fixed. The flat beam has a smaller pinch enhancement and a smaller coefficient for beamstrahlung, and therefore corresponds to a larger N than the round beam. The disruption parameter, shown in Fig. 4.2, follows the behavior of N. Near 8 GHz, the flat beam has $D > 10$, which is probably too large to maintain stability in the final pinch.

The repetition rate is shown in Fig. 4.3. It increases roughly linearly with frequency, as expected from Eq. (2.21). In that equation, if σ_z is replaced by 0.01 λ, the equation yields $f\lambda \simeq$ const. for $R = 1$. For $R \gg 1$, using $F(R) \simeq 1.3/R$ (in the asymptotic limit), the equation gives $fR\lambda \simeq$ const. This $fR\lambda$ scaling implies that flat beams can be used to maintain a reasonable repetition rate at high rf frequencies. This result rigorously holds, however, only in the limit where L_T, δ_{BS}, and H_δ are fixed.

The scaling of N and f determine the behavior of the average beam power, $\overline{P}_b$, as shown in Fig. 4.4. $\overline{P}_b$ varies from 0.70 to 1.33 MW over the frequency range shown in the figure. The total beam efficiency is shown in Fig. 4.5. η_{bT} scales approximately as ω^2, as seen in Eqs. (2.17) and (2.18), if N and E_o are independent of ω. The values of η_{bT} are in the

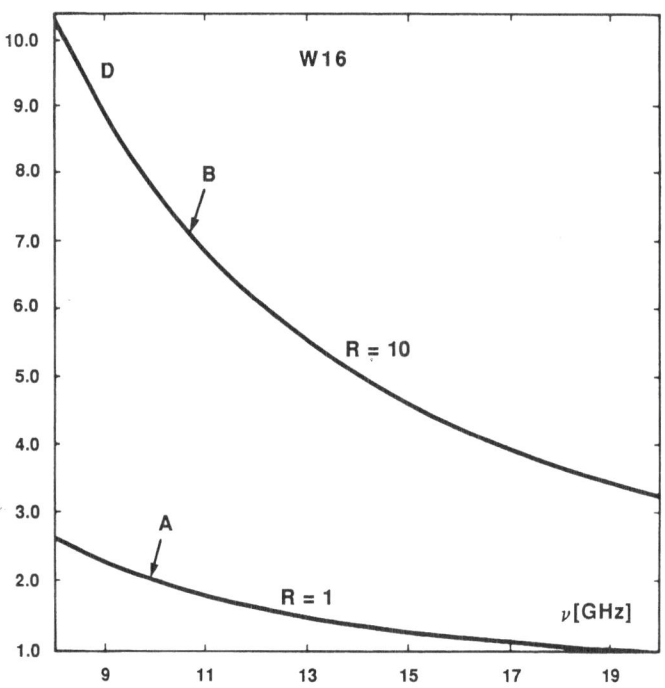

FIG. 4.2. D vs. ν for beamstrahlung-limited scaling scenario.

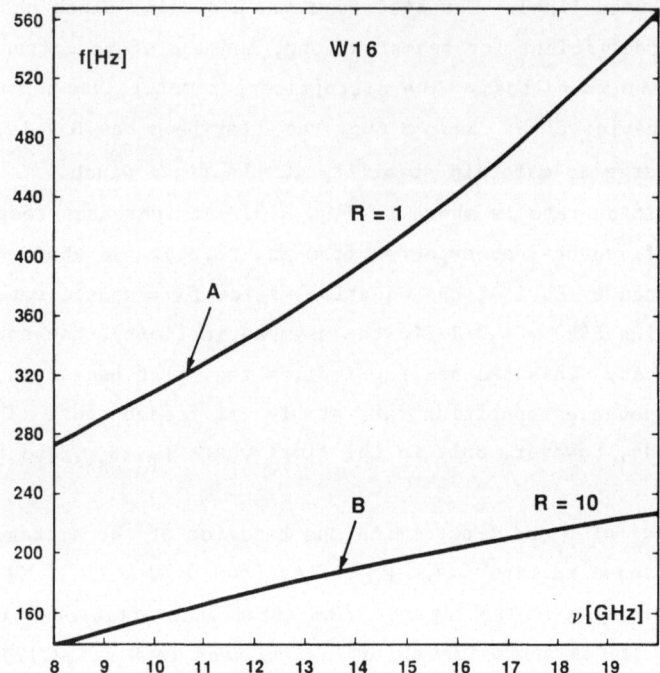

FIG. 4.3. f vs. ν for beamstrahlung-limited scaling scenario.

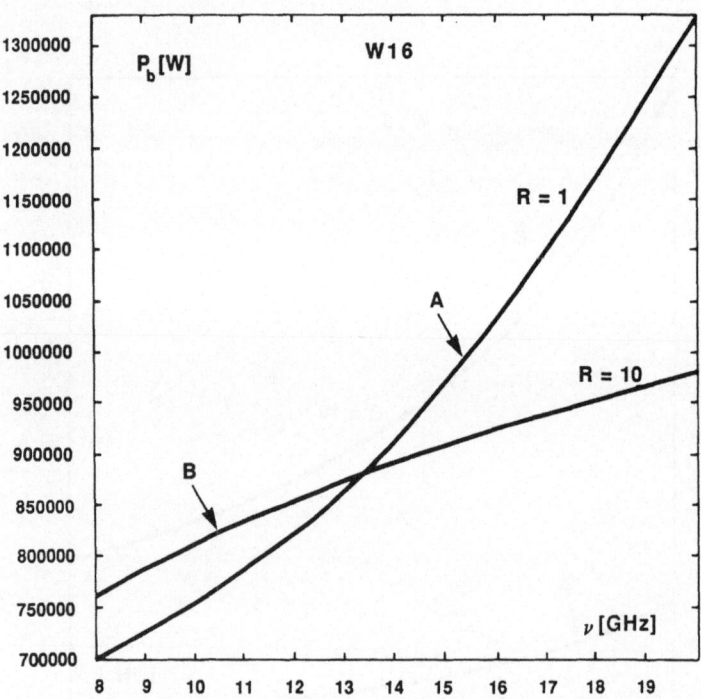

FIG. 4.4. $\overline{P}_b$ vs. ν for beamstrahlung-limited scaling scenario.

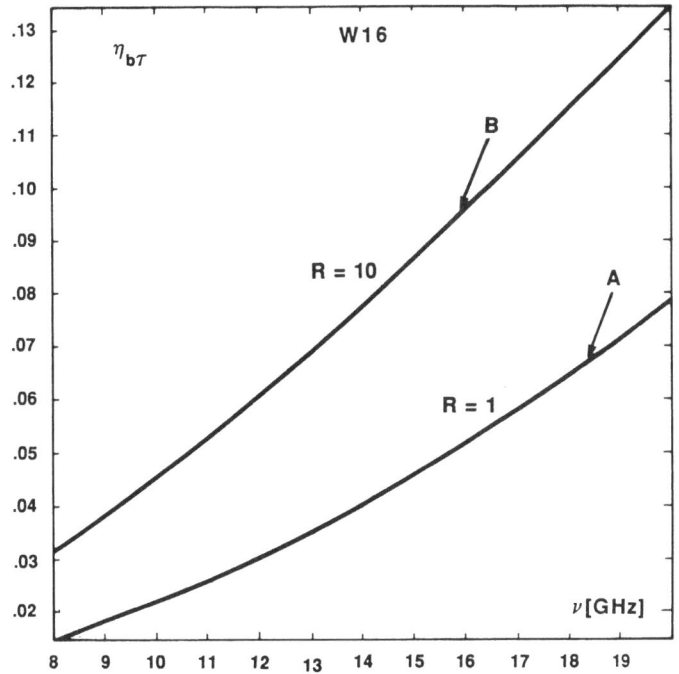

FIG. 4.5. η_{bT} vs. ν for beamstrahlung-limited scaling scenario.

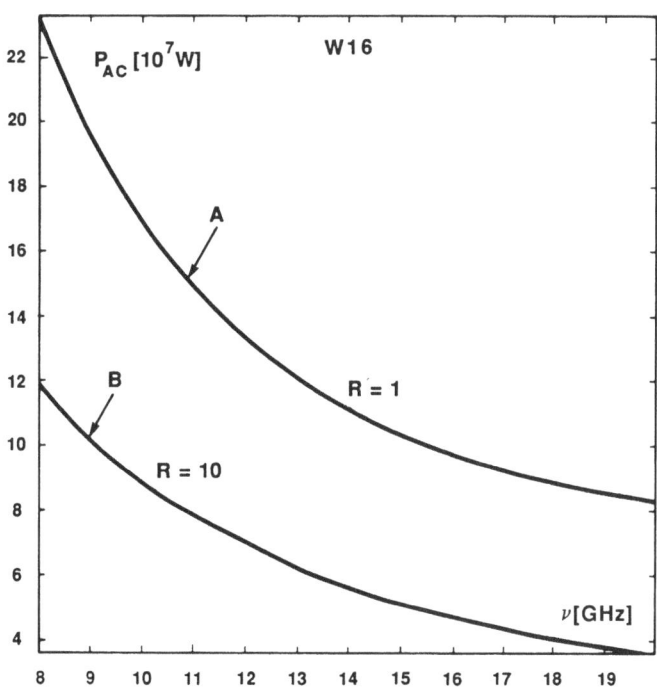

FIG. 4.6. $\overline{P}_{AC}$ vs. ν for beamstrahlung-limited scaling scenario.

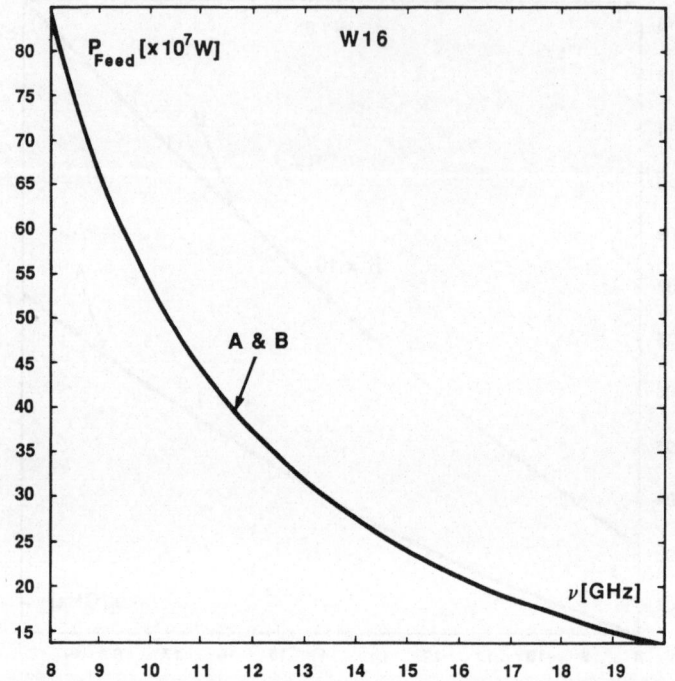

FIG. 4.7. $\hat{P}_{feed}$ vs. ν for beamstrahlung-limited scaling scenario.

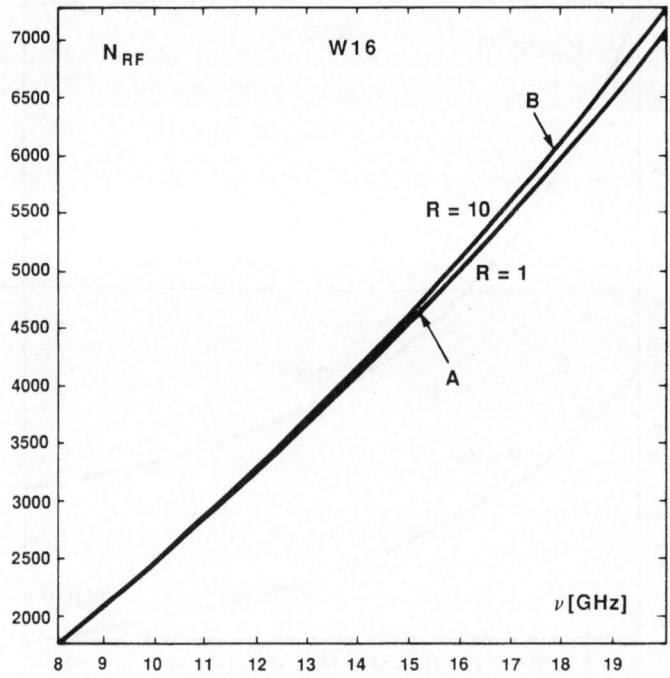

FIG. 4.8. N_{RF} vs. ν for beamstrahlung-limited scaling scenario.

range of a few percent, and are enhanced by the assumption of five bunches per rf pulse. The total accelerator efficiency is therefore of order < 1% to 2%.

The efficiency described above leads to the ac power curve seen in Fig. 4.6. The curves lie mainly in the range of 100-200 MW/linac for round beams, with lower ac power requirements for the flat beams.

The peak rf power per feed is shown in Fig. 4.7. It decreases as ω^{-2}, as expected from Eq. (2.23), since the rf amplitude is held fixed, and therefore w_s decreases as the rf frequency is increased. $\hat{P}_{feed}$ decreases from nearly 850 MW/feed at 8 GHz to under 150 MW/feed at 20 GHz. This scaling is compatible with the performance characteristics of most rf sources.

The number of rf feeds per linac is shown in Fig. 4.8 and varies from 2000-7000 feeds over the frequency range shown in the figure. The expected scaling, $N_{RF} \sim 1/t_F \sim \omega^{3/2}$, is satisfied.

4.1.2 Dipole Wake Field Effects for a Collider Constrained by Beamstrahlung

Using the bunch parameters N and σ_z from the scaling model, we have integrated Eqs. (3.5) to (3.7) to study wake effects in the linac. For four different operating frequencies 8, 10, 11.5 (4 × 2.856), and 20 GHz we have found that value of initial displacement, x*, of the bunch which will give a factor of two growth in normalized emittance at the end of the linac.

The following assumptions are made: the bunch is launched with an initial energy of 5 GeV and with zero angular injection error. The bunch is placed at the value of rf phase (θ_o) which minimizes its energy spread and is launched with that minimum value of energy spread. The beam is launched initially matched to the focusing system with an initial normalized emittance of 10^{-6} m-rad. The smooth focusing approximation is used (chromaticity = − 1) with the betatron wavelength initially set at 5 m and allowed to increase as (energy)$^{1/2}$. The length of the accelerator, L, is adjusted in each case to yield a final energy of 500 GeV. In all cases the aperture a of the SLAC-type structure is taken to be 1.32 times the frequency scaled, nominal SLAC value, i.e. a = 1.32 (2.856 GHz/ν) 1.163 cm.

Results for a collider constrained by beamstrahlung are given in Table 4.2. Since N is roughly fixed, x* decreases rapidly with frequency, as expected. These values of x* imply very stringent conditions on injection jitter. The use of Landau damping[1] can relax these conditions significantly, although its use in the case including random magnet misalignments requires further study.[1]

Table 4.2. Collider constrained by beamstrahlung.

ν (GHz)	N (10^9)	σ_z (mm)	θ_o (^{0})	σ_E/E (%)	L^+ (m)	x* (μ)
8	6.3	0.375	0.8	0.26	2923	9.3
10	5.9	0.300	1.2	0.26	2926	9.0
11.5	5.8	0.260	1.5	0.25	2929	8.5
20	5.7	0.150	4.6	0.24	2960	1.5

L^+ = Length from 5 GeV to 500 GeV

L^+ = Length from 5 GeV to 500 GeV

Note that longitudinal wake effects are fairly mild, resulting in an increase in L of only ~ 4% from 8-20 GHz and virtually no change in the minimum attainable energy spread which, at ~ 0.25% is acceptable for purposes of the final focus.

4.2 Results for the Collider Constrained by Beamstrahlung and Number of Particles per Bunch

4.2.1 Results of the Scaling Model

The scaling scenario described in the previous section results in the number of particles per bunch being nearly independent of the rf frequency. This feature of the model may be difficult to achieve for several reasons. First, since σ_z/λ is assumed fixed, this scaling requires that the bunches be compressed to higher density as λ is decreased. Second, the wake fields become stronger with increasing rf frequency, scaling as ω^2 for the monopole wake field and as ω^3 for the dipole wake field, as seen from Eq. (2.10). It may therefore be necessary to reduce the number of particles as some power of λ in order to mitigate these effects, as was discussed in Section 1.

On the other hand, as the number per bunch decreases, with specified luminosity, the repetition frequency increases, thereby driving the average beam power up. Since reducing N also reduces the beam efficiency, the ac power required will be higher than in the constant beamstrahlung scenario. For these reasons, the scaling of N has been limited to N ~ λ, which is slower than the scaling required to offset the wake field effects. As a figure of merit that is a measure of the phase-space density, one can use the number of particles N (per bunch) divided by the wavelength λ and the normalized emittance ε_N, i.e. $N/\lambda\varepsilon_N$. For SLC the design figures are $N = 5 \times 10^{10}$, $\lambda = 10.5$ cm, and $\epsilon_N = 3 \times 10^{-5}$ m-rad, hence

$$(N/\lambda\epsilon_N)_{SLC} = 1.6 \times 10^{14} \text{ cm}^{-1} \text{ (m-rad)}^{-1} \ .$$

The particular case considered here has $N/\lambda = 1.9 \times 10^{11}$ m^{-1}, which corresponds to $N = 2 \times 10^{10}$ at the SLAC frequency. The figure of merit, with $\epsilon_N = 3 \times 10^{-6}$ m-rad, is then

$$N/\lambda\epsilon_N = 6.4 \times 10^{14} \text{ cm}^{-1} \text{ (m-rad)}^{-1} \ ,$$

which is $\approx 4 \times (N/\lambda\epsilon_N)_{SLC}$. For a round beam this curve intersects the beamstrahlung line ($\delta_{BS} = 0.3$, $N_0 \sim$ const) at a frequency of about 9.5 GHz. Again the "A" and "B" curves correspond to $R = 1$ and $R = 10$, respectively, with $R\sigma_y^2 =$ const. Unlike the previous example, the scaling of the interaction-point quantities with rf frequency in this example is due to the specified variations in both N and σ_z; both are assumed to vary linearly with λ.

The variation of the beamstrahlung energy loss with frequency is shown in Fig. 4.9, and mainly reflects the decrease of N with λ. The round-beam result lies above 30% for $\nu < 9.5$ GHz. The disruption parameter, shown in Fig. 4.10, scales as $D \sim N\sigma_z \sim \omega^{-2}$.

The repetition frequency scales as $f \sim (N^2 H_D)^{-1}$, and therefore increases rapidly with frequency as shown in Fig. 4.11. The resulting beam power, in Fig. 4.12, also increases with frequency, since $Nf \sim (NH_D)^{-1}$. It is interesting that large R does not help the repetition rate and the power in this scenario, as it did in the previous one. There, fixing the beamstrahlung lead to $fR\lambda =$ const. as an approximate scaling law. Here, specifying N causes the repetition frequency to be determined entirely from the luminosity. Since $\sigma_y^2 R$ has been assumed fixed, the R dependence of f comes only from H_D (cf. Fig. 4.1). From Eq. (4.1), with $\sigma_y^2 R$ fixed, D will be twice as large for $R \gg 1$ than it is at $R = 1$, thereby compensating for the reduction in the pinch enhancement factor at large R. The result is that both f and $\overline{P}_b$ are only weakly affected by R.

The total beam efficiency, shown in Fig. 4.13, is reduced by the loss of particles at high frequency in this scenario, and only achieves a maximum value of $\sim 4\%$, corresponding to a total accelerator efficiency of $< 1\%$ at all frequencies shown in the figure. This result leads to the requirement for ac power shown in Fig. 4.14. The flat beam curve lies above 400 MW/linac at all the frequencies shown in the figure. The round

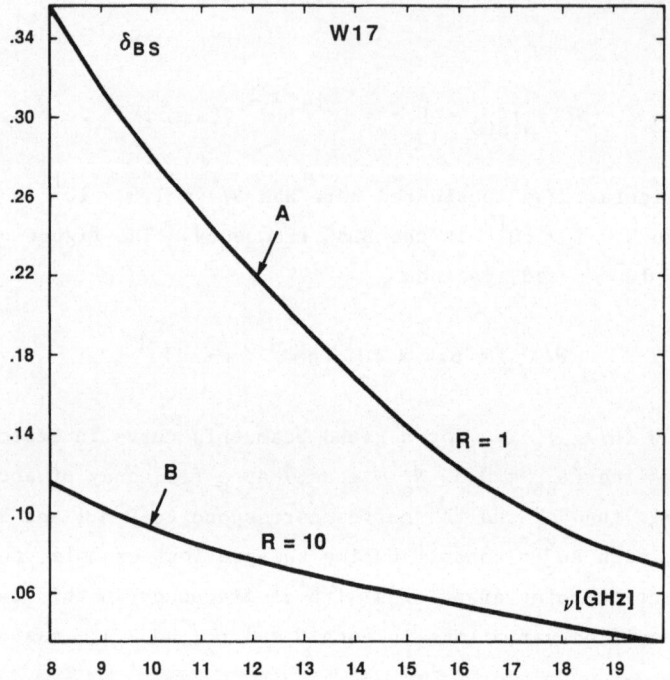

FIG. 4.9. δ_{BS} vs. ν for density-limited scaling scenario.

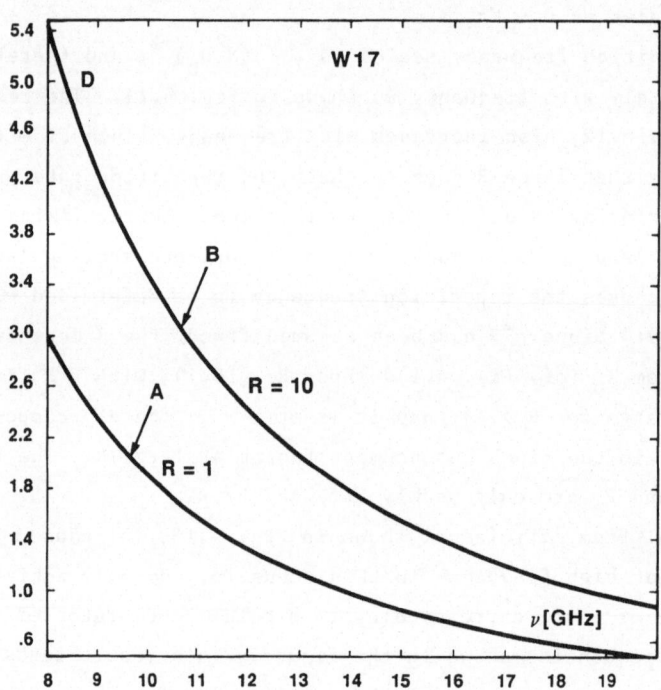

FIG. 4.10. D vs. ν for density-limited scaling scenario.

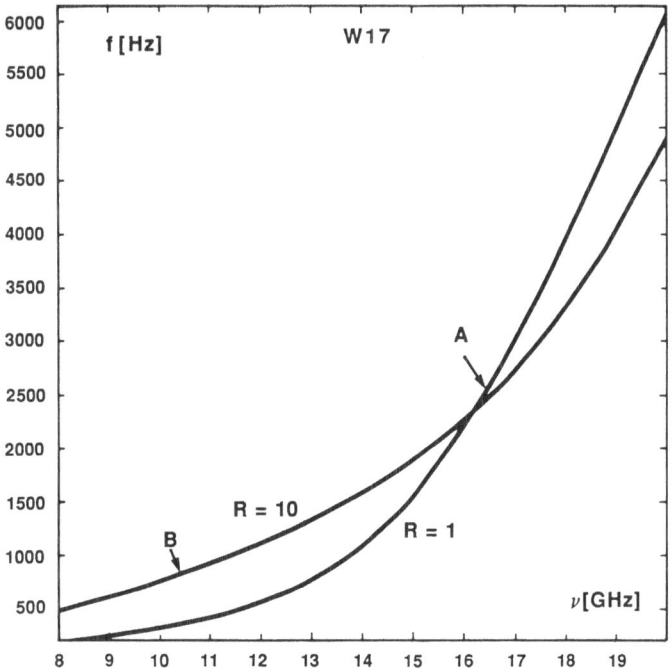

FIG. 4.11. f vs. ν for density-limited scaling scenario.

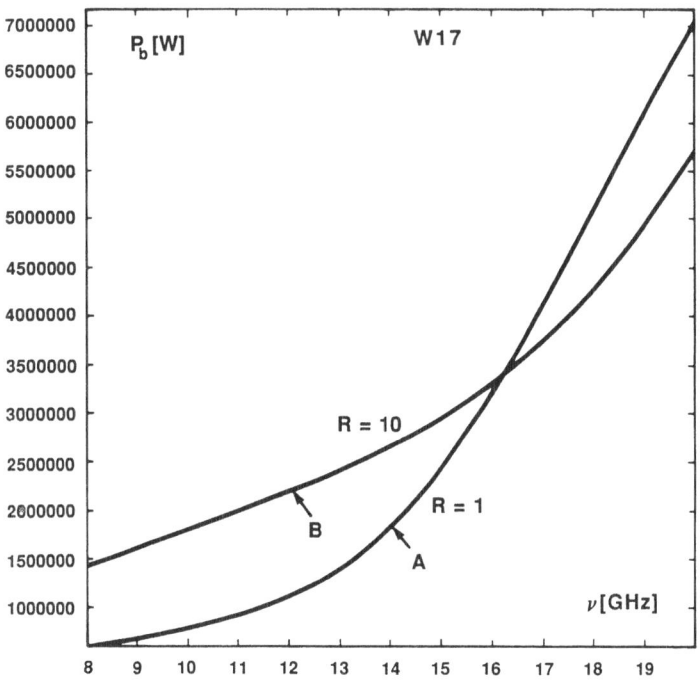

FIG. 4.12. $\overline{P}_b$ vs. ν for density-limited scaling scenario.

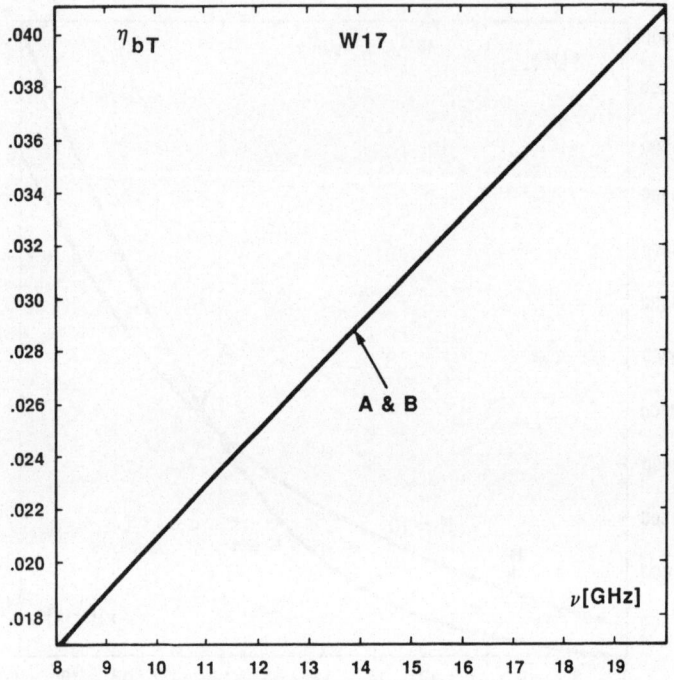

FIG. 4.13. η_{bT} vs. ν for density-limited scaling scenario.

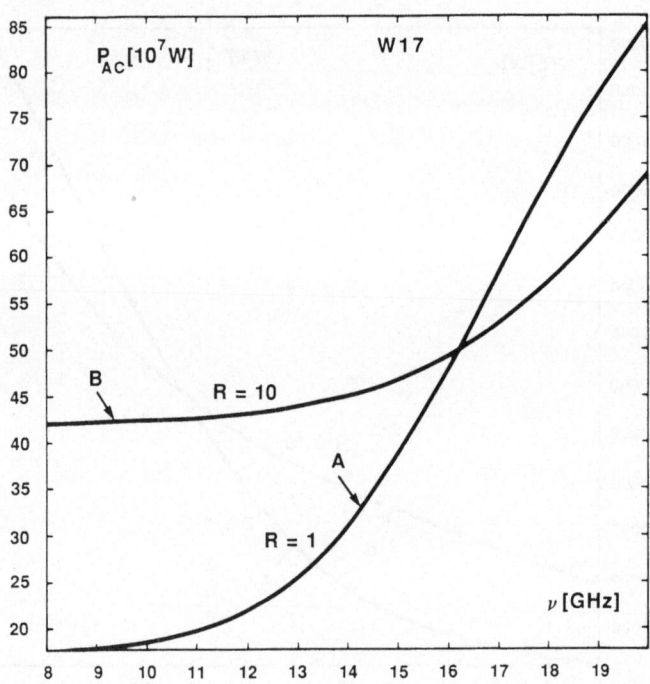

FIG. 4.14. $\overline{P}_{AC}$ vs. ν for density-limited scaling scenario.

beam, on the other hand, has a narrow frequency band in which the power
lies below 200 MW/linac. In the region, 9.5 GHz $<$ ν $<$ 11 GHz, the ac power
is below 200 MW and the beamstrahlung is less than 30%.

The results for peak rf power per feed and for the number of rf feeds
per linac are the same in this scenario as in the previous one, since the
same structure has been assumed with the same rf amplitude.

4.2.2 Dipole Wake Field Effects for a Collider Constrained by Number of Particles per Bunch

Numerical calculations similar to those described in Section 4.1.2
were carried out for the collider constrained by N. The results are given
in Table 4.3. Note that for the 8 GHz case, energy loss to beamstrahlung
is unacceptably large (Fig. 4.9) and for the 20 GHz case the ac power

Table 4.3. Collider constrained by N.

f (GHz)	N (10^9)	σ_z (mm)	θ_o (^{0})	σ_E/E (%)	L^+ (m)	x^* (μ)
8	7.1	0.375	0.9	0.26	2924	9.2
10	5.7	0.300	1.1	0.26	2926	9.1
11.5	5.0	0.260	1.3	0.26	2927	8.7
20	2.9	0.150	2.3	0.24	2936	4.4

consumption is unacceptably large (Fig. 4.14); these cases are included in
Table 4.3 only for illustration.

Figures 4.15 and 4.16 give, for the 11.5 GHz case, plots of the bunch
centroid displacement and the normalized bunch emittance versus distance
along the accelerator. The centroid displacement exhibits complicated
behavior, illustrating the competing effects of focusing, wake fields, and
adiabatic damping. The normalized emittance grows monotonically, though
somewhat faster at the low energy end, where wake field effects are
strongest.

Reference

1. V. Balakin, et al., Proc. 12th Int. Conf. on High Energy Accelerators,
   Fermilab (1983). K.L.F. Bane, IEEE Trans. Nucl Sci. NS-32, 2389
   (1985).

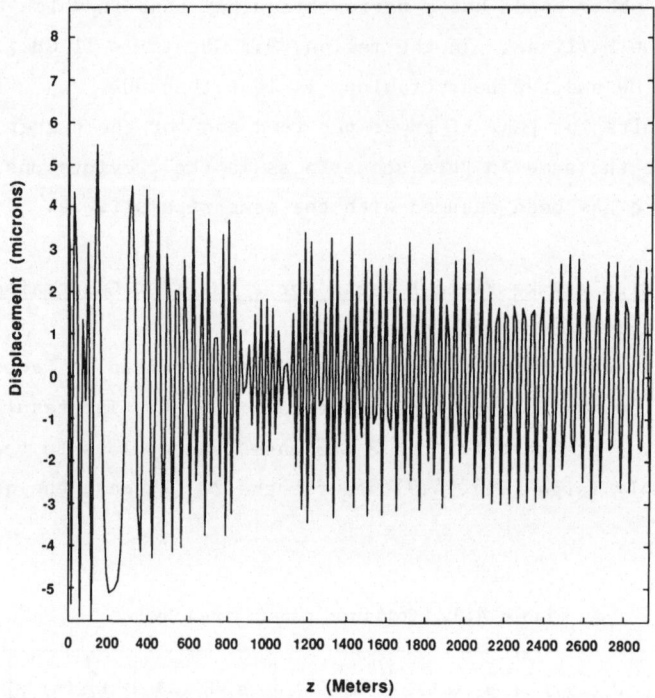

FIG. 4.15. Displacement of the bunch centroid versus distance along the linac for the density-limited scenario at 11.5 GHz.

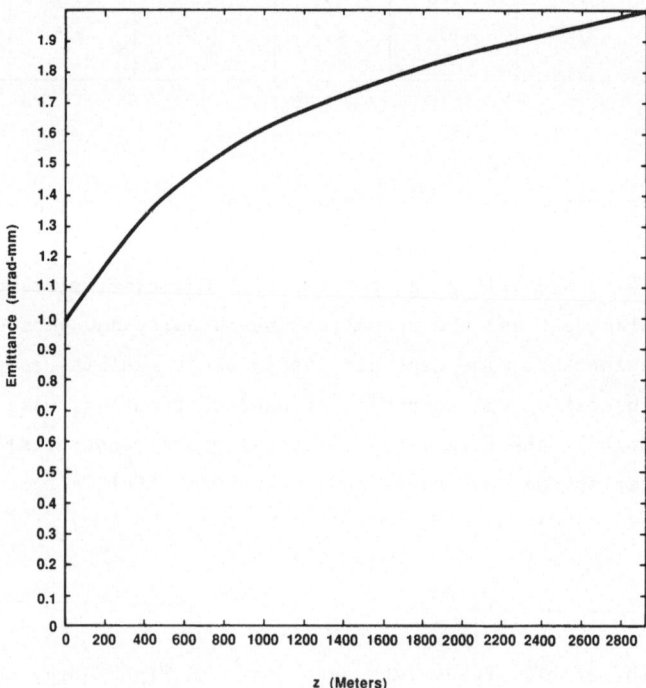

FIG. 4.16. Normalized emittance of the bunch versus distance along the linac for the density-limited scenario at 11.5 GHz.

5. THE GYROKLYSTRON AS A LINAC MICROWAVE SOURCE

The results of the accelerator scaling study place stringent requirements on the microwave source. In particular, the peak power requirement far exceeds the current state-of-the-art for microwave amplifiers[1] and a new approach is needed. The gyroklystron[2-4] has the potential to satisfy all the requirements and is the subject of this section. In Section 5.1 the gyroklystron principle is discussed. In Section 5.2 we detail the design of the University of Maryland's 30 MW, 10 GHz gyroklystron which is currently under construction. We introduce gyroklystron scaling laws in Section 5.3 and use them to predict the gyroklystron's ability to meet supercollider requirements in Section 5.4. In the final section we compare gyroklystrons to klystrons.

5.1 General Considerations

The typical gyroklystron beam is annular, with each electron on a helical orbit in a uniform applied magnetic field (see Fig. 5.1). Ideally, most of the beam energy is in the perpendicular motion and the axial velocity spread is small. Particles are bunched in phase by the negative-mass effect and energy is extracted via the synchronous interaction of the bunched beam with an electromagnetic wave. Bunching occurs because the electron cyclotron frequency depends on electron energy:

$$\omega_{ce} = \frac{eB_o}{\gamma_o m_o}$$

where B_o is the magnitude of the applied field, e and m_o are the charge magnitude and rest mass of an electron, and γ_o is the total electron energy normalized to the rest energy. Consequently, electrons that lose energy rotate more rapidly and electrons that gain energy rotate more slowly.

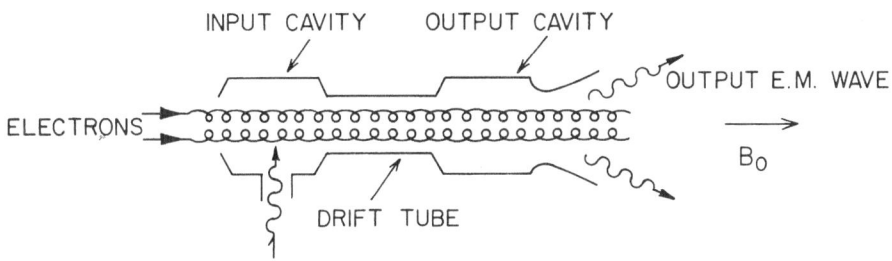

FIG. 5.1.   Schematic of a two cavity gyroklystron.

The typical gyroklystron circuit consists of a collection of microwave cavities separated by drift sections cutoff to electromagnetic waves at the desired frequency. The first cavity is excited with a mode that has a large $E_\perp$ field in the vicinity of the beam. The electromagnetic wave frequency is close to the cyclotron frequency so that most electrons experience a non-trivial energy exchange. That is, $d\gamma/dt \propto \langle v_\perp \cdot E_\perp \rangle$ is significant in the cavity for most electrons. Those electrons that lost energy in the cavity will advance in phase in the drift section and those that gained energy will retard in phase. The drift section lengths are chosen so that most particles are in phase when entering the last cavity, enabling efficient energy extraction (see Fig. 5.2).

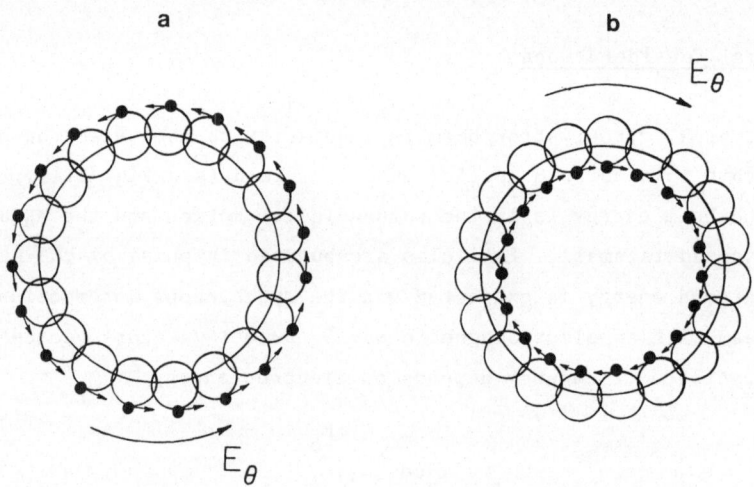

FIG. 5.2. Cross-section of the annular beam showing electrons bunched in phase in their cyclotron orbits. Each dark circle represents an electron bunch and the two figures show the situation a half period apart.

5.2 The University of Maryland's Gyroklystron Experiment

At the University of Maryland we are currently constructing an X-band gyroklystron which is predicted to achieve a peak power of 30-50 MW in 1-2 μs pulses with an efficiency of 45% and a gain of 63 dB. The major subsystems are depicted in Fig. 5.3, and the nominal operating parameters are listed in Table 5.1. A line-type modulator provides the required voltage pulse and the spiraling beam is produced by an electron gun of the Magnetron Injection type. The beam interacts with a four-cavity circuit operating in the lowest circular electric mode (TE_{01}^{o}) and is collected in the beam dump. Details of these subsystems are discussed below.

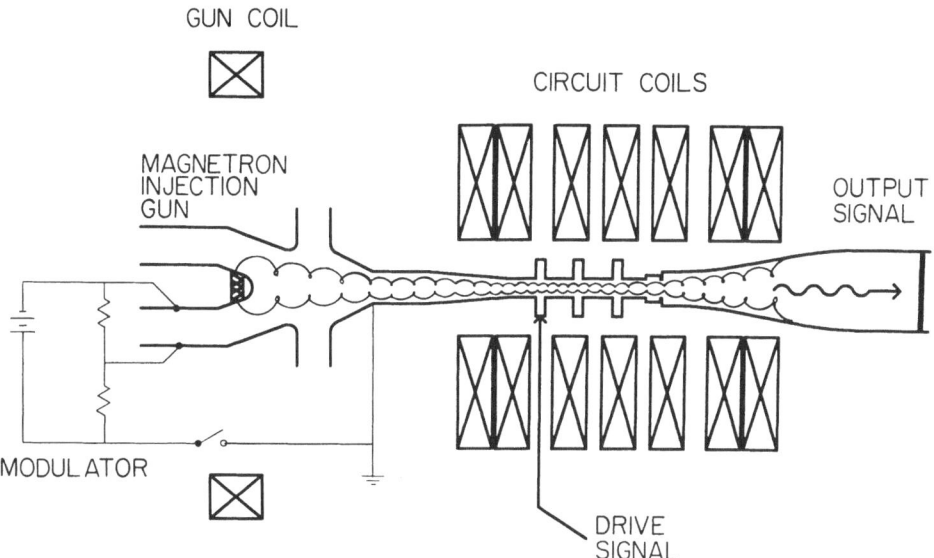

FIG. 5.3. Simplified schematic of the University of Maryland gyroklystron.

TABLE 5.1. The beam parameters.

Beam Power P_o	80 MW
Beam Voltage V_o	500 kV
Magnetic Field B_o	0.555 T
Velocity Ratio $v_\perp/v_z$	1.5
Guiding Center Radius r_g	0.79 cm
Larmor Radius r_L	0.43 cm
Guiding Center Spread Δr_g	< 0.43 cm

5.2.1 The Modulator

The 500 kV voltage pulse is generated by a line-type modulator, which
is the lumped-element equivalent of a transmission line pulser. Our
modulator is similar in design to those used by the SLAC klystrons[3] and is
shown schematically in Fig. 5.4. Four pulse-forming networks (PFN´s) in
parallel are resonantly charged to 46 kV using a 0.68 H charging
inductor. This charging cycle is triggered using a spark gap, and derives
its energy from a 10 μF storage bank. The bank is continually recharged to
25 kV from the main power suppply at approximately 180 mA. Shot-to-shot
voltage regulation of better than 1% is achieved by interrupting the
charging cycle when the PFN voltage has reached a suitable level. To

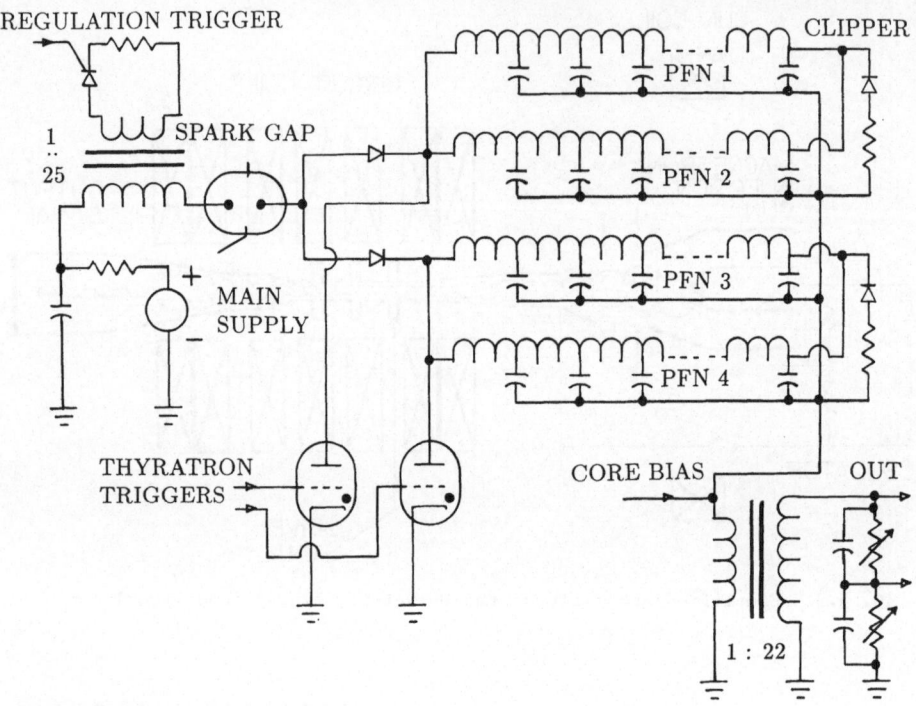

FIG. 5.4. Simplified modulator schematic.

accomplish this, an SCR connects a small resistance to a low voltage secondary winding on the charging inductor.

The PFN´s have 8–12 stages to provide pulses of 1–2 µs. While the PFN capacitors are fixed at 0.014 µf, the inductors can be varied between 0.5 and 3.5 µH by choosing a suitable tap point and by inserting a copper tuning slug. The PFN´s are switched through two thyratrons into a 1:22 pulse transformer which provides the required potential of 500 kV and current of 400 A. Peak-to-peak ripple of better than 1% is achieved by loading down the parasitic oscillations in the pulse transformer and by adjusting the PFN inductors. A dc core bias current of 80 A on the transformer primary is required to achieve the desired maximum pulse length. Approximately half of the available current is shunted through a compensated resistive divider to provide a more linear load and an intermediate voltage for a modulation anode in the electron gun. The maximum repetition rate with the current power supply is 4 Hz.

The modulator also incorporates extensive fault sensing. Short circuits, improper charges, and open circuit conditions are detected by various sensors and appropriate action is taken to prevent damage to either the modulator or the electron gun. An end-of-line clipper is connected to the PFN's to prevent reverse voltage from appearing on the thyratrons after a short circuit.

A simplified modulator with resistive charging, a spark gap trigger, and a purely resistive load has been tested. This configuration was capable of demonstrating all the modulator requirements except for repetition rate and shot-to-shot repeatability. Successful operation at 500 kV, 400 A was achieved with 1.5 μs flat-top pulses. A typical experimental output trace is shown in Fig. 5.5. A flat-top ripple of less than 1% was achieved with inductance tuning and a simple loading network.

5.2.2 The Electron Gun

The standard configuration for the electron gun is shown in Fig. 5.6. This configuration is called a double anode Magnetron Injection Gun (MIG). The MIG generates a rotating beam as follows. A thermionic emitter strip produces free electrons. The electrons are accelerated in the predominantly radial electric field generated by the cathode-control anode potential difference. The cathode is immersed in a magnetic field, so the electrons begin to rotate. The electrons slowly drift away from the cathode into a region where a strong axial electric field generated by the control anode-main anode potential difference accelerates electrons to their full energy. Finally, the electrons traverse an adiabatic magnetic compresion region in which parallel energy is converted to perpendicular energy as the magnetic field increases to its final value.

An analytic model[6] has been developed which can be used to predict MIG characteristics. The model approximates the cathode-control anode system as a coaxial system with effective "slant lengths" to account for the conical tapers. The Hull cutoff condition predicts the electron motion in the coaxial system and conservation laws translate those properties to the interaction region. The resultant adiabatic trade-off equations are summarized below:[7]

$$ f_m = (\mu r_c/r_L)^2 \tag{5.1} $$

$$ (d_{ac}/r_c) = d_f \mathfrak{n}/\cos\phi_c \tag{5.2} $$

$$ (\ell_s/r_c) = I_o/(2\pi J_c r_c^2) \tag{5.3} $$

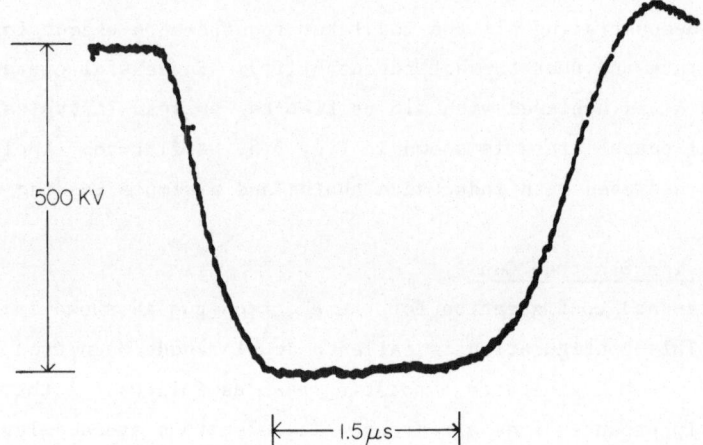

- RESISTIVE CHARGING
- SPARK-GAP TRIGGER
- PURE RESISTIVE LOAD

500 KV

←— 1.5 μs —→

FIG. 5.5. A typical modulator voltage pulse.

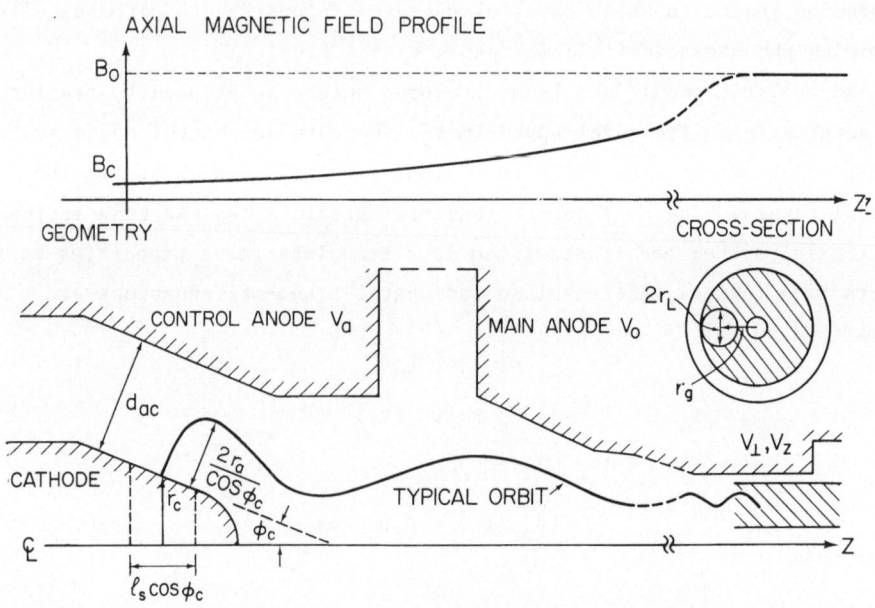

AXIAL MAGNETIC FIELD PROFILE

B_0

B_c

Z

GEOMETRY

CROSS-SECTION

CONTROL ANODE V_a

MAIN ANODE V_0

$2r_L$

r_g

d_{ac}

$V_\perp, V_z$

CATHODE

$\frac{2r_a}{\cos\phi_c}$

TYPICAL ORBIT

r_c

ϕ_c

Z

$\ell_s \cos\phi_c$

FIG. 5.6. Schematic of a double anode Magnetron Injection Gun (MIG).

$$[eV_a/(mc^2)] = \frac{\ell n(1 + d_f\mu)}{\ell n(1 + 2\mu)} \left\{\left[1 + 4\frac{(1 + \mu)^2}{(1 + 2\mu)^2}\frac{(\gamma_o^2 - 1)}{f_m \cos^2\phi_c}(\frac{\alpha^2}{1 + \alpha^2})\right]^{1/2} - 1\right\}$$

(5.4)

$$E_c = \frac{V_a \cos\phi_c}{r_c \ell n(1 + d_f\mu)}$$

(5.5)

$$(\Delta r_g/r_L) = \frac{(\ell_s/r_c) \sin\phi_c}{\mu(1 + \mu^2)^{1/2}}$$

(5.6)

$$(J_c/J_L) = J_c\left\{\frac{k_1}{2k_2}\left[(1 + 4 V_a k_2/k_1^2)^{1/2} - 1\right]\right\}^{-3/2}$$

(5.7)

where

$$k_1 = 5.6872 \times 10^3 [r_c \ell n(1 + d_f\mu)/\cos\phi_c]^{4/3} [1 + \frac{2}{15}\ell n(1 + d_f\mu)$$

$$+ \frac{11}{450}\ell n^2(1 + d_f\mu) + \frac{437}{111375}\ell n^3(1 + d_f\mu) + \ldots]$$

and

$$k_2 = 2.9356 \times 10^2 [r_c \ell n(1 + d_f\mu)/\cos\phi_c]^{8/3} [\frac{1}{14} + \frac{6}{175}\ell n(1 + d_f\mu) + \ldots] \; .$$

where

$$\mu = [(r_g/r_L)^2 - 1]^{-1/2}$$

and

α is the ratio of perpendicular to parallel velocity

r_c is the emitter strip average radius

r_L is the average Larmor radius

r_g is the average guiding center radius

Δr_g is the spread in guiding center radii

ℓ_s is the emitter strip length

d_{ac} is the cathode-control anode gap

d_f is the number of Larmor radii across the cathode–control anode gap

ϕ_c is the cathode half–angle

$f_m = B_o/B_c$ is the magnetic compression ratio

I_o is the total current

J_c is the emitter strip current density

J_L is the approximate space–charge limiting current density

E_c is the electric field on the emitter strip

V_a is the control anode voltage

and

V_o is the main anode voltage

Equations (5.1)–(5.4) provide design data for the MIG system, while Eqs. (5.5)–(5.7) provide constraints on possible designs. Thus, the latter equations are used to obtain the feasible design regions and the former are subsequently used as input to computer simulations. The allowed range for Δr_g depends on the microwave circuit design. The electric field is limited by breakdown considerations, and is usually kept below 50 kV/cm. This constraint sets a lower limit on the allowed range for r_c. Setting an upper limit on the tolerable value for J_c/J_L results in an upper limit on the allowed range for r_c. Thus, acceptable designs must be found in the region between the minimum and maximum allowable values for r_c.

The beam parameters in Table 5.1 form the starting point of the MIG design. The design tolerances for the MIG system are shown in Table 5.2. The peak electric field requirement comes from empirical gun data.[8] We emphasize that this electric field is acceptable only for short pulses ($<$ 2 μs) and would have to be considerably lower in a CW gun (further complicating the design). The numerical simulations indicate that the peak field typically appears on or near the tip of the cathode nose and that its magnitude is approximately twice that of the emitter strip.

TABLE 5.2. Design tolerances.

Peak Electric Field	100 kV/cm
Minimum Current Range	120–200 A
Maximum Axial Velocity Spread (160 A)	7%
Maximum Beam Position	0.0135 m

The velocity spread requirement results from a code that predicts beam-to-microwave energy conversion efficiency.[2] Those code simulations show that a reasonable efficiency can be achieved with a 100 MW beam at an axial velocity spread of Δv_z = 10% and that the efficiency drops rapidly thereafter. Consequently, we arbitrarily define the usable (or operating) current range to be the range of currents having Δv_z < 10%. Because of the high space-charge factor, this range does not extend down to 0 A. The velocity spread is minimized at 160 A so that the usable current range can be maximized while still achieving an acceptable spread at 200 A.

The magnetic field profile is shown in Fig. 5.7(a). It is generated by a set of seven water-cooled pancake coils in the microwave circuit region and supplemented by a large gun coil placed over the cathode. The length of the compression region, defined as the distance between the emitter strip center and the entrance to the microwave circuit, is 48.25 cm. This is a reasonable compromise between the conflicting requirements of a physically short system and an approximately adiabatic compression.

The resultant gun design (from computer simulations) is shown in Fig. 5.7(b), the key dimensions are listed in Table 5.3, and the simulated beam properties at 160 A are displayed in Table 5.4.

The dependence of axial velocity spread on current is summarized in Fig. 5.8, where it can be seen that the operating current range requirement is achieved. This dependence results from the effects of beam self-fields on the non-laminar beam motion.[7] A velocity ratio of α = 1.5 is maintained

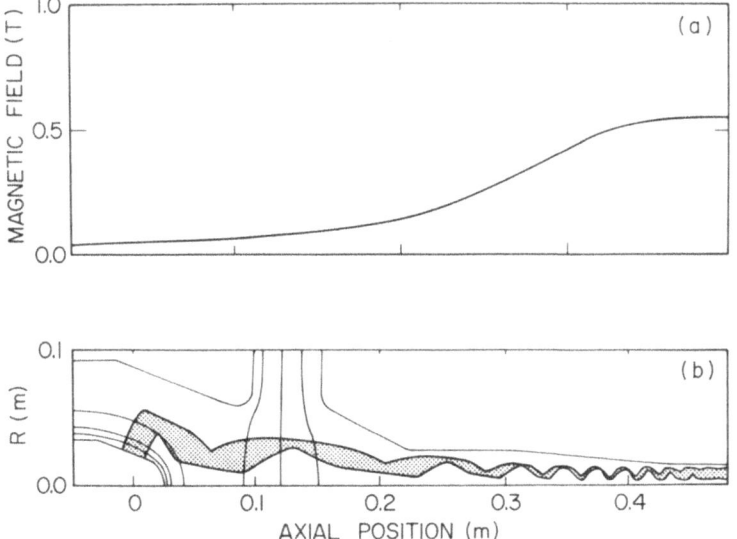

FIG. 5.7. Simulated beam for the University of Maryland MIG.

TABLE 5.3. 100 MW MIG electrode specifications.

Cathode Radius r_c	0.0228 m
Emitter Strip Width ℓ_s	0.020 m
Cathode-Control Anode Gap d_{ac}	0.0613 m
Control Anode-Accelerating Anode Gap	0.0550 m
Cathode Half Angle ϕ_c	20^0
Compression Ratio f_m	12
Control Anode Voltage V_a	143 kV
Emission Current Density J_c (Uniform)	5.61 A/cm^2
Total Current	160 A

TABLE 5.4. Simulated beam properties at 160 A.

Average Velocity Ratio	1.522
Average $\beta_{\perp o}$	0.7175
Average β_{zo}	0.4735
Spread in β_{zo}	5.586%
Average Guiding Center Radius	0.00781 m
Maximum Beam Position	0.012456 m
Fraction of Space Charge Limit	40%
Peak Electric Fields	
Cathode	91 kV/cm
Control Anode	77 kV/cm
Accelerating Anode	84 kV/cm

for the data points shown in the figure; this is achieved by adjusting the control anode voltage.

5.2.3 The Microwave Circuit

The microwave circuit consists of four right circular cavities, two nonlinear waveguide transitions, and a half-wavelength output window. An input signal is injected into the first cavity, the second and third cavities enhance beam bunching and hence gain, and the final cavity has a coupling iris in the endwall to launch the output signal into the waveguide.

The design for the gyroklystron cavities is shown in Fig. 5.9(a) and

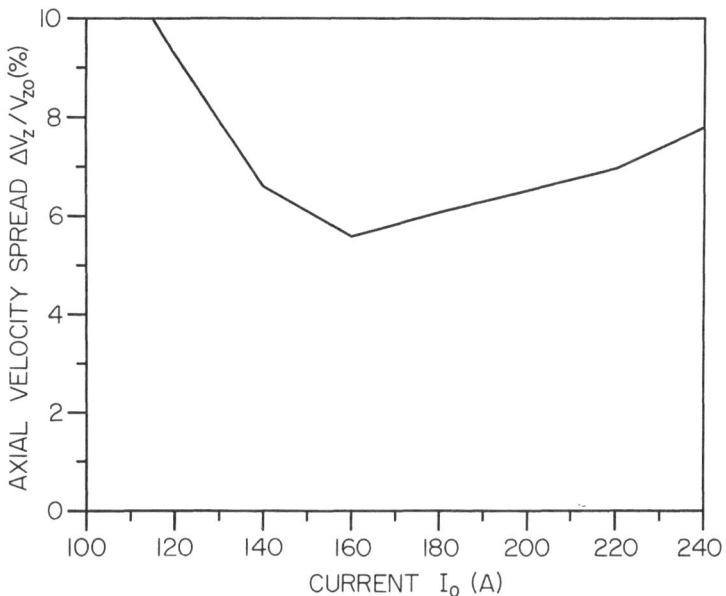

FIG. 5.8. The dependence of axial velocity spread on beam current for the University of Maryland MIG.

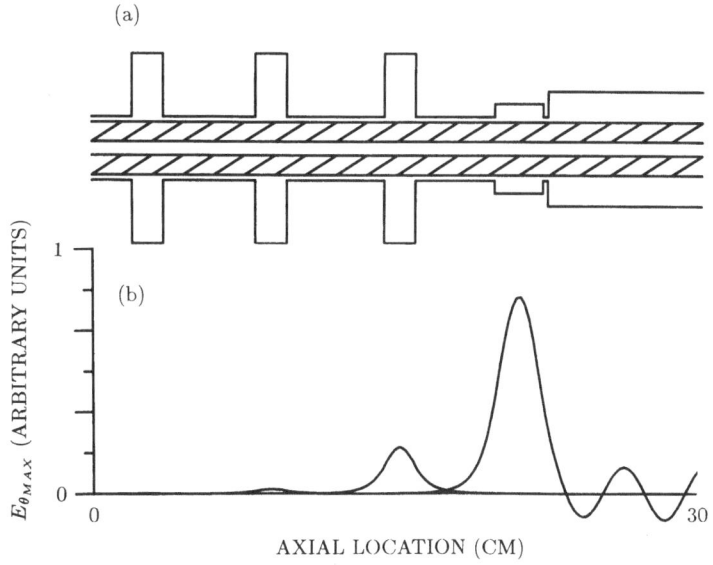

FIG. 5.9. The gyroklystron cavity configuration (a) and field profile (b).

the relevant dimensions are listed in Table 5.5. The optimized design is found with the aid of a partially self-consistent steady-state code[2] that uses the beam parameters from the gun simulation. Particle trajectories are numerically integrated through the cold cavity electromagnetic fields, which consist of only circular electric (TE_{on}^{o}) modes. The relative amplitudes of the circular harmonics are computed using a mode expansion in the cavities and drift tubes. The cavity Q's are such that 160 A is 75% of the start oscillation current in the input and buncher cavities and 101% of the start oscillation current in the output cavity. Detuning the final two cavities enhances efficiency at the expense of gain.

TABLE 5.5. Optimized gyroklystron circuit dimensions.

Cavity No.	Q	f_{res} (GHz)	Length (cm)	Radius (cm)
1	235	10.000	1.53	4.50
2	235	10.000	1.53	4.50
3	240	9.975	1.53	4.50
4	180	9.995	2.38	2.11
Drift Tube				
1			4.50	1.50
2			5.40	1.50
3			3.90	1.50
Coupling Iris			0.336	1.50

Figure 5.9(b) displays the maximum E_θ at the beam center as a function of position in the microwave circuit. The phase shift between cavities is not indicated in the figure. The electric field in the first cavity is too small to be seen; the alternating field to the right of the fourth cavity represents the outgoing wave of the endfire system. Although the field amplitudes decrease rapidly in the drift regions, indicating adequate isolation between cavities, the finite drift tube fields drastically affect circuit performance. Contour plots of saturated efficiency as a function of B_o and $\Delta v_z/v_z$ are displayed in Fig. 5.10. In a perfectly cold beam with no velocity spread, the predicted optimum efficiency is ~ 50%. This falls

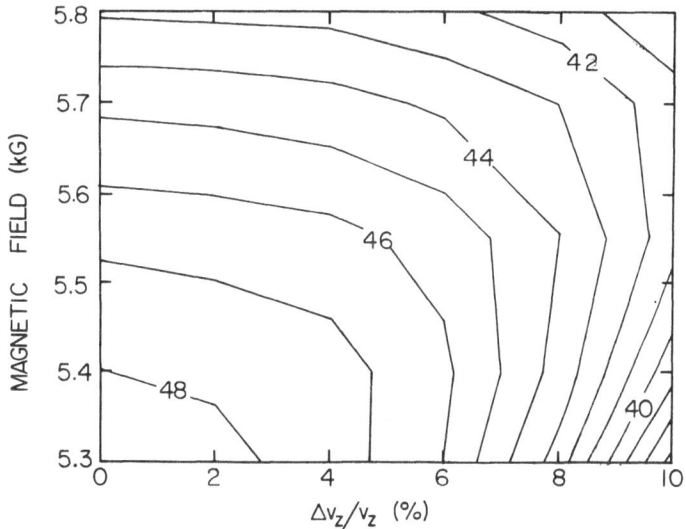

FIG. 5.10. Gyroklystron efficiency as a function of magnetic field and axial velocity spread.

to 45.5% when $\Delta v_z/v_z$ = 6.4%, because the spread in velocity prevents tight phase bunching of the electrons. This deleterious effect of velocity spread was minimized in the design by keeping the drift tubes short, which required relatively large cavity fields. The corresponding large-signal gain is ~ 63 dB.

The output cavity Q is essentially diffractive and is sufficiently larger than the minimum Q so that a simple coupling iris can be used for the endfire system. Cold testing has confirmed both the frequency and Q predicted by the mode expansion code. The Q's of the input and buncher cavities are resistive; a cavity with a ring of carbon-impregnated alumino-silicate placed against the outer wall, approximately 3 mm thick, achieved the desired Q in cold-testing. Microwave power is injected into the input cavity through a resonant coupler. The drift tubes are not cutoff to the TE_{11}^o, TE_{21}^o, and TE_{01}^o modes, but are constructed from alternating rings of metal and microwave absorber to introduce losses and stabilize the regions.

A short taper connects the coupling iris of the output cavity to the water cooled, stainless steel beam dump. The taper is nonlinear to minimize mode conversion of the microwave signal. The beam dump radius is 3.81 cm and its length is ~ 35 cm, insuring an instantaneous temperature rise of no more than 20° C/pulse. A dumping magnet allows no electrons to pass the beam dump area. A second nonlinear taper takes the waveguide up to the 6.35 cm radius required for the half-wavelength, beryllia output window. A 20 hole, 48 dB directional coupler is used with a diode to sample the output pulse shape and estimate the power. A nonresonant waterload measures the full average power.

5.3 Gyroklystron Scaling Laws

Because the current state-of-the-art in microwave amplifiers falls short of predicted supercollider requirements, the question of how to increase their performance is of paramount importance. In this section, we will discuss the critical question of peak power capability. In Section 5.3.1, we will explore increasing the peak power of a gyroklystron at a given frequency. In Section 5.3.2, the question of scaling with frequency is considered.

Both the MIG and the microwave circuit must be considered in the scaling process. In the low-voltage, small Larmor radius ($r_L \ll r_g$) limit, the adiabatic trade-off equations yield analytic scaling laws. To achieve the scaling, it is assumed that the MIG is operated at the maximum permissible values of electric field, cathode loading, and guiding center spread. These scaling laws are compared to the actual trade-off solutions in the figures that follow.

5.3.1 Scaling with Power

An approximation for the relative potential depression in the gyroklystron beam is given by[7]

$$\Delta V/V_o \approx 430 \ G \ \left\{\frac{\alpha \ I_o \ (kA)}{\omega_{ce} \ (GHz) \ r_g \ (cm) \ V_o \ (kV)}\right\} \ .$$

At a given frequency, beam power can be increased by raising the current until a potential depression limit is reached. To increase the power beyond this point, either the beam voltage or the guiding center radius must be increased.

A significant increase in the guiding center radius must be accompanied by a change of operating mode. Furthermore, since the drift tube radius must be larger than the outer beam radius, the drift tubes will be able to support electromagnetic waves and could cause stability problems. The drift tube problem may be solved in part by using a coaxial microwave circuit. If a suitable circuit configuration can be found, then the MIG trade-off equations predict that the electron gun can provide a current proportional to r_g. Thus, the MIG and the circuit are well matched and the beam power can be scaled as $P_o \propto r_g$. Scaling the current MIG design (from 100 MW) with r_g is shown in Fig. 5.11. The circles represent the actual trade-off solutions and agree well with the scaling approximation.

In the microwave circuit, peak power scales with voltage as $P_o \propto V_o^2$. In the MIG, however,

$$P_o \propto V_o^2 [1 + (1 + \frac{V_o}{m_o c^2})^{-1}] ,$$

which means that the relative potential depression in the circuit region will slowly decrease with increasing voltage. Furthermore, empirical electron gun studies predict that the peak electric field must decrease as beam voltage increases,[8] further reducing the MIG power capabilities. Scaling the current MIG design (from 100 MW) with V_o is shown in Fig. 5.12. The actual trade-off solutions reveal that the scaling law is a slight overestimate at these large voltages.

If the limits on potential depression and MIG electric field have not been reached, then beam power can be increased because $I_o \propto E_c$. However, $J_c \propto E_c^{5/3}$, so the limits in thermionic emission technology can soon be reached. Scaling the current design with electric field is shown in Fig. 5.13. Again the scaling approximation gives a slight overestimate.

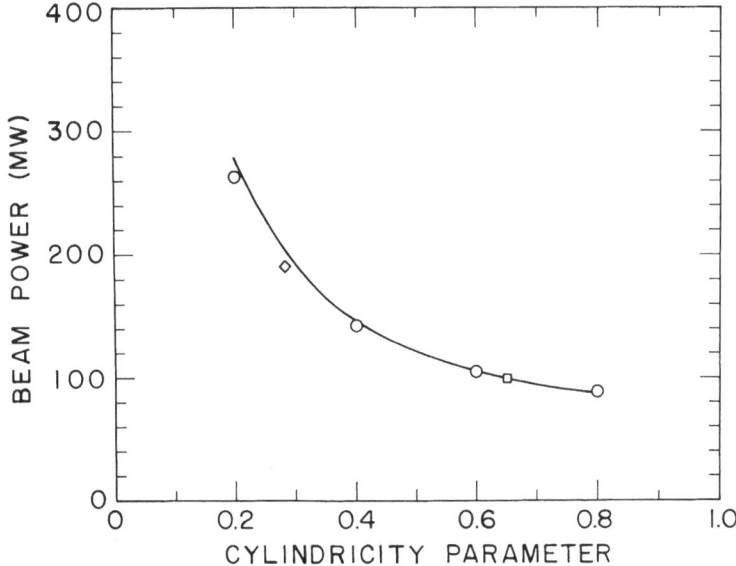

FIG. 5.11. MIG scaling with beam radius from the University of Maryland design.

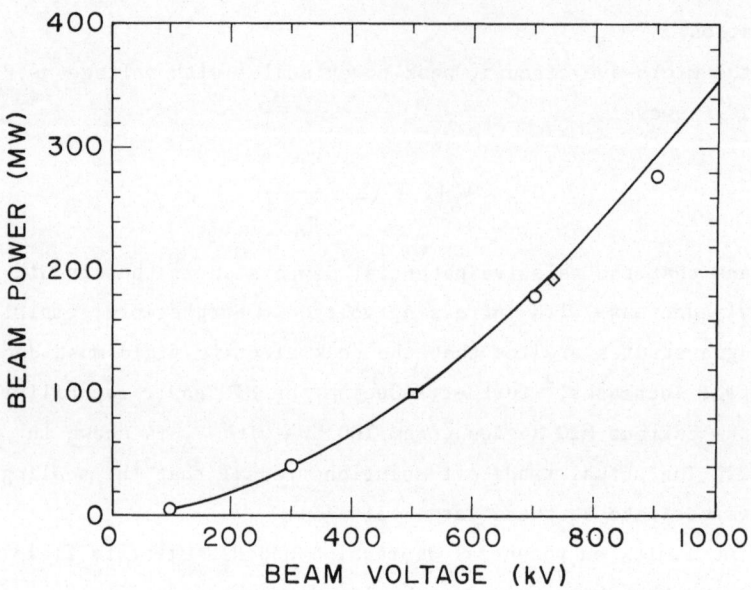

FIG. 5.12. MIG scaling with beam voltage from the University of Maryland design.

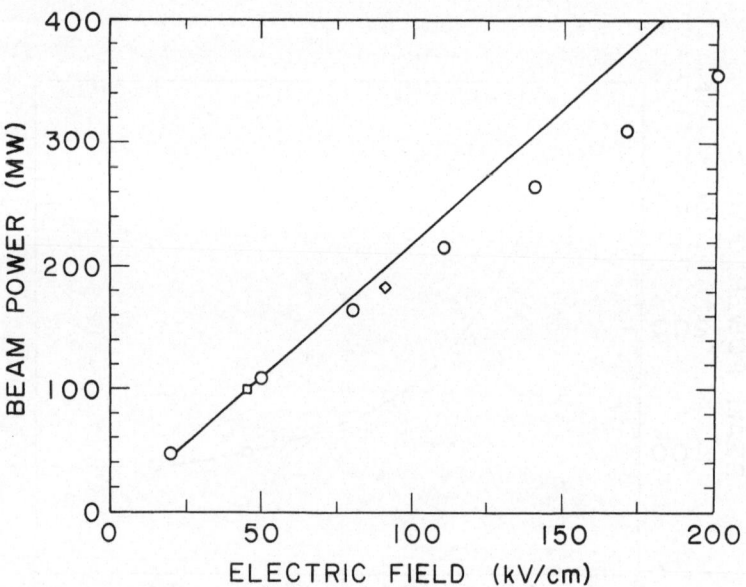

FIG. 5.13. MIG scaling with electric field from the University of Maryland design.

5.3.2 Scaling with Frequency

The simplified trade-off equations reveal that $P_o \propto \omega_{ce}^{-1}$ for the MIG configuration. This is confirmed in Fig. 5.14 where the current design is scaled with frequency. Because the peak power per feed in the accelerator is proportional to $\lambda^2 \propto \omega_{ce}^{-2}$, it appears as though the technology gap can be closed simply by going to higher frequencies. However, average power soon becomes the dominant microwave tube problem at high frequencies since it is proportional to $\omega_{ce}^{-5/2}$ [e.g., see Ref. 2].

5.4 Design Features of a Gyroklystron for an X-Band Supercollider

In this section, we will predict the features of a gyroklystron that would be capable of powering a 500 GeV on 500 GeV collider. The predictions will be based on the accelerator scaling studies, the experience gained from the 30 MW experiment, and the gyroklystron scaling laws.

The first assumption is that the accelerator will be at 10 GHz, where the peak power per feed requirement is $\hat{P}_{feed}$ = 545 MW. At a minimum, we will assume two feeds/gyroklystron, and if we assume a 45% efficient tube, then the required gyroklystron beam power is P_o = 2.42 GW. It is not feasible to increase the beam power by a factor of 24 even by a combination of all three techniques, and so another new approach is needed.

This new approach is called the Binary Power Multiplier (BPM) and was developed by Z. D. Farkas[9] at SLAC. A simplified BPM schematic is shown in Fig. 5.15. Two high gain amplifiers, labeled GK_1 and GK_2, are required. Each amplifier is connected to the input signal through a 180^0 (digital) phase shifter. The amplifier outputs are connected to a hybrid coupler with the following property: if the output signals are in phase the total signal appears in the first arm, and if the output signals are 180^0 out of phase, the total signal appears in the second arm. For the first half of the pulse both amplifiers are in phase, and in the second half of the pulse the amplifiers are out of phase. A delay line whose length corresponds to half a pulse width is attached to the first hybrid arm so that a pulse with twice the power and half the width appears at the end of the BPM.

With the proper waveguide mode, a BPM has the potential to halve the microwave pulse length and double the pulse power with good efficiency.[9] Furthermore, BPM's can be placed in series for further power enhancement. A three-stage system, with an estimated efficiency of 80%, could reduce the gyroklystron beam power requirements to 380 MW while increasing the pulse length requirements from 0.14 μs to 1.1 μs.

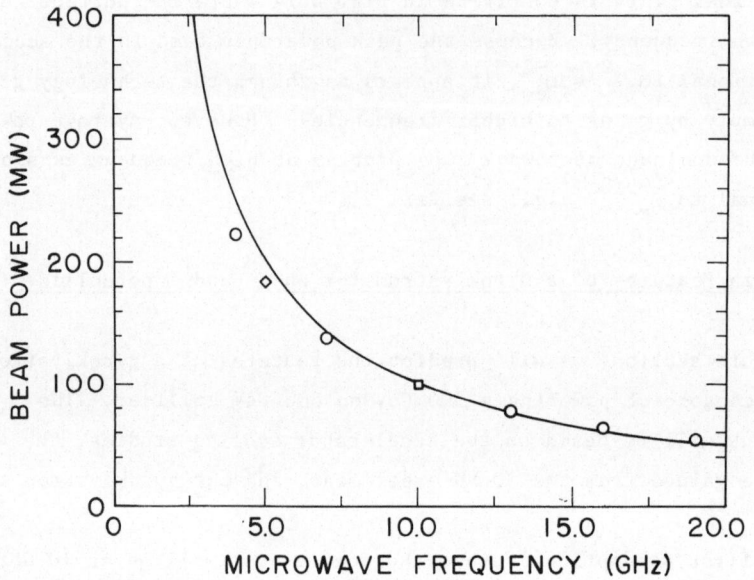

FIG. 5.14. MIG scaling with frequency from the University of Maryland design.

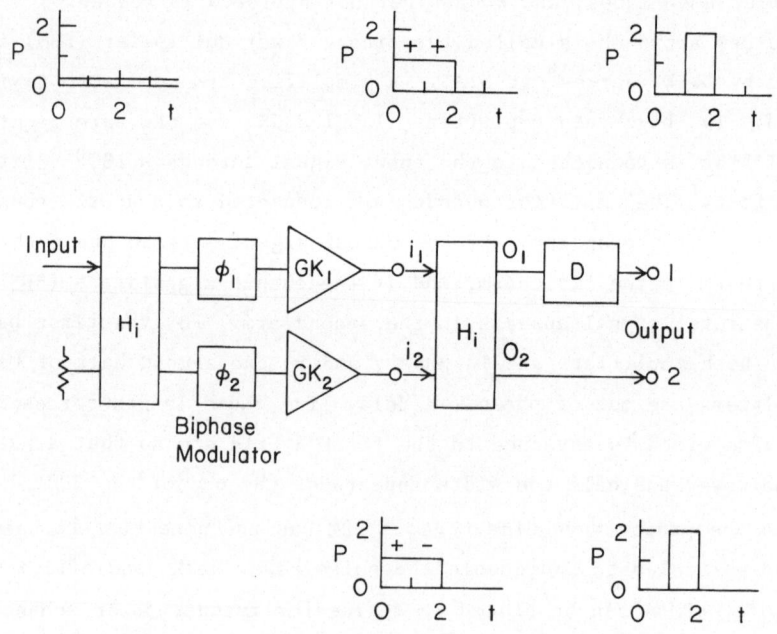

FIG. 5.15. A single stage Binary Power Multiplier.

One scaling possibility would be to fix E_c, double the guiding center radius, increase the beam voltage to 800 keV, and increase the beam current to 500 A. A coaxial circuit would be used with inner and outer drift tube radii of 0.9 cm and 2.24 cm, respectively. The drift tube lengths would be increased to 5.2 cm to provide comparable isolation between cavities. The coaxial drift sections would not be cutoff to the TE_{11}^o, TE_{21}^o, and TE_{31}^o modes, but the inner conductor would be constructed as a circular electric mode filter to stabilize those modes. In Fig. 5.16, an estimate of the 400 MW system is superimposed on the current gyroklystron design.

A MIG has been designed for the 400 MW system; again the starting point was the adiabatic trade-off equations. The electrode configuration is shown in Fig. 5.17. The cathode half-angle was changed to 30^0 to shorten the cathode nose. The peak electric fields were maintained well below 90 kV/cm on all surfaces, and the cathode loading at 500 A was approximately 37% of the computer-estimated space-charge limiting current. The dependence of axial velocity spread on current is shown in Fig. 5.18. The 10% (and under) range is maintained from 325 A to 575 A. A summary of the MIG specifications is given in Table 5.6.

5.5 Comparison with Klystrons

It is instructive at this point to compare the expected performance of the gyroklystron and the klystron. Table 5.7 lists specifications for four tubes. Klystron A is the 150 MW S-band SLAC klystron that has been built

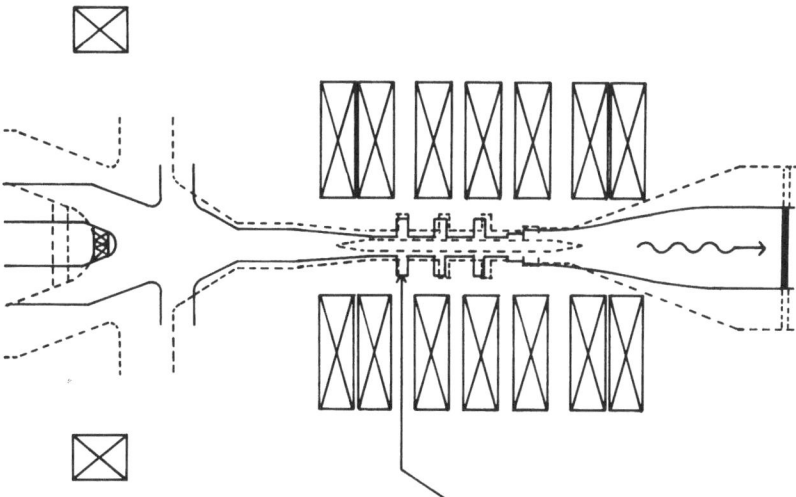

FIG. 5.16. Estimate of 400 MW (beam power) gyroklystron configuration. The solid line represents the current University of Maryland design and the dashed line indicates the expected geometry for a coaxial gyroklystron.

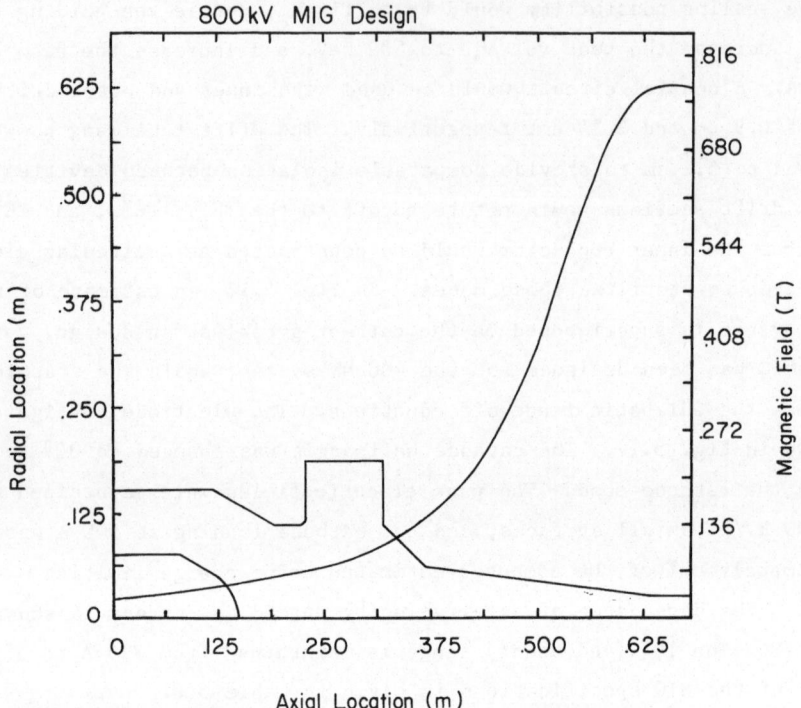

FIG. 5.17. The 400 MW MIG configuration.

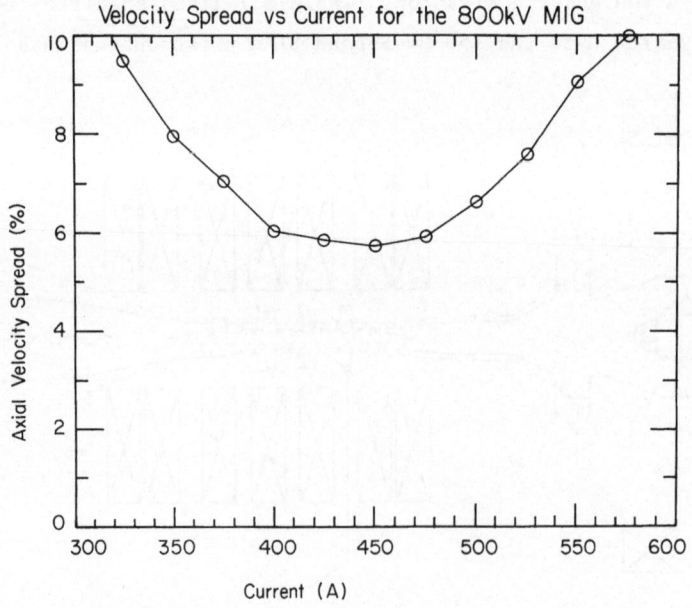

FIG. 5.18. The dependence of axial velocity spread on beam current for the 400 MW MIG.

TABLE 5.6. 400 MW MIG specifications.

Cathode Radius r_c	0.0632 m
Emitter Strip Width ℓ_s	0.0274 m
Cathode-Control Anode Gap	0.0864 m
Control Anode-Main Anode Gap	0.0900 m
Cathode Half-Angle ϕ_c	30^0
Compression Ratio f_m	17.7
Control Anode Voltage V_a	270 kV
Emission Current Density J_c	4.59 A/cm^2
Total Current	500 A
Average Velocity Ratio	1.52
Average β_{zo}	0.506
Spread in β_{zo}	6.63%
Average Guiding Center Radius	0.0156 m
Peak Electric Field	< 90 kV/cm

TABLE 5.7. Comparison of gyroklystrons and klystrons.

Parameters	Klystron		Gyroklystron	
	A	B	A	B
Frequency (GHz)	2.87	10.0	10.0	10.0
Peak Power (MW)	150	100	30	180*
Pulse Width (μs)	1	2	2	1.1
Efficiency (%)	51	65	45	45*
Gain (dB)	59	45-50	63	60*
Voltage (kV)	450	546	500	800
Current (A)	603	282	200	500
Cathode Loading (A/cm^2)	8.2	12.0	5.6	4.6
Drift Tube Radius (cm)	1.87	0.65	1.50	2.24
Peak Electric Field (kV/cm)	220	320-400	90	80

*Based on current gyroklystron design.

and tested,[1] and Klystron B is the SLAC proposal for a conventional X-band klystron.[10] Gyroklystron A is the current Maryland experiment and Gyroklystron B is the proposed gyroklystron of Section 5.4.

The main klystron advantage is the higher efficiency. This is an intrinsic feature because the gyroklystron can readily tap only perpendicular energy while the klystron draws its power from the total beam energy. While the gyroklystron achieves its power with overmoded cavities, the klystron achieves its power with extremely high electric fields. Even if the X-band klystron experiment is successful, that tube will inherently have a higher fault rate (than the gyroklystron) due to the large electric field. Finally, if the gyroklystron electric field constraint is relaxed to the S-band klystron level, then a 400 MW beam could be achieved without increasing the guiding center radius.

References

1. T. G. Lee, G. T. Konrad, Y. Okazaki, M. Watanabe, and H. Yonezawa, IEEE Trans. Plasma Sci. PS-13, 545 (1985).

2. K. R. Chu, V. L. Granatstein, P. E. Latham, W. Lawson, and C. D. Striffler, IEEE Trans. Plasma Sci. PS-13, 424 (1985).

3. J. McAdoo, W. M. Bollen, A. McCurdy, V. L. Granatstein, and R. K. Parker, Int. J. Electron. 61, 1025 (1986).

4. M. Caplan, J. Neilson, A. Salop, H. Jory, 1985 IEDM Tech. Digest, p. 528 (1985).

5. The Stanford Two Mile Accelerator, edited by R. B. Neal (W. A. Benjamin, Inc., New York, 1968), p. 411.

6. J. M. Baird and W. Lawson, Int. J. Electron. 61, 953 (1986).

7. W. Lawson, J. Calame, V. L. Granatstein, G. S. Park, C. D. Striffler, and J. Neilson, Int. J. Electron. 61, 969 (1986).

8. A. Staprans, 1986 High Voltage Workshop, Monterey, California.

9. Z. D. Farkas, IEEE Trans. Microwave Theory Tech. MTT-34, 1036 (1986).

10. Private communication.

6. SUMMARY

A general review of the design concept of a future e^+e^- linear collider using a SLAC-type linear accelerator was presented. The constraints imposed on luminosity and collider performance by interaction-point and accelerator beam physics were discussed. Power requirements and overall efficiency are strongly affected by these constraints. Two crucial parameters are the number of particles per bunch, N, and the rf frequency ($\nu = \omega/2\pi$). If beamstrahlung in the interaction point were the only constraint, N would remain approximately constant with frequency. Average rf power as well as peak power per linac feed section then decrease with frequency, and the highest frequency possible (from manufacturing considerations for the accelerating structures) should be chosen.

On the other hand, it is argued that accelerator beam physics considerations will not permit keeping N constant as wavelength and bunch size are decreased. At high frequencies, the linear bunch density, which scales as $N \sim \lambda \sim \omega^{-1}$, or transverse dipole wake field effects, which scale as λ^3, are most likely the limiting factors.

A computer model incorporating the interaction-point and beam physics scaling laws as well as linac parameters and power requirements was used to study a 500 GeV on 500 GeV collider of fixed length (2 × 3 km = 6 km) and fixed luminosity (10^{33} $cm^{-2}s^{-1}$). Such a collider is being considered by SLAC. The results of the computer studies confirm the general expectations mentioned above. We considered only the cases N = const. and $N \sim \lambda$ in the frequency range 8 to 20 GHz.

In the first case (δ_{BS} = 0.3), the number of particles per bunch for a round beam is $N \sim 6 \times 10^9$. In the second case ($N \sim \lambda$), we chose $N = 2 \times 10^{10}$ at SLAC frequency, and the corresponding figure of merit for the phase-space density, $N/\lambda\varepsilon_N$, is four times greater than the design value for SLC. For δ_{BS} = 0.3 (N = const), average power decreases with increasing frequency as expected. For the $N \sim \lambda$ case considered, on the other hand, we find that only a small frequency range (9.5 to 11.0 GHz) yields acceptable performance parameters. Below 9.5 GHz the collider operation is limited by beamstrahlung, and N would have to be kept constant (reducing the luminosity). Above 11 GHz the average total power exceeds 200 MW for each of the two collider linacs and rises rapidly with frequency; furthermore, dipole wake field effects become increasingly strong.

Of major concern for such a future collider scenario is the microwave source for the linac where very high peak power (hundreds of megawatts per feed) is required, which is beyond the state-of-the-art of existing rf source technology such as the klystrons being used at SLAC in S-band. We examined the scaling laws and expected performance parameters of the gyroklystron presently being developed at the University of Maryland with a frequency of 10 GHz. The peak power per feed required at 10 GHz for both scenarios studied with our computer model (N ≈ const. and $N \sim \lambda$) is 545 MW. Our studies show that we can achieve this power level by scaling the current gyroklystron design from 30 MW to 180 MW output power and using a three-stage binary power multiplier.

RF SYSTEM OF A SYNCHROTRON FOR PROTONS AND HEAVY IONS

Dieter Böhne

Gesellschaft für Schwerionenforschung
Darmstadt, West Germany

INTRODUCTION

In this paper the potential and the constraints of producing many kilovolts of rf accelerating voltage for synchrotrons in a cumbersome broad frequency range are reviewed from the electrical engineering standpoint. Particle dynamics aspects are not touched. Those issues are covered broadly in the proceedings of CERN and FNAL accelerator schools. The beam loading aspects are not touched either, they are treated in a separate paper by G. Rees in this lecture series. This paper elaborates on numbers and limits which determine cost and complexity of the rf system. For convenience of the author, most data and material are taken from SIS 18 synchrotron, presently under construction at GSI. Comparisons are made with other machines (Fig. 2). The numbers for other existing synchrotrons are found in the "Catalogue of High Energy Accelerators", a supplement to the Proceedings of the 1980 High Energy Accelerator Conference. However, this official compilation does not contain data of the no more existing synchrotrons which have played a splendid role in the 35 year-old history of rf acceleration in synchrotrons. Those machines were: a) the Cosmotron at BNL, b) the Princeton Pennsylvania Accelerator, c) the Zero Gradient Synchrotron at ANL, d) the Nimrod at Rutherford Laboratory, and the first proton synchrotron in Europe in Birmingham, which came 1 year later than the Cosmotron.

Citations on all these forerunners are made in the text at appropriate instances. There have been quite a lot of proposals written for machines, which then never were built. The reasons for that are reported as being of scientific policy nature and not because of technical difficulties. The rf developments associated with these proposals, however, contributed significantly to the progress in this field.

THE MACHINE CYCLE

Before we concentrate on the rf system we briefly review a few characteristics of an accelerating cycle of a synchrotron, Fig. 1. At a low field level of the bending dipoles, the preaccelerated beam is injected, preferably with many turns. Due to the poor reproducibility of low fields, because of remanent fields and eddy currents, this initial field level should be comfortably high, thus implying a reasonable preacceleration of the beam by a linear accelerator. This level is about 0.1 Tesla for the SIS 18. The Unilac was upgraded to provide a generous injection energy of 11.3 MeV/u. Older machines, which had a small Van-de-Graaf preaccelerator, had an about 10 times smaller field level. As we will see later, a high injection particle velocity results in a favourable reduction in frequency swing of the accelerating voltage.

After the short injection time, 4 to 100 µs for the SIS 18, the magnet current starts to rise due to the switch-on of the power supply voltage, 13 kV in case of the SIS 18 for the ring dipoles. One is always interested in making the ramp-up and ramp-down time very short. A fast field rise $\dot{B}$ requires a large voltage and hence a high instantaneous mains power at the end of the ramp. This power amounts to 35 MW for the SIS and allows for a $\dot{B}$ of 10 T/sec. In larger machines one cannot afford this fast field rise, though the magnet power is a few times higher. For the CERN PS the peak power is about 41 MW, the field rise is 2 T/sec and the total cycle time is around 2 seconds. A much faster field rise can be achieved in smaller machines by draining and restoring the large magnet power into a condenser battery. This results in a sinusoidal shape of the magnet cycle on the expense of a flexible flat top for slow extraction. Those machines are used as boosters for larger machines (KEK, Japan, 20 Hz; ANL, USA, 30 Hz; FNAL, USA, 15 Hz) or for production machines like the SNS at Rutherford Laboratory. The latter has a $\dot{B} = 70$ T/sec at the steepest point of the 50 Hz machine cycle. We will need those $\dot{B}$ numbers later.

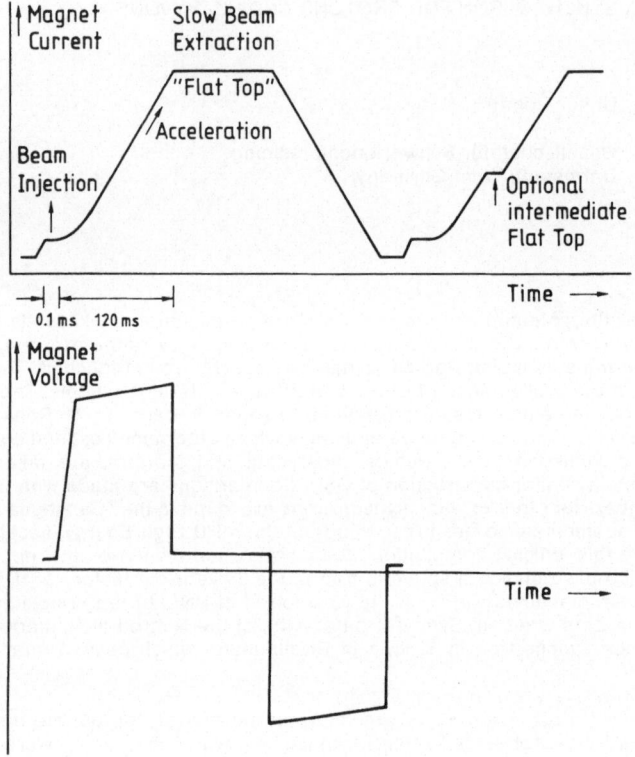

Fig. 1. The magnet cycle of the SIS 18 synchrotron.

The implication of a high $\dot{B}$ is the necessity of a non-conductive vacuum chamber in the magnets, to prevent eddy current problems. In the SIS, where a stainless steel chamber could not be avoided due to extreme ultra-high vacuum requirements, the $\dot{B}$ on the ramp was limited to 10 T/sec. Still, this value cannot be tolerated at low fields, excessive eddy currents would distort the field properties. The field rise is smoothly turned on. The latter procedure is also done in other machines, because particle dynamics reasons require a smooth turn-on of the rf voltage, the amplitude of which is proportional to $\dot{B}$.

During the steep field ramp the full rf voltage is acting on the beam and the frequency is raised proportionally to the increasing particle velocity. At the end of the ramp, when the bending field has reached the iron saturation level, the rf voltage can be switched off and the beam starts to lose its bunch structure and a nearly dc beam current can slowly be extracted during the flat top time. Or, if bunch structure in the external beam is preferred, the rf is left on during extraction. A phase jump between rf and bunches prevents further acceleration. It is not necessary that the field rise is linear as in Fig. 1, in fact in machines with resonant magnet power supplies it is definitively not. However, in machines with separate power supplies for the dipoles and quadrupoles, as in the SIS, the field values in both types of magnets must always be proportional in the 10^{-3} range, and this is easier for a linear current rise.

THE FREQUENCY RANGE OF THE ACCELERATING VOLTAGE

We first consider the revolution frequency for a particle or a bunch of particles. It is simply geometrically determined by the particle velocity v over the ring circumference. It is convenient to introduce β = v/c because this value indicates how close the particles approach the velocity of light and β occur in many formulas

$$f_r = \frac{c}{2\pi R} \cdot \beta \tag{1}$$

For large proton machines, β at ejection is extremely close to 1, for small heavy-ion machines it is not. At injection the β depends on the generosity of the injector linac or even of an intermediate booster synchrotron. Those machines are characterized by their final particle energy T in MeV or MeV/nucleon and we must relate β to those units:

$$\beta = \frac{\sqrt{\gamma^2-1}}{\gamma}, \qquad \gamma = 1 + \frac{T}{T_o} \tag{2}$$

γ is the so-called relativistic factor and relates the kinetic energy T to the rest energy of the proton T_o = 931.5 MeV. Table 1 summarizes the revolution frequency for 3 different machines at injection and extraction. The values h and f_{rf} will be explained later.

Table 1. Revolution Frequencies of 3 Different Machines

	SIS	CERN PS	CERN SPS
R in m	34.4	100	1100
T in MeV	11.4	800	10000
f_r at injection in MHz	0.212	0.402	0.0432
f_r at extraction in MHz	1.365	0.477	0.0433
h	4	20	4620
f_{rf} in MHz	0.85-5.5	8.04-9.54	200

From Table 1 the following is obvious: a) the revolution frequency for large machines becomes quite low, eventually uncomfortably low for a technical acceleration station; b) machines with low injection energy require a large frequency swing during acceleration, which is also undesirable.

The frequency swing in the SIS is abourt 1:6.5, much higher than in the CERN PS in the old days, when the 800 keV booster did not exist and the injection energy from the linac was 50 MeV. It must be mentioned that among the old generation of synchrotrons, where the injected beam came from a 4 MeV Van-de-Graaff, the frequency ratio could be as high as 1:30 and a ratio of 1:10 was quite usual. That means, that the old synchrotrons 20 - 35 years ago had the more difficult rf system to master. This trend still goes on: for reasons of pushing up the space charge limit of the ring, the injection energy should be high, and with modern technology the cost per MeV of linear accelerators can be decently reduced. In case of the proposed European Hadron Facility, a two-stage synchrotron of 30 MeV (in size like the CERN PS), an injector linac of 1200 MeV is envisaged, leaving a frequency swing of the first ring of only 1:1.11.

It is important to mention that the actual operating frequency f_{rf} of an rf accelerating station can be higher by a factor h than the revolution frequency f_r. The h is called the harmonic number and it must be an integer figure. The number of bunches in the machine, which are formed out of the dc beam during the capture process, correpond to the same number h. What value is then taken for h, is left to the designers' choice, according to what he considers a comfortable frequency f_{rf}. There is an upper bound in small machines: the longitudinal spacing of the bunches must leave a beam free gap for the rise time of the kicker magnet for the fast beam extraction. This time cannot be made much shorter than 0.1 µsec presently. The bunches should also be spaced largely enough so that they do not "see" each other. Otherwise undesirable bunch oscillations could occur due to coupling effects by space charge forces. There are other particle dynamic reasons for the choice of h, which are not strong either. There is certainly the trend that large machines have considerably higher harmonic numbers than smaller machines. This is evident from Table 1. In the CERN SPS an h = 4620 was chosen, resulting in a 200 MHz rf system. In the FNAL main ring, which has essentially the same size as the SPS (-10%), an h = 1113 was chosen resulting in an accelerating frequency of 53 MHz. It is more precise that h determines the number of rf buckets. One can prevent by the injector that all are filled with particles. It looks like that the future hadron facilities adopt this frequency range of about 50 MHz.

There is, however, a classical frequency domain of 0.5 - 10 MHz. In all but 3 cases of the about 20 proton synchrotrons that have been built so far this frequency range was chosen. The first machine, the Cosmotron at BNL, worked at 0.36 - 4.2 MHz. For a new machine one tends to take advantage of existing experience. The above cited frequency domain is also the range for medium wave and short wave broadcasting and communication transmitters, for which industrial competence and well engineered rf power components are available. And if a machine is costwise not dominated by the rf system, a further optimisation of the frequency range was not deemed to be imperative. For all those arguments a frequency of 0.85 to 5.5 MHz was chosen for the SIS, even when the dipole bending power B • ρ was moved up and down between 12 and 100 Tesla • meters in consecutive proposals and finally B • ρ was settled at the value of 18 T•m for the machine under construction. The harmonic number in the SIS is now 4.

We finally must come to the frequency program itself, the circumstances that gave the synchrotron its name. The cost saving from the synchrotron principle, especially in large machines, is due to its small magnet aperture, which should be used by the beam and should not be used up by a radial displacement of the beam orbit. That implies that a highly precise relation between the revolution frequency, hence also the accelerating frequency, to the particle velocity v is maintained. Since v, β or T are technically not instantaneoulsy measurable, we must resort to the instantaneous magnetic field B(t) and assume that the design bending radius is maintained. The latter can be assured by electrical beam position measurements and a fine correction of f_{rf} is derived from this signal. For very low beam intensities, which might occur in heavy-ion machines or in case of acceleration of secondary particles, this signal disappears in the noise and one really has to rely on a presumably centered orbit, determined by experience with higher intensity beams. We now must rewrite equation (2) and introduce the B field via the momentum p = eBρ with the inconvenience that the particle mass in p = m • v is a relativistic variable and hence dependent on v again. The value β in equation (2) then becomes

$$\beta = \frac{B}{\sqrt{\left|\frac{T_0}{\rho ec}\right|^2 + B^2}} = \frac{B}{\sqrt{\left|\frac{3.1}{\rho}\right|^2 + B^2}} \tag{3}$$

The field B is given in Tesla and the bending radius ρ in m.

For the SIS with a ρ = 10 m and an injection field of B = 0.1 T (for charge over mass ratio = 0.5) the B² in the denominator can be ignored and the β and hence the acceleration frequency starts linearly with B. The frequency program is shown in Fig. 8. However, during the whole acceleration cycle the instantaneous B must be measured and converted according to the complicated formula (3) by a function calculator into the instantaneous accelerating frequency. In view of the high accuracy required, the B does not follow close enough the current curve of Fig. 1 due to nonreproducible remanent fields, eddy current and saturation effects. B(t) is determined by integrating the voltage signal induced in a probe loop, which is proportional to $\dot{B}$(t) and the starting field B_0 at injection must be added ("peaking strip" method).

THE ACCELERATING VOLTAGE AMPLITUDE

It is technically not difficult to generate a master oscillator signal in a way that the required synchronisation is respected. The basic low level circuitry is still the same today as it was three decades before. However, the low level circuitry is the domain, where clever things can be done to the beam with a few transistors. It is the high power end of the amplifier chain, where interaction with the beam takes place, which still requires a substantial engineering effort. Before we describe the hardware, numbers should be given for the precious kilovolts. Equation (4) relates the required energy gain per turn to the quantities $\dot{B}$, ρ, R, used before. There might be a difference between the bending radius ρ and the mean machine radius R because of the straight sections occupied by focusing magnets, injection and ejection elements, rf stations, beam diagnosis boxes and, as a reserve, space for future beam correction and stabilizing items. Fig. 2 outlines the situation schematically. The CERN PS has an R = 100 m and an ρ = 70 m, whereas in the SIS R = 34.4 m and ρ = 10 m.

$$\hat{V} \cdot \sin \varphi_S = 2\pi R \cdot \rho \cdot \dot{B} \tag{4}$$

The left side denotes the voltage gain per turn, $\hat{V}$ being the peak sum voltage, reduced somewhat by sin φ_S in its acceleration action. For reasons of stable particle motion the particles are not allowed to cross the accelerating voltage at the peak of the sine wave, rather than shortly

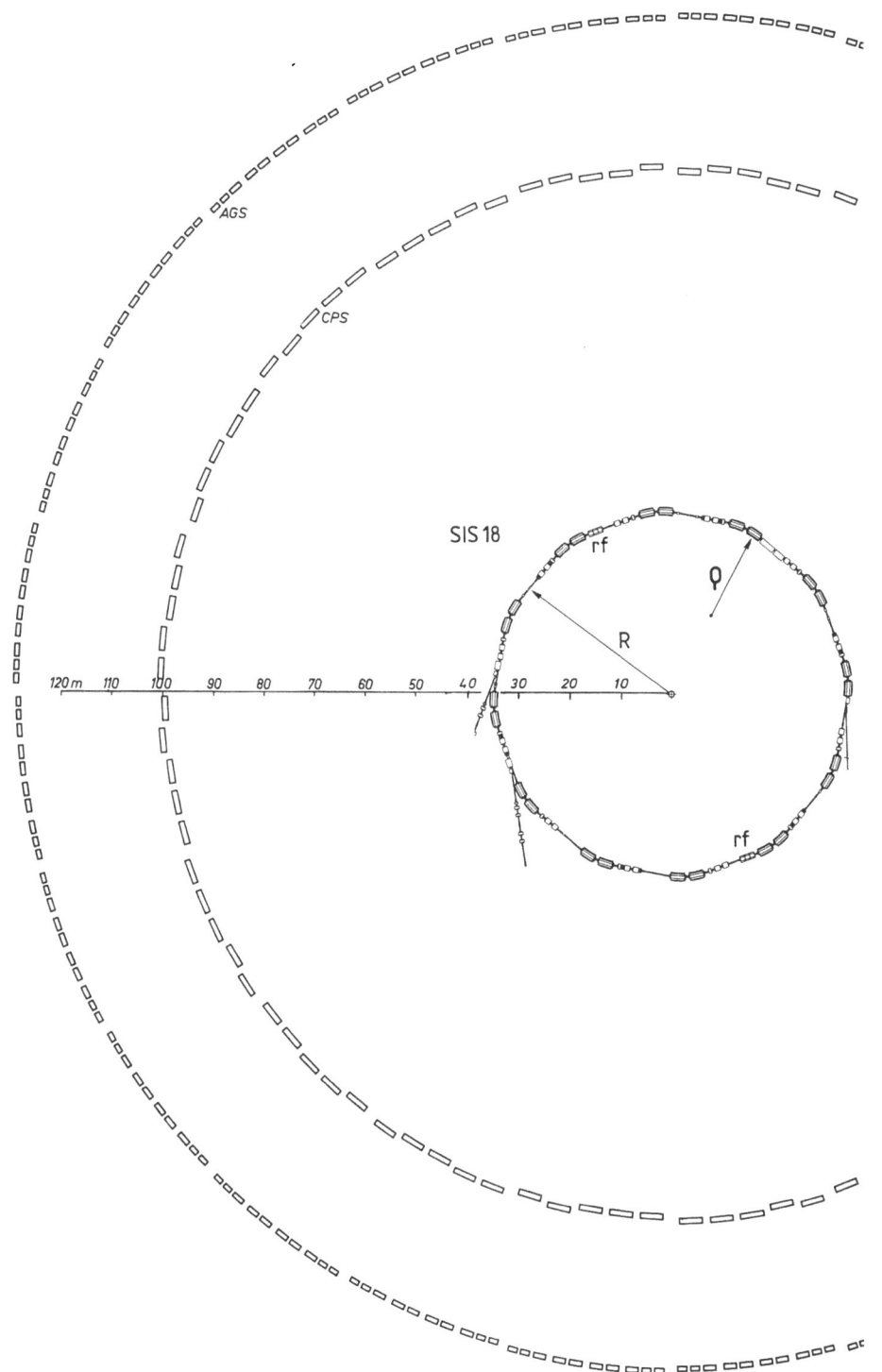

Fig. 2. Comparison of 3 synchrotrons: The Brookhaven AGS, the CERN PS and the SIS 18 at GSI. R is the ring radius and ρ is the bending radius of the dipole magnets.

before. φ_S is manipulated during the cycle for various reasons, which we will not explain here. We take a sin φ_S of about 0.7 during the accelerating cycle throughout.

The right side of equ. (4) can qualitatively be illustrated as follows: $2\pi R$ is the ring circumference. The larger the ring, for instance due to straight sections, the less is the revolution frequency and the occasions, where the particles can experience the voltage per time unit. The $\rho \cdot \dot{B}$ can be rewritten as $d\rho B/dt$, the momentum increase per time unit, and it is plausible that a faster acceleration needs more voltage installation. Another explanation is of a more Maxwellian nature: ignoring the straight sections, $\pi R \cdot \rho$ is the area of the ring through which the varying flux $\dot{B}$ passes, and on the circumference of this area the voltage $\hat{V}$ is then available like in a single secondary winding of a transformer. Table 2 lists some approximate numbers for the already mentioned machines:

Table 2. Total Accelerating Voltage for 3 Different Machines

	SIS 18	CERN PS	CERN SPS
$\dot{B}$ in T/sec	10	2	0.6
R and ρ in m	34.4/10	100/70	350/335
$\hat{V}$ sin φ_S in kV	21.6	88	1390

According to the chosen φ_S the $\hat{V}$ is certainly higher than the figures in the lower line. For the SIS the $\hat{V} = 32$ kV. If the chosen $\dot{B}$ seems to go down with the ring radius, this is not only because the upgoing rf voltage demand would be excessive, it is also because of the expenses for the magnet power supplies, which go up with $\dot{B}$. Small machines, for which the injected beam current is severely space charge limited, tend to pick up in average beam intensity by a higher cycling rate and hence by a higher $\dot{B}$.

We conclude that many tens of kilovolts are required and we can anticipate that this voltage cannot be generated in one single accelerating station today, except the frequency swing is negligible like in the SPS or in the FNAL main ring. Since 2/3 of the cost of an accelerating station is not voltage-dependent, there has always been the challenge for rf engineers, working on a new machine, to push up the voltage of a single unit to the technical limits of the time.

THE TECHNICAL CONCEPTION OF AN ACCELERATION STATION

Before we come to rf in its pure sense, we first consider a quasi dc acceleration. In one now highly developed species of linear accelerators, the induction accelerator, a voltage pulse of 250 kV is applied on a single winding of a transformer core, in fact on hundreds of them. The same voltage occurs on an imaginary secondary winding, namely the particle beam orbit passing through the core window. The vacuum chamber is a ceramic cylinder in this location. Fig. 3 shows this schematically. We apply a voltage pulse of the same shape as on the bending magnets of Fig. 1, the current - unfortunately in this case - starts to rise. A constant voltage is now acting on the beam during the accelerating cycle, which is ended by the onset of core saturation. With a B_{max} of 1.5 T and a core cross section of 1 m^2 we obtain 1.5 V · sec or 15 V for a 0.1 sec accelerating cycle. Though the beam goes about 10^5 times across this voltage, the final energy gain would only be 1.5 MeV. This value is independent of whether the cycle is short with a higher voltage and less turns or vice versa. Certainly, we have spent only a 15 V primary power supply, but still a sizeable transformer core. This kind of acceleration is called betatron acceleration and is used in the Saturne II synchrotron for very small energy corrections during the beam extraction time.

One sees immediately how we can do better: we apply very short voltage pulses which now can be very high and which are no more related to the length of the magnet cycle. The uncomfortable polarity reversal, which is now unavoidable during the cycle, implies some particle free space in the circulating beam. If we now conceive a clever scheme, so that the voltage pulses always hit the beam segment, we have the rf acceleration in a synchrotron.

The hardware in the first proton synchrotron, the Cosmotron at Brookhaven, was exactly as shown in Fig. 3: an rf power amplifier was connected to a single primary winding of a large transformer core and a voltage of 2 kV and of a frequency of 0.36 to 4.2 MHz was applied. A material problem became evident: High permeability core material was mandatory in order to have at the lower frequency end an impedance of at least $\omega L = 50$ Ω, otherwise the magnetisation current could not be delivered by the amplifier tube. At the high frequency end, where the ωL problem has disappeared, the eddy current losses of the core presented again a power limit. At the time of the Cosmotron development, end of the forties, ferrites were just

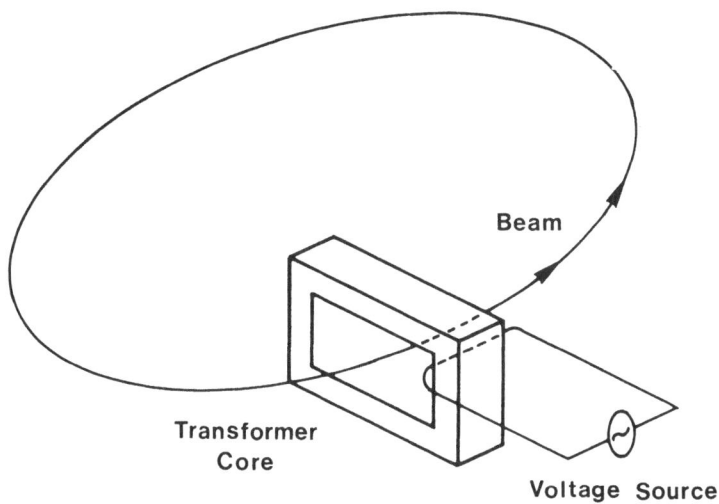

Fig. 3. Beam acceleration by a transformer coupled pulse or rf voltage.

appearing on the market in small samples and essentially showed the required properties: a high μ_r and low rf losses. It took a long way for the ferrite manufacturers to arrive at nearly the same favourable magnetic and dielectric properties on larger pieces, and the Cosmotron transformer was assembled from hundreds of bricks. The unique effort of the Cosmotron engineers and the accidental willingness of the industry in this circumstance is a remarkable event in accelerator history.

During the transformer testing on the full scale item, which was 1.3 x 1.6 x 1.6 m large, nasty resonances occured in connection with the stray capacitance of the primary winding, gap capacitance, amplifier leads and anode capacitance, which just appeared uncomfortably in the envisaged frequency range. The shunt resonance, which otherwise exhibits a desirably high transformer impedance, caused a phase jump and had to be damped away.

The Cosmotron worked with this transformer during all its life. But the designers concluded that one should work in future on this shunt resonance frequency and should adjust this frequency by changing the ferrite permeability by some means of dynamic dc biasing. This was done in all but three proton synchrotrons thereafter. But a new material problem came up: μ_r and loss properties as a function of dc saturation. The impedance problem at the low frequency end disappeared, because reasonable amounts of capacitors could be added to the gap capacitance, and the size of the core could be made smaller. But the nasty stray resonances still appear in devices of today. They present the most serious and time consuming problems and reappear in every new generation of accelerating stations, when small changes in electrical or mechanical characteristics have been applied compared to the design of a previous device.

In one case, in the Saturne I synchrotron, where originally a very large frequency swing was required, the allowance of additional gap capacitance was used to introduce a rotating condenser, which determined the variable frequency in a programmed way and only fine adjustments were done by the inductance control. In the Princeton Pennsylvania accelerator where the frequency swing was 1:30, two accelerating stations were used, one for the lower and one for the upper frequency range. A very clear method of dealing with an extreme range of revolution frequencies is used now in the Brookhaven AGS in case of heavy-ion acceleration. The magnet cycle is stopped when the rf system has approached its upperr tuning limit. The beam stays at this energy and loses its bunch structure. This is indicated as an intermediate flat top in Fig. 1 on the right hand cycle curve. The frequency tuning is reset to its lower edge and rebunching starts on a higher harmonic number and then the magnet ramp and further acceleration continues. This apparently simple method can be used with all modern machines. It could not in older machines, because the magnet power supplies did not allow for this flexibility in ramp shaping.

It is worthwhile to mention that though we operate our accelerating gaps on the shunt resonant frequency, the device is still a 1:1 transformer as it originally was. The name "accelerating cavity", that is used commonly, does not pay tribute to this fact. Since the plate voltage of a power tube is of the order of 8 to 12 kV, we now have the essential number per cavity and the rest is just the multiplication of units and cost. There are means of applying a step-up ratio of 1:2, but what is mostly done is the paralleling of two or four cavities per amplifier unit.

Before we turn to the chapter on ferrite properties, we review the variety of cavity designs; they are not related to the ferrite properties. There is no continuity visible in the preference of cavity concepts. They reflect the choice of the designer, whether he prefers to deal with spurious resonances or with the problem of 10 kA in fast acting bias power supply. It is indeed the ambiguity of having simultaneously the dc bias field and the rf magnetic field in the ferrite core and a sophisticated decoupling must be conceived.

Fig. 4a shows the classical symmetric "reentrant" cavity, properly speaking it is a $\lambda/2$ coaxial cavity. The cavity volume is loaded by ferrite rings to shorten the length and to provide

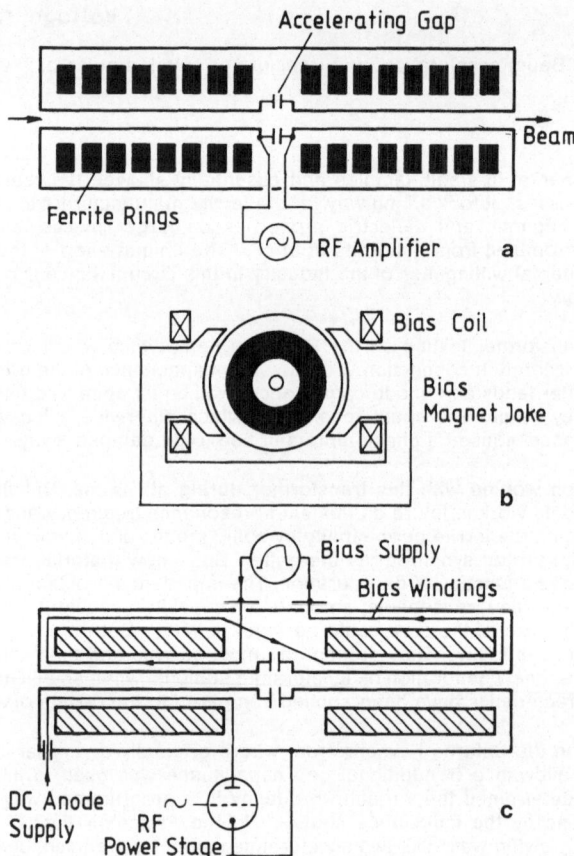

Fig. 4. Ferrite loaded $\lambda/2$ acceleration cavity and two methods of applying the polarization field.

the opportunity of inductance tuning. Such a cavity is about 2 m long and contains about 60 rings. The ring shape became quite standardized: 50 cm outer, 25 cm inner diameter and 2.5 cm thick. The inner diameter is sometimes chosen to be larger when the beam tube size and additional wire conductors it requires, or even because of the consideration of avoiding rf flux enhancement and overheating on the inner diameter. Space is left between the rings for copper cooling plates or a cooling fluent, which floods the whole cavity. The accelerating gap is insulated by a ceramic cylinder and most times shunted by additional capacitors. The rf power amplifier is a push-pull stage of a few tens of kilowatts. This power is absorbed by the ferrite and must be cooled away. Ferrite is quite temperature sensitive in its mechanical and magnetical properties. No saturating bias provisions are yet visible in Fig. 4a. This is sketched in a perpendicular view in beam direction in Fig. 4b. A laminated iron core with its bias coils, runs along each ferrite section. In this design a clear decoupling of rf and bias field is achieved and the number of coil turns is freely selectable according to the bias power supply design. This concept was used in the first generation rf stages of the CERN PS. The second generation now is entirely different. Except for the cost of the large magnet and some wasted ampereturns in the large air gap, no obvious disadvantage can be reported. In fact, most recent cavity designs at Los Alamos, reuse this external polarization, but for a much smaller ferrite volume.

A further step in the development led to the concept of Fig. 4c. This has become one of the 3 choices of today. The amplifier rf current flows only through one core half and the plate dc current as well, thus avoiding the difficult anode choke and coupling condensers to the gap. The second half core is excited by the "figure of eight" winding, the gap voltage is now twice the anode voltage. This figure of eight coupling was already applied in the original CERN PS cavity. The most important feature of this winding is the fact that it can simultaneously be used as saturation coil for the ferrite. The induced rf voltages cancel in the whole "figure of eight" winding of Fig. 4c, and only a modest rf current transporting the ferrite loss power to the right hand core is flowing across the by-pass capacitors on the bias terminals of the cavity.

One is tempted to apply more than one bias winding, which should not change the coupling properties, but reduces the bias current requirements decently. This was done in the Saturne II cavity and in the SIS cavity. In the latter, six bus bars, equally distributed on the circumference, are running back and fore in the air gap inside and outside the ferrite rings and are crossing in the gap region. This large number of rf current carrying conductors with their various stray capacitances gave rise to many spurious resonances for harmonics of the accelerating frequency with associated voltage arc-overs and insulator overheating. The problems were cured finally, but it took a long time. The SIS cavity is now running with a safe 16 kV gap voltage, an rf power consumption of 40 kW and a bias current of 10 - 800 A. It should be mentioned that the acceleration gap in Fig. 4c can be placed at one end, as well. This removes the high voltage environment of the gap from the obstructed lead crossing zone in the middle of both cores. This concept was realized in the MIMAS rf system at Saturne and will be used for the storage ring cavity at GSI.

We now turn to a second cavity concept, being entirely asymmetric and having the gaps on one end, as well. The principle provides also rf voltage cancellation in the bias current path, but there is one turn only, the inner and outer conductors of the cavity and a cavity overpass line serving as current leads. Fig. 5 illustrates this alternative. The arrows indicate the bias current flow: from the left outer conductor back through the inner conductor to the gap, from there-on the lead in the overpass is on rf potential, then continues to the right hand gap, back through the inner conductor to the right outer conductor where the rf potential now cancels out, and back to the power supply. For the rf both gaps are connected in parallel via the bias overpass. The rf amplifier can be connected to any point of this potential. The advantage of this design is the clear coaxial geometry for the rf current, except for the overpass lead, which can be multiplied. A high filling factor can be achieved since the figure of eight winding with all its resonating problems is avoided. Some effort must be devoted to a careful design of the blocking condensors. The single turn polarization field requires a high current. This cavity concept was chosen for the new rf system for the CERN PS and the AGS at Brookhaven, the KEK booster and the SNS synchrotron at Rutherford Laboratory.

The third principle of rf acceleration in a synchrotron is no more from the transformer type. It rather came out of the tool box of low frequency linear accelerators. Its name is "drift tube" principle. It was first used in the Birmingham synchrotron and the Bevatron at LBL around 1953. It reappeared in the Princeton Pennsylvania Accelerator around 1960 and then celebrated a remarkable come-back in both the Booster and the Main Ring at FNAL in 1971. Its advantage is the ideal decoupling of the rf field and the bias field and its suitability for quite high gap voltages, no more closely related to the amplifier characteristics. It is particularly suited for application in a high frequency range and-preferably under circumstances of a low frequency swing. All these features comply with the parameters for future proton synchrotrons.

Fig. 6 shows schematically the design of the drift tube cavity. The beam bunch is accelerated in the left hand gap, traverses the drift tube well shielded, whilst the polarity of the rf reverses, and then the bunch is accelerated in the right hand gap again, where the polarity is now in the right sense. The resonant circuit is formed by the drift tube capacitance and the ferrite loaded inductance below. There is much freedom in the design of this inductor, it is decently away from the beam. There can be a few of such devices in parallel, and the important feature is that the bias current, indicated by arrows, passes now well decoupled from the rf current through the center conductor of this coaxial device. There is the free choice of winding numbers. There are no geometrical constraints for a tap on the inner conductor for the rf power tube connection. The whole device, including the ferrites, could be operated under vacuum, the voltage at the rf feed-through being relaxed. But the devices built so far still have ceramic cylinders across the gap and this component presents a substantial problem area when the voltage approaches the 100 kV level.

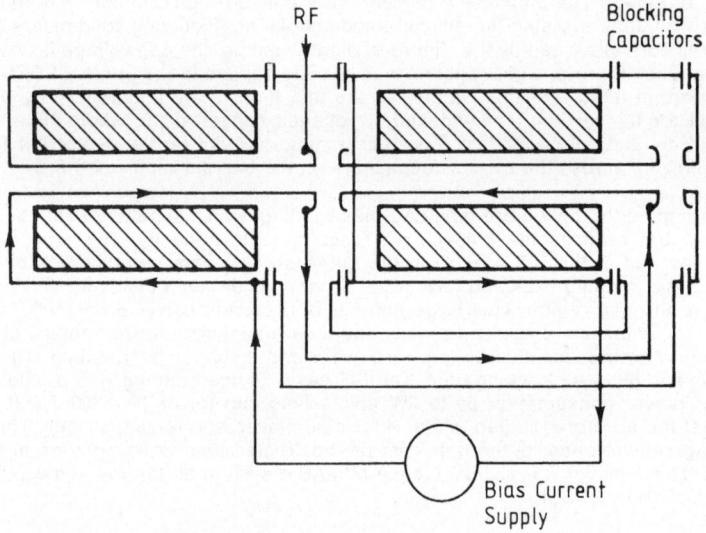

Fig. 5. Double λ/4 cavity with the polarizing current flowing on the cavity walls.

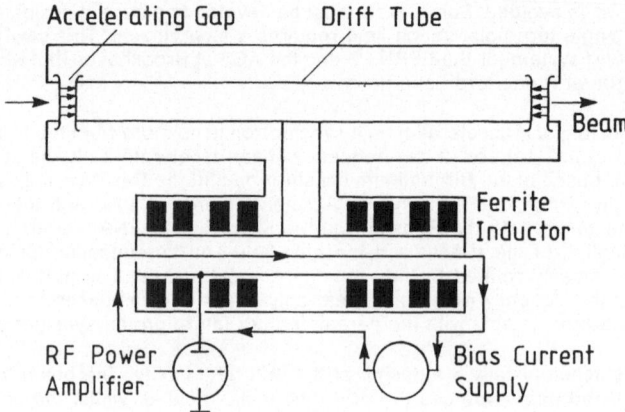

Fig. 6. Drift tube accelerating cavity with a ferrite loaded tuning inductor.

There is one serious constraint associated with the drift tube principle: it requires a high harmonic number. Otherwise its efficiency drops off drastically. This results from the inherent relationship between drift tube length and bunch spacing, which should be one half. The length should be $L = \beta \cdot \lambda/2$, β being the particle velocity over the velocity of light and λ the wavelength of the accelerating voltage. This relation does not vary during the cycle; if β increases, λ decreases proportionally. If the above mentioned condition is not fulfilled, the following formula denotes how efficient the drift tube voltage still is used, T is called the transit-time factor.

$$T = 2 \sin \frac{Lh}{2R} \tag{5}$$

Assuming a drift tube length of 2.5 m, the values for T for the three machines are the following:

	SIS	CERN PS	FNAL Booster
T =	0.29	0.49	1.96

The result is obvious: the FNAL Booster was designed with its high harmonic number of 84 for drift tube acceleration, the other two machines were not. A figure of $T = 1$ still would be acceptable, but this is a matter of compromising between efficiency loss and simplicity of the accelerating device.

THE FERRITE PROPERTIES

The ferromagnetic property, that means the enhancement of the magnetic flux in a given magnetic field, inherently implies that the flux saturates at some point, when the field is raised. For tuning a cavity it is desirable that the ferromagnetic property, the relative permeability μ_r changes smoothly with the increasing field, and not abruptly, as the word saturation would suggest.

Fig. 7a shows the hysteresis curve of a "soft" ferrite material A and a "hard" ferrite material E. We concentrate on the curve of material E because it is more illustrative. However, material A is used for low frequency cavities because it allows for a larger μ_r variation. The polarizing dc field H is swept between 0 and 3 A/mm, and the reversible μ_r, which is the slope of the hysteresis curve, varies as indicated in Fig. 7b from about 100 to about 3. Ferroxcube is a trade name of the Philips Company, and the type No. 4 was developed for synchrotron cavities. The frequency of the cavity varies like the square root of the μ_r ratio. From Fig. 7a one can conclude that the μ_r ratio would be much higher when the field would be changed from the crossing point of the upper curve segment with the $B = 0$ axis to the saturation value on the right. This would then give a μ_r variation as the dashed curve indicates in Fig. 7b. (Actually this curve belongs to the situation, where the cycle would start from complete demagnetisation. But it has the same shape, when the cycle starts from the left intersection of the B curve with the $B = 0$ axis.) This kind of field sweep was applied in the Princeton Pennsylvania accelerator, but never again. Perhaps the necessity of a bipolar power supply was judged to be complicated, but actually it is not.

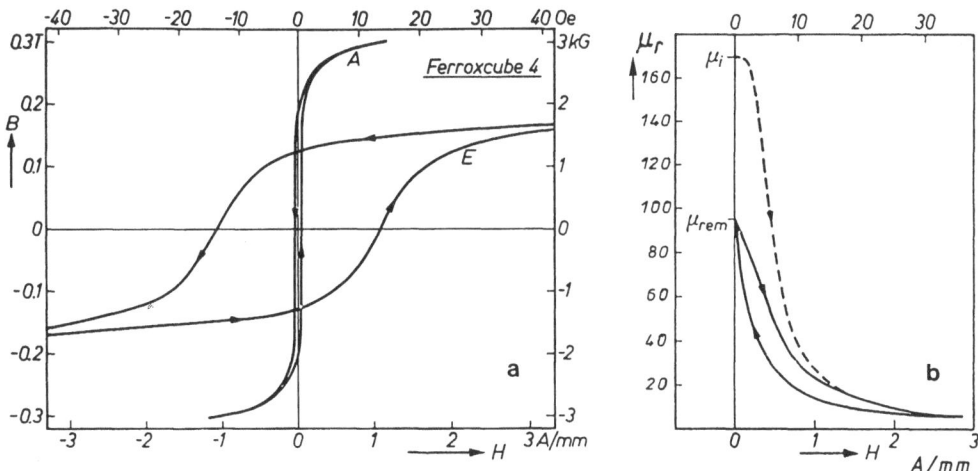

Fig. 7. Hysteresis curve and μ_r characteristics for two types of ferrite material, Ferroxcube 4.

In Fig. 8 a complete machine cycle of the SIS is shown with the linearly raising dipole field B, the corresponding rf frequency curve F_{HF} and the drastic change of the associated μ_{Δ} characteristic. The development of the bias power supply, which delivers the current of 0 - 800 A in 40 msec in a closely programmed way, was nearly as laborious as the development of the rf hardware.

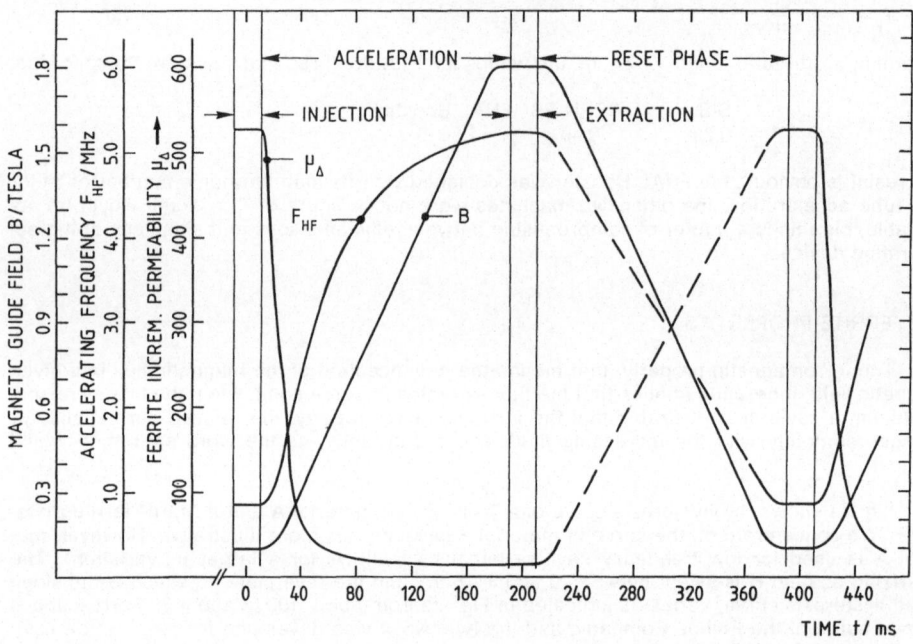

Fig. 8. Machine cycle of the SIS 18 with dipole field B, accelerating frequency F_{HF} and the ferrite permeability μ_{Δ}.

The ferrite manufacturer offers a couple of standard qualities with μ_r values ranging from 10 to 600. The frequency dependence of μ_r is inverse to the μ_r numbers. This is tolerable, since at the high frequency end the μ_r is biased down anyway. Fig. 9a shows the μ_r versus bias field and Fig. 8b shows the frequency dependence. The figures have been taken from Philips publications. The accelerator builder makes his selction of the μ_r range and should not go too far in Fig. 9b in his frequency choice. Then the manufacturer can develop a particular material, with some chance of optimization in respect to lower power loss.

The inherent rf power losses of the ferrite, induced by friction during polarization and reversal of magnetized domains inside the material, is one of the two parameters, about which the rf engineer feels uncomfortable for two reasons. First, because of the effort of producing large amounts of rf power in a wide band amplifier and because of the necessity of complicated cooling structures inside the cavity. Secondly, and this is more severe, there exists an upper limit of about 0.3 to 1 Wats per cm³ for the tolerable power density in the ferrite. Beyond this limit the electrical and mechanical properties deteriorate drastically. Ferrite is a poor heat conductor.

The second uneasy parameter is the onset of the "high loss effect", which might occur below the power loss threshold. This effect occurs as follows: after the rf power is switched on, a few milliseconds thereafter the cavity voltage drops immediately to about one half of its initial value and a strong noise occurs on the rf signal. Only speculations exist about the physics of this effect ("spin wave resonance") and the ferrite manufacturer can never give data about it. This effect depends on material casualities, on the bias cycle, the rf flux and apparently on everything. It can easily be measured in the static case, at a fixed frequency. But it tends to disappear at high cycling rates, when the frequency is rapidly changed. The determination of the onset of this effect is always left to the customer, who usually has in the testing phase not the final dynamic power supply either. The manufacturer does not know what to do against this effect, or he does not tell. There are unacceptable differences, about a factor of two, between otherwise comparable commercial products. Even the reproducibility is not existent: if a ferrite quality delivered 10 years ago is specified again, the product quality in respect to the high loss effect actually is unpredictable.

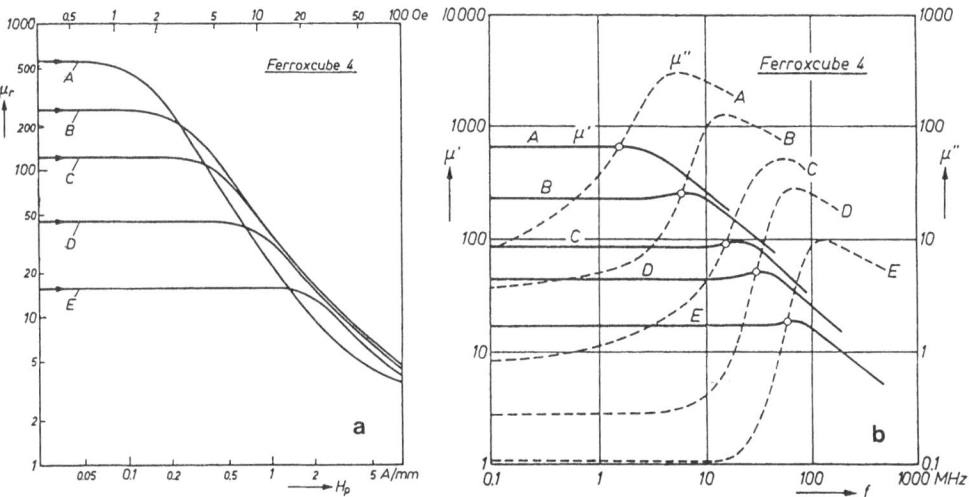

Fig. 9. μ_r as a function of bias field and frequency.

For the SIS circumstances the onset of this high loss effect was observed at 15 mT· MHz, giving a voltage per ferrite ring of 260 V. This limit occurs in the broad mid of the bias cycle, not at the ends. This consideration, and not the power losses, determined the maximum gap voltage of 16 kV for 64 rings. All one can do about both effects, the power loss and the "high loss effect", is to increase the volume of ferrites, or trivially the number of rf stations.

Now coming to actual numbers of the rf power, which finally has to be matched to one or the other of the above mentioned thresholds.

The shunt resistance R_p of a parallel resonant circuit is $R_p = Q\omega L$ and it should be high, because the power loss is $P = \hat{U}^2/2R_p$. Q is the ratio between the ohmic shunt resistance and the frequency dependent impedance of either the inductance or the capacitance. ω is $2\pi f$ the resonant frequency and L the inductance of the circuit. The Q value is quite abstract, but easily measurable and used in many formulas. By the way, Q is equal to μ'/μ'' in Fig. 9b.

If we expand $R_p = Q\omega L$ a little further, we obtain

$$R_p = Q\omega L = \mu_r \cdot Q \cdot \omega \cdot L_o \qquad (6)$$

L_o is the inductance of the empty cavity without ferrite and purely geometry dependent:

$$L_o = \frac{\mu_o}{2\pi} \cdot (\ln D_a/D_i) \cdot l \qquad (7)$$

D_a is the outer, D_i the inner conductor diameter of the cavity and l is the cavity length.

The number $\mu_r Q\omega$ in Equ. (6) has become a quality measure for the ferrite, it is depicted in fig. 10, but it remains very abstract again. From the curves of fig. 10, which belong to negligible rf amplitudes, it could be concluded that the shunt resistance R_p would go up with frequency. But this is not the case, because the $\mu_r Q\omega$ value goes down with the bias field. The $\mu_r Q\omega$ is therefore an approximate constant over the frequency cycle and its value equals the low frequency values of Fig. 10.

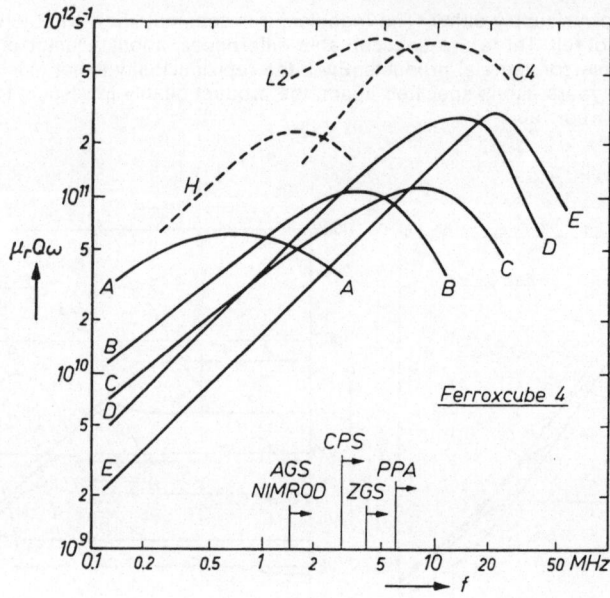

Fig. 10. The $\mu_r Q\omega$ characteristic of different ferrite materials.

We assume for the moment a given geometry of two meters in length and ferrite rings of 50/25 cm diameter, filling up the volume, L_o becomes 0.28 µH and from Fig. 10 we take a $\mu_r Q\omega$ product of 10^{10}. The shunt resistor R_p becomes then 2.8 kΩ. This is the order of magnitude which is typical for synchrotron cavities. In linear accelerator cavities R_p is of the order of a few MΩ. With the gap voltage of 16 kV in the case of the SIS, the rf power requirement becomes 45 kW. The qualities A to E in Fig. 10 are catalogue materials. The qualities H, L2 and C4 are optimized according to the specifications from various accelerator projects.

RF FOR FUTURE ACCELERATORS

It is likely that there will be built a few small synchrotrons like the SIS and several storage rings of still smaller size. For these projects the design options outlined in this paper can be applied and no further developments are needed, even not for the ferrites.

Large proton colliders will certainly have classical high-energy synchrotrons as injectors. In the main ring the very efficient linear accelerator cavities can be used, because the beam is highly relativistic and the frequency remains constant during a further acceleration. The envisaged superconducting magnets limit the $\dot{B}$ severely, thus the demand of rf voltage is not excessive.

Completely new designs will be needed for the Hadron Facilities because of the envisaged high repetition rate of about 25 Hz. There a sum voltage of about two MV will be needed and one certainly will try to avoid a number of 100 rf cavities of the classical design. At Los Alamos, in the context of developments for LAMPF II, more than 100 kV have been obtained in a completely new cavity concept with quite small amounts of ferrite with very low power loss.

ACKNOWLEDGEMENT

The author owes thanks to Klaus Kaspar, GSI, for many discussions and for providing the literature collection, data and figures of the SIS rf system.

BASIC FEATURES OF SUPERCONDUCTING RF CAVITIES

H. Piel

Fachbereich Physik der
Bergischen Universität – Gesamthochschule Wuppertal
5600 Wuppertal, West Germany

1. INTRODUCTION

In 1986 we celebrated the 75th anniversary of the discovery of super-conductivity by Heike Kammerling Onnes [1] and the same year saw the discovery of a new class of high T_c Superconductors by Bednorz and Müller[2]. The transition temperature for superconducting materials is now approaching 100 K[3]. It is an exciting time for basic research in superconductivity.

Already the application of the technology of rf superconductivity to particle accelerators, which we want to discuss in this lecture and the following one by H. Lengeler, has by itself some history. It started in 1965 with the accelerating of electrons in a superconducting lead-plated resonator at Stanford[4]. In the beginning of the 70's, construction of the Stanford Superconducting Recyclotron[5], the Illinois Microtron using a superconducting accelerating section[6], and the CERN-Karlsruhe s.c. Particle Seperator[7] was started. In 1974 a superconducting resonator successfully accelerated an electron beam to 4 GeV in the Cornell Synchrotron[8] and in 1976 the construction of the Argonne s.c. Heavy Ion Post Accelerator[9] was begun. In 1977 the first Free Electron Laser was operated using the high brightness beam of the Stanford Superconducting Accelerator[10]. Several of these devices have now been operated for many thousands of hours reliably and under routine conditions. It was shown that the drastic reduction of the rf surface resistance in s.c. cavities could be achieved even in complex resonators. The early expectations, however, to reach the very high electric accelerating or deflecting fields promised by the elementary theory of superconductors in radio frequency fields, were not fulfilled.

Although the accelerating fields of 2 to 3 MV/m achieved in the first operating s.c. accelerators were about 10 to 15 times lower than expected from BCS-theory, the early results at X-band[11] and recent experiments at L- and S-band frequencies[12,45], show that there are not other fundamental limitations. Research and development work in rf superconductivity is therefore still rewarding.

This lecture tries to give an introduction to the fundamental features of superconducting cavities and is organized as follows: In the following section the concept of coupled resonators, which form an accelerator module, and important quantities like the rf surface resistance, the cavity Q and the shunt impedance, are introduced. Section 3 discusses the fundamentals of rf superconductivity. A short introduction to superconductivity is given. The surface resistance of a superconductor in an rf field is explained in the framework of a two-fluid model and the critical rf magnetic surface field is introduced. In section four the importance of anomalous losses in s.c. cavities is outlined. The diagnostic method, microscopic defects, thermal stability and high purity niobium, as well as the progress in electron field emission studies, are described. Section five deals with high T_c superconductors and their use in rf cavities. Niobium cavities with a Nb_3Sn surface are discussed, and first experiments on the microwave properties of the new metal oxide superconductors are described explicitely.

Section one through four follow very closely and update a seminar given by the author at the CERN Accelerator School in Oxford 1985[14].

For additional reading on the subject the references 13,15 and 16 are suggested.

2. SOME CAVITY FUNDAMENTALS

2.1. Coupled cavities

The heart of each high energy accelerator is the rf accelerating section which generally is composed of a number of accelerating modules each of which is a chain of coupled rf resonators. For educational purposes we want to assume that such a module is a string of weakly-coupled pill-box cavities as shown in Fig. 1a, each of which is excited in the TM_{010}-mode. This mode has a longitudinal electric field on the axis of the cavity, which is surrounded by the circular field lines of the magnetic field and reaches its maximum at the cylindrical wall of the cavity. The accelerating module of Fig. 1a is a chain of coupled oscillators very much like the coupled pendula shown in Fig. 1b. The resonant frequency of the free pendulum corresponds to the resonant frequency ($\omega_0 = 2\pi f_0$)

$$\omega_0 = 2.405 \ c/a$$
c = velocity of light
a = radius of pill-box cavity $\qquad\qquad (1)$

of the TM_{010}-mode of the pill-box cavity. The coupling spring between the pendula is equivalent to the coupling electric flux through the small iris openings connecting the individual cavities.

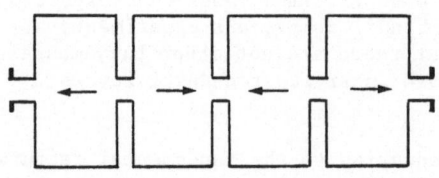

Fig. 1a. Chain of weakly-coupled pill-box cavities representing an accelerating module.

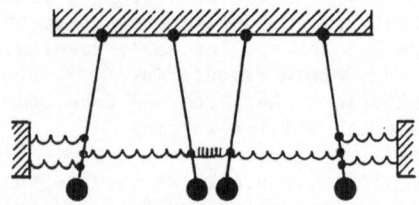

Fig. 1b. Chain of a coupled pendula as a mechanical analogue to Fig. 1a.

In the classical normal conducting linear accelerators such a module consists of many cavities and is generally operated in a travelling wave mode. The rf power is coupled into the first cavity of the string, travels down the structure, and is adsorbed strongly by the rf losses in the cavity walls. In superconducting accelerating modules these losses are reduced by many orders of magnitude, and a travelling wave operating mode is inappropriate. A superconducting accelerator module is therefore operated as a chain of N coupled resonators. Such a module is then excited in one of its N eigenmodes. By solving the characteristic equations of such a coupled oscillator system one obtains for the resonant frequencies ω_q of the eigenmodes and the axial electric field E_q (q,t) of the n-th cavity the following relations

$$\omega_q^2 = \omega_o^2(1 + K(1-\cos\alpha_q)) \tag{2}$$

$$E_n(q,t) = E_o\sin(\frac{2n-1}{2}\,\alpha_q)\,\cos\omega_q t \tag{3}$$

E_o = maximum axial electric field

$$\alpha_q = q \cdot \frac{\pi}{N} \qquad\qquad q = 1,2\,..N$$

N = number of coupled cavities

K = coupling factor between cavities.

The "mode angle" α_q is often used to denominate one specific number of the N eigenmodes. In the so called π-mode (Q = N or $\alpha_q = \pi$), the accelerating module oscillates in its highest frequency and, normalized to the acc. field, has the smallest rf losses in its walls. This is the reason why the π-mode is a favourite mode of operation for accelerating modules. In this mode the accelerating fields are equal in magnitude and opposite in direction in each pair of cavities as shown in Fig. 1a and seen from Eq. (3). A velocity of light electron which enters the first cavity at time 0 will enter the second cavity after a time $\tau = d/c$. If this time equals half the rf period (π/ω_N), then the electron will receive a maximum of acceleration in the accelerating module. The length d of one cavity of the module is then equal to $\pi c/\omega_N$ where ω_N equals the π-mode frequency of the module according to Eq. (2). A disadvantage of the π-mode is its sensitivity to mechanical tuning errors of the individual cells of a module which scales with N^2. The average accelerating field E_a referred to frequently in this seminar is given as $E_a = V/1$, where V is the voltage gain of the electron after traversing an accelerating module of length $1 = N \cdot d$. E_a is directly proportional to E_o.

2.2 Surface resistance, cavity Q and shunt impedance

In the normal conducting cavities fabricated from high conductivity copper the electromagnetic field penetrates into the cavity wall by the skin depth δ with

$$\delta = (2/\mu\sigma\omega)^2 \tag{4}$$

σ = electrical conductivity (for copper at room temperature $5.80\cdot10^7\Omega m$)

μ = magnetic permeability of the cavity wall.

At 500 MHz this skin depth is about 3.0 μm. The rf losses per unit surface area P_s produced in this thin layer can be expressed as

$$P_s = \frac{1}{2} R_s H_s^2 \qquad (5)$$

where H_s is the magnetic surface field and R_s is the surface resistance. R_s has the dimension of Ohms and for a normal conducting cavity is given by

$$R_s = (\frac{\mu\omega}{2\sigma})^{1/2} = \frac{1}{\sigma\delta} . \qquad (6)$$

This gives at 500 MHz a surface resistance of 5·8 mΩ. In very pure metals, σ which is proportional to the mean free path 1 of the conduction electrons, can be increased by more than four orders of magnitude if the conductor is cooled to the temperature of liquid helium. The rf surface resistance, however, decreases only by a factor of about five. This behaviour is not explained by (6) and is due to the anomalous skin effect which has to be considered when 1 becomes comparable to the classical skin depth δ. In the limit of 1>>δ the surface resistance is given by[17]:

$$R_s = \frac{8}{9} (\frac{\sqrt{3}}{16\pi} \frac{\mu_o^2\omega^2 1}{\sigma})^{1/3} . \qquad (7)$$

R_s scales like $\omega^{2/3}$ and becomes independent of ℓ. It is therefore of no benefit to cool a normal conducting cavity to low temperatures.

The quality factor Q of a cavity is directly related to its surface resistance. Q is defined as the ratio of the energy stored in a cavity to the energy lost per rf period. Energy can be transferred to the particle beam; it can be radiated out of the cavity through openings or antennas; and it is dissipated and converted to heat by the rf losses in the cavity wall. If only the losses P_o from the unavoidable Joule heating of the cavity wall are taken into account, one arrives at the unloaded Q of a cavity:

$$Q_o = \frac{\omega W}{P_o} . \qquad (8)$$

The stored energy W is proportional to the cavity volume and to the square of the average accelerating field. P_o scales with the surface resistance, the area of the cavity wall, and is also proportional to E_a^2. It therefore can be shown that (8) reduces to

$$Q_o = G/R_s \qquad (9)$$

where G is the so called geometry constant of the cavity. It is independent of the cavity frequency and for resonators like the ones shown in Fig. 1a and 5 is approximately 270 to 300 Ω.

If P is the rf power per unit length necessary to maintain an accelerating field E_s in an unloaded cavity, then the shunt impedance (per unit length) r of the accelerator cavity is defined by

$$P_o = E_a^2/r. \qquad (10)$$

The shunt impedance is proportional to Q_o. The specific shunt impedance is defined as the ratio r/Q_o. For a single cell cavity, shaped like the cells of the accelerator structure in Fig. 5, $r/Q_o \cdot d$ is about 150 Ω. The length d of one cell in an accelerator structure for highly relativistic electrons operated in the π-mode is $\pi c/\omega$, as already mentioned in section 2.1. One therefore obtains from Eqs. (9) and (10)

$$P_o \simeq \pi c R_s E_a^2/150 \cdot \omega \cdot G. \qquad (11)$$

From this one concludes that the rf power necessary to maintain a given accelerating field per unit length is, for normal conducting cavities, proportional to $\omega^{-1/2}$. High frequencies are therefore favoured for the operation
of normal conducting linear accelerators. At 500 MHz, which is a frequency
typical for storage ring cavities, Eq. (11) gives for copper at room temperature and for E_a = 5 MV/m a dissipated rf power of 1.0 MW/m.

This very high power dissipated in normal conducting accelerator cavities is the main reason for the interest in rf superconductivity.

3. SUPERCONDUCTING CAVITIES

3.1. Short introduction of superconductivity

It is well known that many metals and alloys become superconducting
below a certain critical temperature T_c, which is characteristic for the
specific material. Only three years after the first liquefaction of helium,
Heike Kammerlingh Onnes[18] found in Leyden 1911 that mercury lost its resistivity completely below 4.15 K. A great many, but not all, elemental
metals become superconducting, and although superconductivity seemed to be
a very basic phenomenon for electrical conductors, it took almost 50 years
before Bardeen, Cooper and Schrieffer[19] could explain the mechanism behind
this phenomenon in their theory, often referred to as the BCS theory. It
would be beyond the scope of this seminar to give an account of this
beautiful theory, but it may be useful to extract some ingredients in order
to explain the two-fluid model of a superconductor given by H. London
already in 1934[20]. This model is very useful for understanding the basic
features of a superconductor in an rf field. If the new high T_c superconductors can also be described by this model is presently unknown, but a
first look at their properties is given in section five.

It is shown in the BCS theory that due to the interaction of the conduction electrons in a metal with the vibration of the atoms in the
lattice, there is a very small net attraction between electrons. As a result
of this conduction electrons can form into pairs, the so called Cooper
pairs. The energy of pairing 2Δ(T) (Δ(T) would be the pairing energy per
electron) is very weak and in the BCS theory given at T=0 to be

$$\Delta(0) = \alpha k T_c \tag{12}$$

$\alpha = 1.76$
k = Boltzmann constant.

Only a very small thermal energy is needed to ionize a Cooper pair
back into two "normal" electrons. At T=0 all conduction electrons are
paired, but at finite temperatures there is always a probability that a
pair is broken up. This probability is given by the Boltzmann factor
$\exp(-\Delta(T)/kT)$ and for the ratio of the densities of normal electrons (n_e)
and Cooper pairs (n_c) we find:

$$n_e/n_c \approx e^{-\frac{\Delta(T)}{kT}}. \tag{13}$$

At temperatures below $T_c/2$, n_c and Δ are very close to their values n_0 and
$\Delta(0)$ at T=0. The "two fluids" therefore are the superfluid of Cooper pairs
of density n_0 and the normal fluid of conduction electrons of density

$$n_e(T) = 2 \cdot n_0 \, e^{-\alpha \frac{T_c}{T}} \tag{14}$$

for $T < T_c/2$.

R.P. Feynman[21] gives a very instructive explanation as to why the fluid of Cooper pairs can carry an electric current without any losses. Contrary to the normal electrons, Cooper pairs are Bose particles. When there are many Bosons in a given state then there is an especially large probability for the other Bosons to go into the same state. So nearly all Cooper pairs will be locked down at the lowest energy in exactly the same state, and it will not be easy to get one of them out of this state. The probability to go into this state is by a factor $\sqrt{n_0}$ higher than into any other state, and n_0 is a very large number. Therefore, all Cooper pairs move in the same quantum state. Resistivity comes from knocking on electrons and transferring energy to the lattice, but this becomes impossible because they are all bound to Bosons.

Cooper pairs can be ionized by electromagnetic radiation if the frequency is high enough. The energy of the photons has to be

$$h\omega = 2\ \Delta(T) \tag{15}$$

which in the case of niobium ($2\Delta(o) = 3.12$ meV) results in a frequency of about 700 GHz.

It should be noted that Cooper pairs are not closely bound like, for example, the nucleus and its electrons in an atom. Cooper pairs are ordered states in momentum space with the two electrons having opposite but equal moments and opposite spins. For our purpose, however, it is qualitatively acceptable, although somewhat superficial, to consider a Cooper pair as a bound state with a rather large extension for which the coherence length ξ is a material constant and ranges typically between 38nm (niobium) and 1600 nm (aluminium). The distance between Cooper pairs is therefore considerably smaller than their "size".

A sufficiently strong magnetic field will destroy superconductivity. The critical value of the applied field is denoted by $H_c(T)$ and exhibits a temperature dependence given by

$$H_c(T) = H_c(o)\ (1-(T/T_c)^2). \tag{16}$$

Meissner and Ochsenfeld[22] found that in a superconductor which is cooled in an external field smaller than H_c, below T_c the magnetic field is completely expelled. The interior of the superconductor is screened by currents which flow in a very thin skin layer. The external magnetic field exponentially decays in this surface layer, and its decay length is called the London penetration depth λ. It ranges between 15 and 110 nm and is material dependent.

There are two classes of superconducting materials denoted as type I and type II superconductors. There is no difference in the fundamental mechanism of superconductivity between them. They differ from each other only by a completely different Meissner effect. A good type I superconductor excludes a magnetic field until superconductivity is destroyed abruptly at H_c, and then the magnetic field penetrates completely. A good type II superconductor expells the field only for relatively weak external fields smaller than H_{c1}. Above H_{c1} the field partially penetrates into the superconductor which remains superconducting. At a much higher H_{c2} field, sometimes 100 kOe or more, the flux penetrates completely and the superconductivity vanishes. The so called thermodynamical critical field is then approximately the geometric mean of the lower and upper critical magnetic field:

$$H_c = (H_{c1} \cdot H_{c2})^{1/2}. \tag{17}$$

3.2 Basic characteristics of a superconducting cavity

3.2.1. The rf resistance of a superconducting surface

In the case of a normal conducting rf resonator the electromagnetic field penetrates by the skin depth into the cavity wall. In a superconducting cavity the equivalent "superconducting skin depth" is approximately equal to the London penetration depth and therefore about two orders of magnitude smaller the δ. In contrast to the zero resistivity for dc electric currents there are losses if the superconductor is exposed to a high frequency field. This can be explained by the two-fluid model. The time varying magnetic surface field, $H_s \cos\omega t$, penetrates into the superconductor and induces in the "s.c.skin depth" an electric field. The amplitude of this field will therefore be proportional to ωH_s. The electric field accelerates the Cooper pairs, which transport this part of the surface current without losses. It will also accelerate the normal electrons, which can interact with the lattice and produce losses according to the anomalous skin effect. The power dissipated in the wall of the s.c. cavity per unit area $P_s{}^t$ (the index t denotes the two fluid model) can therefore be expressed as

$$P_s{}^t \sim n_e(T)\ \omega^2\ H_s{}^2. \tag{18}$$

Using Eq. (14) one arrives at

$$P_s{}^t \sim n_o \omega^2\ e^{-\alpha \frac{T_c}{T}}\ H_s{}^2 \tag{19}$$

$$T < T_c/2.$$

Comparing Eqs. (5) and (19) one gets for the surface resistance in the two-fluid model for frequencies well below the ionization limit and for $T < T_c/2$:

$$R_s{}^t = A\omega^2\ e^{-\alpha \frac{T_c}{T}} \tag{20}$$

where A may depend on material parameters like λ, ξ, ℓ and v_F. For frequencies below 10 GHz and for $T < T_c/2$ the experimental data are in fact described well by the relation

$$R_s = A(\lambda, \xi, \ell, v_F)\ \frac{\omega^2}{T}\ e^{-\frac{T_c}{T}} + R_{res}. \tag{21}$$

The residual resistance R_{res}, which is temperature independent and not related to the superconducting surface, is easily separated. The first term in Eq. (21), which is often referred to as the BCS resistance, agrees remarkably well with the result of the two-fluid model.

Expressions for the surface resistance which are based on the BCS theory have been derived by Mattis and Bardeen[23] and Abrikosov, Gorkov and Khalatnikov[24]. Computations of the surface resistance based on these rather complex expressions have been performed by Halbritter[25] and Turneaure[26]. A further refinement of the BCS theory in regard to rf superconductivity has been achieved by R. Blaschke[27] by including the anisotropy of the pairing energy which is induced by the anisotropy of a crystal lattice. This modification removed a long existing discrepancy between experiment and theory in respect to the frequency dependence of the surface resistance. The quadratic dependence reflected by the two-fluid model has to go into a $\omega^{2/3}$ behaviour as the frequency approaches the ionisation limit.

Figure 2 compares experimental data on the surface resistance of niobium at 4.2 K by U.Klein[28] and G.Müller[29] with the computational results of Blaschke. The agreement is excellent. The two-fluid model describes the frequency dependence below 10 GHz quite well but cannot account for the change of slope at very high frequencies.

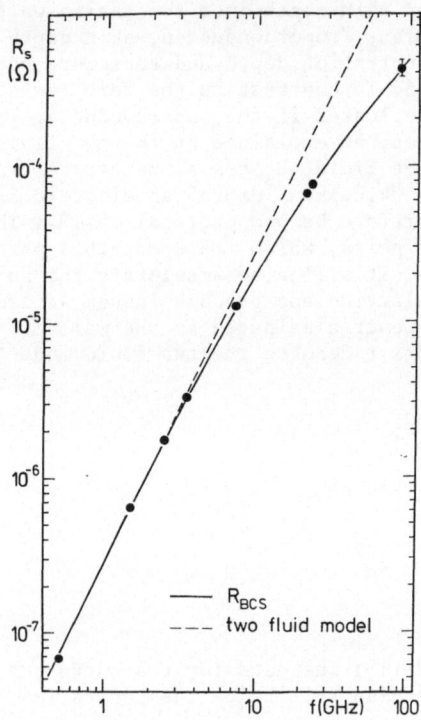

Fig. 2.   Frequency dependence of the surface
resistance of niobium at 4.2 K.

Figure 3 shows the temperature dependence of the surface resistance of a single cell niobium cavity at a frequency of 3 GHz. The exponential temperature dependence explained by the two-fluid model is nicely demonstrated as well as the existence of a residual resistance which is characterized by its temperature independence. Extracting α from the data of Fig. 3 and many other experiments at other frequencies, one finds α very near to 1.85 for frequencies below 10 GHz. This value is close to the prediction of the BCS theory.

At 500 MHz and at a temperature of 4.2 K the BCS surface resistance of niobium is 70 nΩ compared to the 5.8 mΩ of copper at room temperature. For an accelerating field of 5 MV/m (example in section 2.2) the dissipated power in a superconducting accelerating module will be only 12 W. This power is adsorbed at 4.2 K and has therefore to be corrected for the Carnot and technical efficiency of a 4.2 K refrigerator. This brings the 12 W to 5.5 kW,which is two hundred times lower than the power dissipated in an equivalent copper structure.

Another important difference between a superconducting and normal conducting cavity becomes apparent if one combines the equations (11) and (21) neglecting the residual resistance R_{res}. One obtains then for the power dissipated in a superconducting cavity:

$$P_{sc} = \frac{A\pi c\omega}{150GT} \ e^{-\ \alpha\ \frac{T_c}{T}} \ E_a^2. \tag{22}$$

152

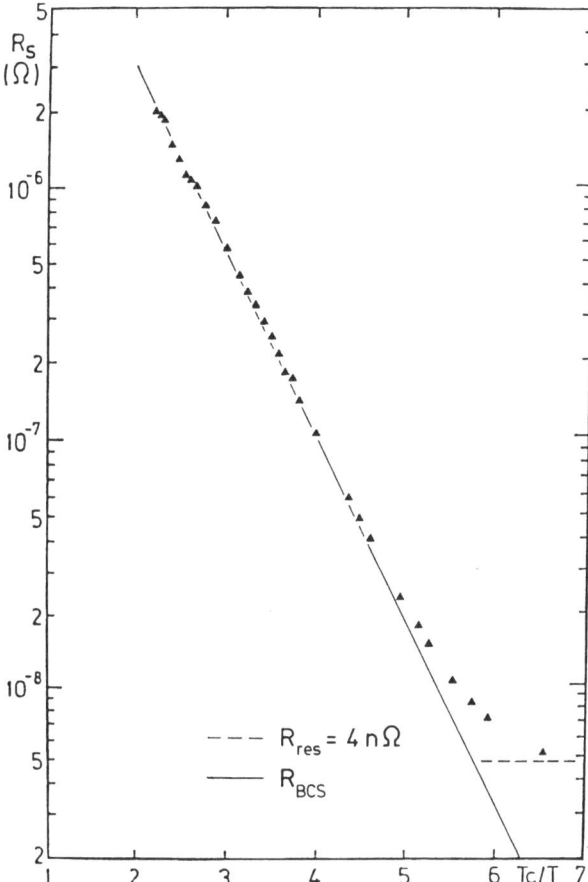

Fig. 3. Temperature dependence of the surface resistance of niobium at 3 GHz.

One sees that, contrary to the case of normal conductivity, low frequencies are preferred in superconducting cavities. Presently the validity of this statement is limited to a surface resistance larger than 50 to 100 nΩ. The reason for this is the residual resistance R_{res} which, to our present knowledge, is not a property of a superconducting surface in an rf field. It is caused by anomalous losses which are described in more detail in section 5. They are critically dependent on the purity of the cavity surface. Chemical etching, electropolishing, rinsing with ultra pure water and methanol, and very high temperature treatment (up to about 1800°C) in an UHV furnace are final preparation steps for superconducting cavities fabricated from niobium. Normal conducting residues left on the cavity surface by these procedures can contribute significantly to the residual resistance. Achieved residual resistance of 1 nΩ or, more typically, 10 nΩ correspond to only about 0.1 to 1 ppm of normal conducting surface area. It is therefore obvious that superconducting cavities have to receive their final surface preparation and assembly in a clean room environment.

3.2.2 Fundamental field limitations

All the considerations given above are valid only if the superconducting cavity is in a true Meissner state. On first sight this can only be the case if the maximum magnetic rf surface field H_s^{max} is smaller than H_c or H_{c1} in a type I or a type II superconductor respectively. This statement, however, may hold only in equilibrium condition and may therefore not apply to microwave cavities. The transition from the superconducting to the normal conducting state is a phase transition. Such a transition needs nucleation centers, and it is therefore possible that there may be a metastable or superheated state before the superconductor returns to its normal conducting state. The maximum field up to which this transition state may persist is called the critical superheating field H_{sh}. In type I superconductors like lead, for example, H_{sh} is higher than H_c. For type II superconductors (Nb$_3$Sn for example) the superconducting state persists beyond H_{c1} but H_{sh} stays below he thermodynamical critical field H_c. Matricon and James[30] have calculated the dependence of H_{sh} on $\kappa = \lambda/\xi$ by solving the Ginzburg Landau equations[31] which are based on a pheonomenological theory of superconductivity. Their results have the limiting form[16]:

$$H_{sh} \simeq 0.75 \ H_c \quad \text{for} \quad \kappa \gg 1 \qquad (23)$$
$$H_{sh} \simeq 1.2 \ H_c \qquad \kappa \simeq 1$$
$$H_{sh} \simeq \frac{1}{\sqrt{\kappa}} \ H_c \qquad \kappa \ll 1.$$

H_{sh} shows a smooth behaviour as κ passes through the interesting value of $1/\sqrt{2}$ which separates type I from type II superconductors.

The persistence of the Meissner state may be very stable in rf fields. This is expected because the nucleation time of lux lines is around 10^{-6} s compared to the 10^{-8} s typical for the rf period of microwave cavities. Experimentally the rf critical field has been studied for type I superconductors like In, Sn and Pb near their critical temperature T_c. The results of these experiments are in agreement with theory[32]. For a typical type II superconductor like Nb$_3$Sn the lower critical field H_{c1} has also been surpassed[33]. One therefore presently assumes that the fundamental limit for H_{rf} is given by the critical superheating field. Table 1 gives some material parameters for Pb, Nb and Nb$_3$Sn. Table 2 gives the best results from present high energy accelerator projects for single cell cavities in the frequency range from 350 MHz to 3 GHz[14].

The maximum magnetic surface field in a cavity excited in the TM$_{010}$-mode is close to its equator and a good rule of thumb is $H_s^{max}/E_a = 45$ Oe/MV/m.

Table 1

Transition temperature and critical fields of the most frequently used materials in rf superconductivity:

Material	T_c [K]	$H_c(0)$ [Oe]	κ	$H_{sh}(o)$ [Oe]	H_s^{exp} at T≈2K [Oe]	E_a^{max} at T≈2K [MV/m]
Pb	7.2	804	0.5	1050	900[32]	22
Nb	9.2	2000	1.5	2400	1590[34]	50
Nb_3Sn	18.2	5400	20	4000	1060[33]	88

Table 2

Best performance of cavities from present high energy accelerator projects

LABORATORY	CERN			KEK	DESY	CORNELL	DARMSTADT/WUPPERTAL	
ACCELERATOR	LEP			TRISTAN	PETRA/HERA	CESR	130 MEV RECYCLOTRON	
MATERIAL	Nb	Nb	Nb on Cu	Nb	Nb	Nb	Nb	Nb_3Sn
FREQUENCY (MHz)	350	500	500	500	1000	1500	3000	3000
TEMPERATURE (K)	4.2	4.2	4.2	4.2	4.2	1.8	1.8	4.2
SINGLE CELL E_a^{max} (MV/m)	10.8	13.0*	10.8	7.6	5.5	22.8**	23.1**	7.2
Q at $E_a \cdot 10^9$	1.8	0.7	0.4	0.6	0.5	2.5	1.2	1.1

*) Cavities fabricated from high thermal conductivity niobium
**) yttrified niobium

It is the ratio which is used in tables 1 and 2 to compute the maximum accelerating field E_a^{max}. The comparison between H_{sh} and H_s^{exp} (compare Eq. (16)) shows that even the latest experimental results do not yet attain the theoretical expectations. This, however, from experimental evidence, is due to the anomalous and point like losses described in section 4. Today we do not know of any fundamental limitation which prevents us from reaching the limiting fields given in Table 1. The high values for the accelerating field promised especially for niobium and Nb3Sn cavities make it worthwhile to continue the experimental efforts.

4.   ANOMALOUS LOSSES

4.1   Temperature mapping and microscopic defects

The origin of field limitations well below H_{sh} and the causes of the residual resistance are the main areas of interest for the research on superconducting cavities. Several diagnostic techniques have been developed to study these questions. In the framework of this seminar only one, namely the"temperature mapping in subcooled helium"[35],[36], will be described.

As each energy loss mechanism will finally lead to an increase of the temperature of the cavity wall, temperature measurements are of prime importance to identify causes for field and Q-limitations. C.Lyneis at Stanford was in 1972 the first to use a chain of rotating carbon resistors mounted a few millimeters from a cavity wall to detect the location of a thermal instability[37].

This method has since been used by many groups working in this field. The carbon thermometers (see Fig. 4) used are 56 or 100 Ω (1/8 W or 1/4 W) Allen Bradley resistors, the bakelite insulation of which is often ground off to increase their sensitivity. Different electric schemes have been used to read the resistance value of the many thermometers generally used on one cavity, either an oscilloscope display or an automatic data acquisition system. During a quench all the energy stored in a cavity is set free and a substantial heat flux develops which leads to film boiling and a marked increase of the temperature of the helium film close to the quench area. This can be detected easily even in superfluid helium and if the resistor is not in contact with the cavity wall. The detection of quench areas is certainly a most useful diagnostic procedure. A temperature map of the surface of a cavity well below the breakdown field, however, will reveal even more information about the nature of high loss areas. Temperature mapping can only be done for bath temperatures above the λ-temperature. The main obstacle for a temperature mapping experiment is the fact that only the temperature of the outside of the cavity wall can be measured which is very effectively cooled by the surrounding liquid helium. In an experiment performed at CERN in 1978[35], it was shown that temperature mapping can be carried out quite well in a subcooled helium bath (favourable subcooled condition: bath temperature slightly above T_λ and bath pressure $\simeq$ 100 mb).In a subcooled bath, bubbles are absent and therefore the microconvection produced by bubbles rising from the heated surface is avoided. This reduces the cooling capability and liquid helium substantially and increases the heat transfer resistance between the niobium surface and the helium.

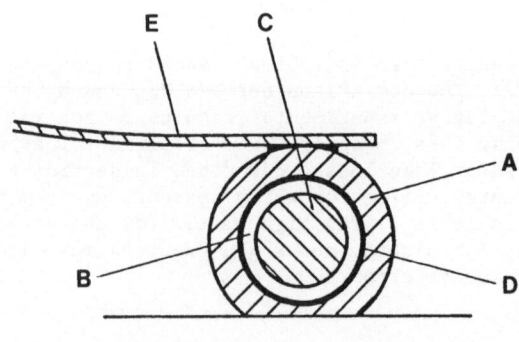

Fig. 4.   Cross section through carbon thermometer used for temperature mapping in subcooled helium
A = copper tube housing
B = bakelite insulation
C = carbon body of resistor
D = gap filled with conduction silver
E = copper beryllium spring

The first set up used for the temperature mapping of a 500 MHz spherical cavity [38] is shown in the photograph of Fig. 5.

 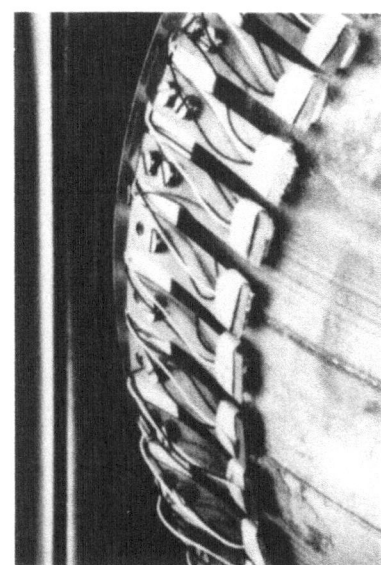

Fig. 5.   General configuration and close up view of the carbon thermometer scanning system[38].

Thirty-nine carbon thermometers (100 Ω, 1/4 W Allen Bradley) slide under spring tension on the cavity wall and can be rotated around the cavity. The resistor voltages and their angular position are read by a computer controlled data acquisition system. Figure 6 shows one of the first 3-dimensional temperature maps of a superconducting 500 MHz niobium cavity operated at an effective accelerating field of 3.2 MV/m[38]. This measurement was done in a subcooled helium bath at a temperature of 2.3 K. On the x-axis the distance along one circle of constant latitude around the spherical cavity is plotted. The y-axis shows the number of carbon thermometers (with resistor 1 corresponding to the top of the cavity and resistor 39 to the bottom of the resonator). The vertical axis displays the temperature increase ΔT detected by the carbon resistor. The residual resistance of this cavity was rather poor ($R_{res} \simeq$ 330 nΩ). It can be attributed to the very non-uniform high-loss area at the top of the cavity. In this early experiment at CERN the clean-room handling was not as well developed as today and already at an accelerating field of 3.2 MV/m one observes strong non-resonant loading.

After these first measurements the technique of temperature mapping was refined considerably[39,40]. The temperature increase of the intermediate helium layer at the outside cavity surface was calibrated against the heat flux density, and the dependence of this calibration on the bath temperature was experimentally determined. The relation between the measured temperature increase and the heat flux density is very dependent on the "hydrodynamics" of the flow of the local convection stream in the subcooled helium bath. All these effects have to be considered carefully. The necessary calibration experiments can be carried out at higher temperatures (for example at a frequency of 3 GHz at 3 K) where the well known BCS losses on the s.c. surface determine solely the heat flux through the cavity wall.

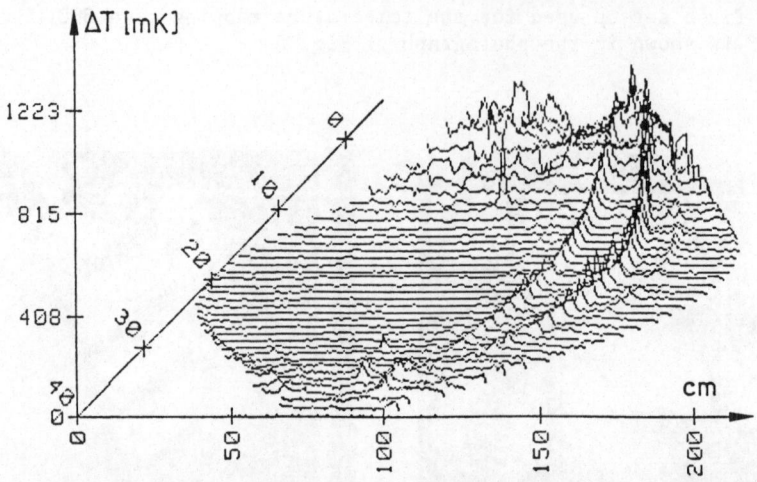

Fig. 6.   Early temperature map of a CERN 500 MHz cavity E_a = 3.2 MV/m
with line like regions of increased temperature due to the
impact of electrons field emitted by point sources[38]

The spatial distribution of the heat flux density on a 20-cell superconducting accelerator module for the Darmstadt Recyclotron is shown in
Fig. 7. This map is typical for the present day diagnostic technique and
for the specific losses observed in a s.c. cavity. Spikes in the heat flux
density are seen on the flat background of uniform losses which are expected from the BCS part of the surface resistance. Similar spikes have already

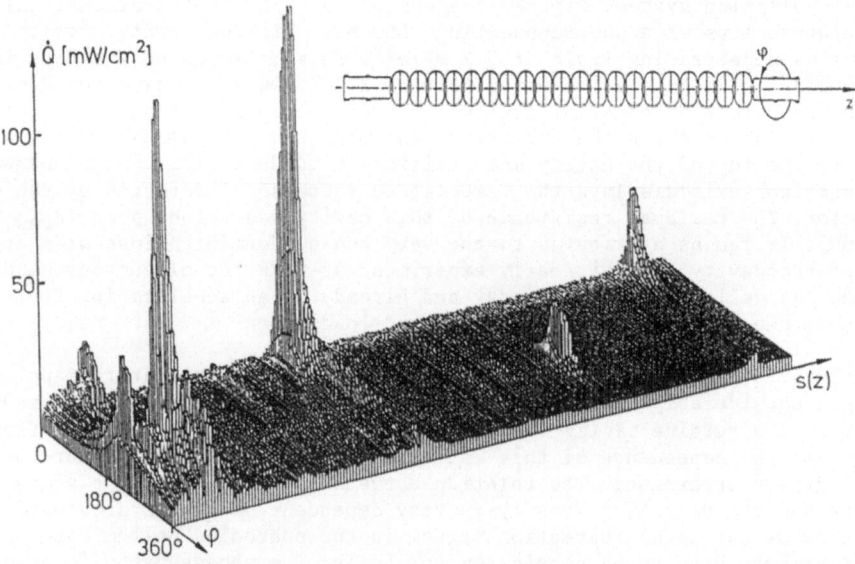

Fig. 7.   Spatial distribution of the heat flux density on a 20-cell
superconducting accelerator module for the Darmstadt 130 MeV
Recyclotron at E_a = 4.8 MV/m.

been observed in the very first temperature maps at CERN. They are produced
by high loss areas on the rf surface which must be smaller than a few millimeters in diameter. They are in fact found to be microscopically small and
in most cases invisible to the naked eye.

In a few but important experiments at CERN[41], such defects were detected
in 3 GHz single-cell cavities by temperature mapping, then cut out of the
cavity and analysed with a scanning electron microscope. Four of the photo-
graphs obtained are displayed in Figs. 8 to 11.

Foreign material inclusions, beads from electron beam welding, holes
in welds and chemical residues were found. During the mounting of the cavity
to the vacuum system, to rf couplers or other parts, or during rapid pump
downs particles can fall onto the s.c. surfaces. They can heat up in the
cavity field to very high temperatures, emit light, cause thermal electron
emission so leading to an excessive heating of their environment and thereby induce quenching. If a quench location is detected during temperature
mapping a later inspection of the cavity often shows a dark spot composed
of a central region and a halo as in Fig. 12. One can assume that this
halo is produced by material from the "dust particle" evaporated during

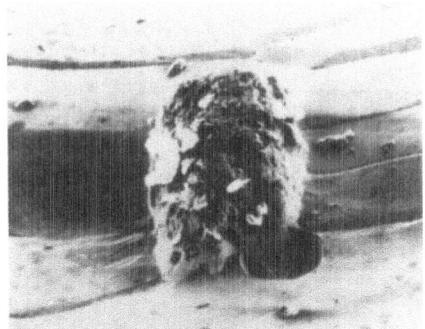

Fig. 8.   Tungsten inclusion on a cavity weld, probably
          embedded during TIG-welding. Quench field E_a
          = 4.5 MV/m.

Fig. 9.   Nb sphere (presumably originated from a welding
          bead). Diameter: 80 μm. Quench field E_a =
          6.8 MV/m.

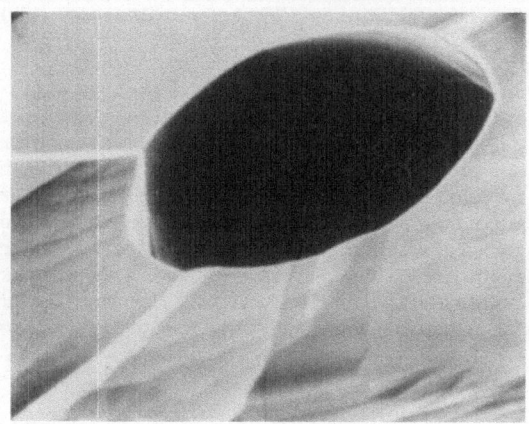

Fig. 10.   Microscopic hole in a weld of a 3 GHz cavity,
causing a thermal instability close to E_a =
8 MV/m

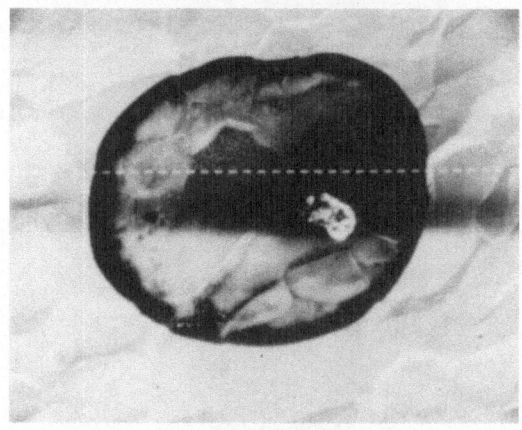

Fig. 11.   Chemical residue (drying mark). Diameter:
400 μm. Quench field E_a = 3.4 MV/m.

high field operation. In cavities mounted horizontally such particles would fall onto the equatorial surface, where the electrical field is small. They would not give rise to electron loading and would initiate quenches only if they were of metallic nature. At CERN it was in fact observed that horizontally mounted cavities reach higher quench fields at the first attempt and also show a much reduced electron loading compared to cavities mounted vertically into the test cryostat[42].

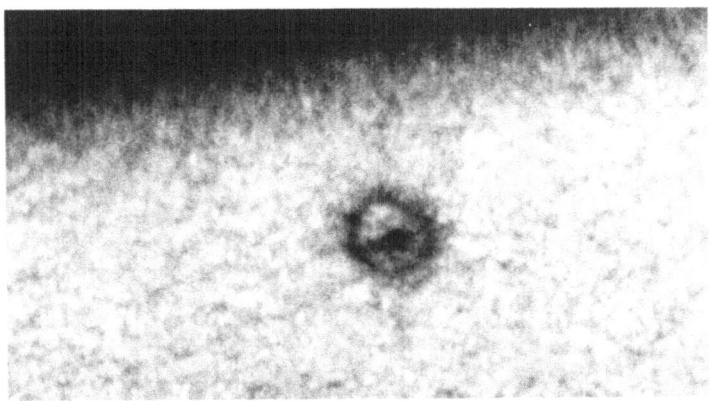

Fig. 12.   Defect halo in a 500 MHz cavity, often observed at quench locations after quenching occurred.
Diameter: ~ 1.5 mm[42]. The defect is located close to the iris (dark region)

All the above observations lead to the adoption of very careful cleaning and mounting procedures for superconducting accelerator cavities. Chemical treatment of the cavities with clean chemicals, the final rinsing procedures carried out with demineralized and dustfiltered water, and the mounting of the cavities to the test facility in clean rooms have improved the reliability with which low residual resistances and high accelerating fields can be presently obtained.

4.2 Thermal instabilities and the virtues of high purity niobium

Defect induced thermal instabilities and electron field emission from point sources (see section 4.3) are the main mechanisms which limit the performance of s.c. cavities. A defect on a cavity surface like the ones shown in Figs. 8 to 11 is heated in the rf field and the dissipated energy is transferred to the helium bath. The temperature gradient produced across the cavity wall may lift the temperature of the defect's environment above the critical temperature of the niobium and a sudden dissipation of the energy stored in the cavity wall will result. The threshold field of such a thermal instability can be increased if the thermal conductivity of niobium can be improved[43]. In standard, commercial reactor-grade niobium the interstitial impurities O,C and N determine the poor thermal conductivity[44]. These impurities can be controlled to a large extent during electron beam melting of the raw niobium and the consecutive manufacturing steps of the sheet material. The residual resistivity ratio (RRR) of niobium is proportional to its electronic thermal conductivity. Typical RRR values of standard niobium range are between 20 and 40. Due to a refinement in production techniques, niobium of RRR values between 80 and 160 has been commercially available since the end of 1983. This advance was achieved mainly by improving the vacuum condition and the procedure during the multiple electron beam melting of the niobium ingots. The progress in cavity performance compared to the status of 1983 can be attributed mainly to this improvement of the

thermal conductivity[45,46]. Not only the obtainable fields have increased,
but also the reliability with which the present design fields of 5 MV/m
can be reached.

An effective procedure to clean niobium of the most critical impurity,
oxygen, is the evaporation of yttrium onto the niobium surface developed
at Cornell[47]. During this process the surfaces of a niobium cavity are
brought into the proximity of an yttrium foil at a pressure of about 10^{-5}
Torr at 1250°C for several hours. A vapor deposited film of several μm
thickness traps the oxygen which diffuses rapidly from the bulk to the
surface. The oxygen enriched surface layer of yttrium is then dissolved
chemically. Starting from standard material (RRR = 30) the RRR value and
thereby the thermal conductivity can be improved by about a factor of
three (depending on the initial oxygen content). Starting from high purity
commercial niobium, RRR values of up to about 500 were obtained at Cornell.
The same technique has been tried experimentally at Cornell[48] and KEK[49]
using much cheaper titanium foils at slightly higher temperatures, with
similar success. The KEK results on a single-cell 500 MHz cavity in
Table 2 were obtained that way.

Figure 13 shows the measured temperature dependence of the thermal
conductivity λ^{50} of niobium samples of different purity. Curve a) represents
the status until 1983 and curve b) shows the quality of niobium which is
now commercially available. Curve c) gives the thermal conductivity of a
niobium sample which was yttrium treated at Cornell.

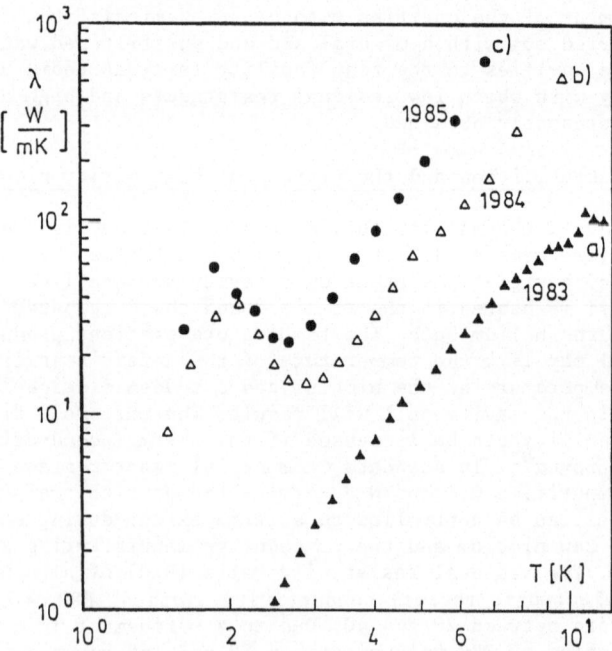

Fig. 13.   The temperature dependence of the thermal conductivity
           of niobium samples of different purity characterized by
           its residual resistivity ratio (RRR)[50].
           a) RRR = 36,   b) RRR = 152,   c) RRR = 360.

A very instructive display is shown in Fig. 14. There the performance of cavities fabricated from niobium of different purity is compared. The measurements were carried out with single-cell, 3 GHz cavities of spherical shape excited in the TM_{010}-mode. Each measurement is made after a new chemical treatment which dissolves more than 20 μm of the cavity surface and therefore creates a completely new surface as far as the shallow penetration depth of the rf field is concerned. Measurements of two laboratories (CERN and Wuppertal) are contained in the data. The dependence of the cavity performance on the purity of the niobium or its RRR is clearly seen.

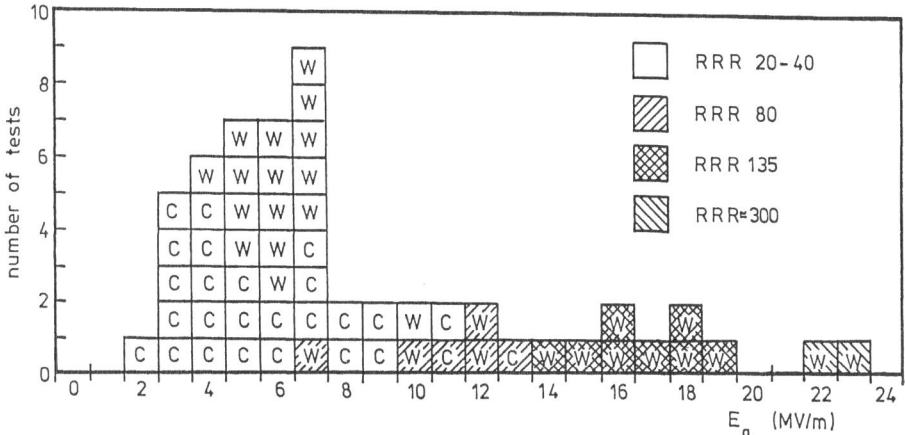

Fig. 14.   Performance of s.c. 3 GHz single-cell cavities fabricated from niobium of different purity[45,51]

The Q_o versus E_a dependence of the very high purity niobium cavity, yttrium treated at Cornell and built and tested at Wuppertal is shown in Fig. 15. This cavity is not limited any more by defect induced thermal instability but its field is limited by electron field emission.

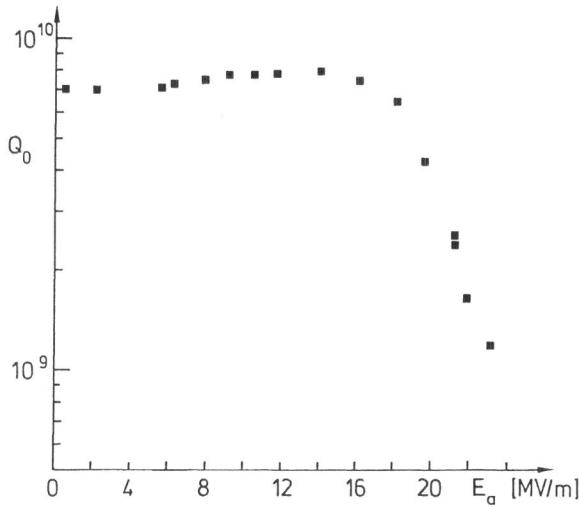

Fig. 15. Q_o versus E_a dependence of a RRR = 300 single-cell cavity which was obtained by the yttrification of an RRR= 80 cavity. The cavity was built and tested at Wuppertal and yttrified at Cornell[50].

163

4.3. Progress in understanding electron loading phenomena

The resonant multiplication of free electron currents (electron multi-pacting) was a very annoying field limitation in practically all supercon-ducting cavities before 1979[14]. This phenomenon was analysed and virtually eliminated by work done at Stanford[52], Genoa[53] and Wuppertal[54] in 1977 to 1979. Cavity shapes[54] and, in special cases, grooving of the cavity surface[55] were proposed, which later proved to suppress multipacting up to the highest fields reached so far.

It is because of this that today all s.c. accelerator structures are of the spherical or elliptical[55] design as can be seen from the examples shown in Figs. 5 and 7. In addition the improving ability to avoid lossy defects on niobium surfaces, and the progress in thermal stability of s.c. cavities, have allowed surface electric fields of more than 25 MV/m at all frequencies suitable for accelerating structures. At such surface fields, field emission induced electron loading is observed and constitutes an important field limitation. Already one of the very first temperature maps obtained at CERN in 1980, as displayed in Fig. 6, gave evidence for the existence of point-like electron sources which emit at anomalously low electric surface fields. The measured emission currents from the point-like sources seen in s.c. resonators do not correspond to predictions by the Fowler-Nordheim theory[56] applied to an ideal niobium surface. The origin of this anomaly is still unknown, but it can be assumed that the field emission in rf cavities is related to the dc field emission from broad area cathodes. At the University of Geneva, experiments are underway to study the field emission properties of niobium samples prepared in a similar way to cavity surfaces[57]. The measurements are carried out in a commercial Vacuum Generators "ESCALAB" UHV System including a scanning electron gun producing a beam of 0.5 μm in diameter, a 157° spherical sector electron analyser, a secondary emission detector and an argon gun. Niobium samples of 1.4 cm diameter can be fixed to a purpose built manipulator which permits the cathode x-y-z-movement necessary for the field emission scans. The anode holder can accommodate several units; for example, a 1 mm diameter flat anode, and a pointed tungsten anode which has been electrolytically etched to a micron size tip radius (Fig. 16).

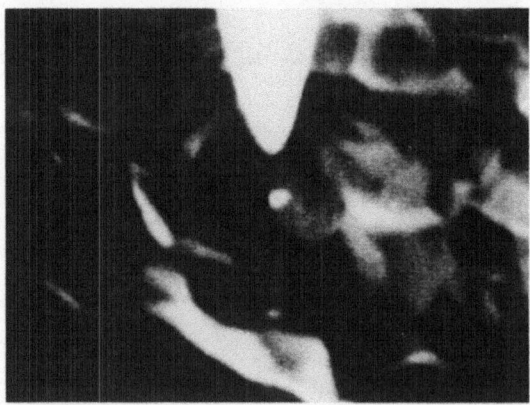

Fig. 16. Tungsten tip of the pointed anode of the "field emission scanning microscope" set up at the University of Geneva together with emitting particle on a niobium surface.

Using this anode a high electric field can be produced on a very small area of the niobium sample. Peak surface fields of 500 MV/m have been measured locally. By moving the cathode the anode is scanned automatically across the sample with a 1 µm setting precision. Figure 17 shows a scanning image of 1 cm² of a niobium surface at different scanning fields and after different treatments of the sample. The scan along each line of one image is carried out at a constant field. When a field emission site is encountered, the electric field is electronically reduced to hold the emission current below a fixed limit. These field reductions result in vertical deflections on the plotted lines. After localizing an emitting site it is investigated with the built-in scanning electron microscope (Fig. 16) and its chemical composition is determined by Auger analysis. So far the most important results of Ø. Fischer's group at the University of Geneva are as follows:

In general one can say that broad area cathodes seem to show the same kind of anomalous field emission as observed in s.c. cavities. In detail the following statements seem to be valid: The emission are most certainly not coming from metallic protrusion with a static electric field enhancement. The emission sites are usually associated with micron-size particles, some of them sitting probably rather loosely on the surface. The elemental composition of these particles is not unique. In a minority of cases no particle was seen down to a resolution of 0.5 µm. The emission from micron-size particles underlines the importance of the clean room techniques applied to the final cavity treatments before assembly. Another result of the Geneva group is shown in Fig. 17.

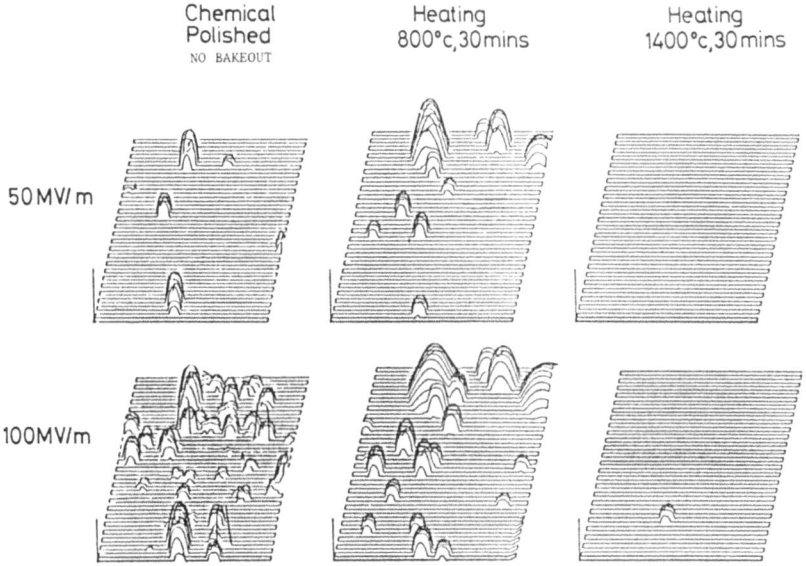

Fig. 17.   Field emitting sites on a niobium surface and their sensi-
tivity to bakeouts.
Upper row: Scanning field 50 MV/m
Lower row: Scanning field 100 MV/m
From left to right: No bakeout, 800°C, 1400°C (30 mins each).
All scans are performed with an anode of 1 mm diameter.

In the ESCALAB System the samples can be moved under UHV conditions to a station where they can be baked out at temperatures up to 2000°C. Figure 17 shows a series of field emission scans of one and the same sample which prior to each scan was baked out for 30 minutes at a given temperature. The number of emitting sites is reduced considerably after bakeout at a temperature of more than 800°C. This interesting observation certainly asks for more studies, but it may already be seen as a hint to apply high temperature firing under UHV and clean room conditions for s.c. cavities to surmount the field emission barrier.

A very recent and quite spectacular result[58] shows that, if a sample which was treated at temperatures of at least 1400°C, and on which the emitters at a given surface field have been reduced as shown in Fig. 17, is heat treated again at .800°C, a very large number of emitters reappear. This is attributed by the Geneva group to microscopic segregations which dissolve in the niobium at high temperatures and which segregate again at temperature specific for the impurity. This hypothesis is supported by the results of a series of experiments which are displayed in Fig. 18.

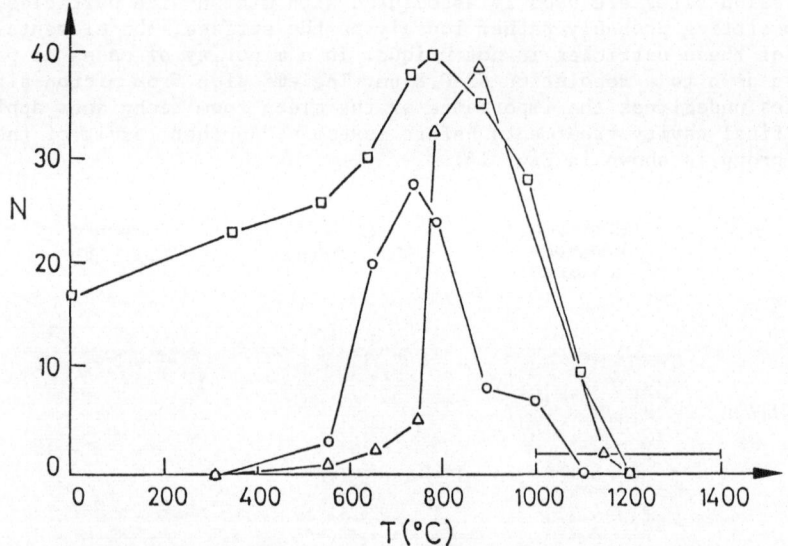

Fig. 18. Number of field emitting sites as a function of the heat treatment temperature for a Nb sample during the first (□) second (Δ) and third (o) thermal cycle. (Sankarramann et al., 1986).

5.   MICROWAVE PROPERTIES OF HIGH T_C SUPERCONDUCTORS

The Q_0 and thereby the shunt impedance of a s.c. accelerating structure increases exponentially with the critical temperature T_C of the superconducting material (see Eq. (2)). Therefore niobium, the element with the highest T_c, is the material most frequently used for s.c. cavities. Considerably higher transition temperatures are reached in superconducting compounds like the A15 materials and especially, and very recently, in the new superconducting perovskites. Superconducting compounds with high critical temperatures have the general tendency to be very brittle and to show a very poor thermal conductivity. Superconducting cavities need only to be covered with a very thin layer (1 to 2μ in thickness) of these materials; therefore, their brittleness and poor thermal conductivity is of no harm to their application in rf superconductivity.

5.1. Nb_3Sn coated accelerator cavities

Among the A-15 materials, characterized by high critical temperatures and critical thermodynamic magnetic fields (H$_c$). Nb$_3$Sn gained early attention. Its T$_c$ of 18.2 K, α of 2.2 and H$_c$ of 5400 Oe make it a promising material for superconducting cavities. A Nb$_3$Sn layer of typically 5 μm is formed on a niobium cavity by the vapor diffusion[59]. The cavity is processed in a vacuum furnace at around 1100°C in a tin atmosphere with a partial pressure of a few 10^{-3} Torr. Resently, work with Nb$_3$Sn resonators has been resumed at Wuppertal[60]. A single-cell and a five-cell cavity (3 GHz) have been covered with a Nb$_3$Sn layer. Fig. 19 shows a scanning electron micrograph of the obtained Nb$_3$Sn surface. The Nb$_3$Sn grains which have grown on the niobium surface resemble a cobble stone paved road of ancient Rome. Their average size is 2-3 μ.

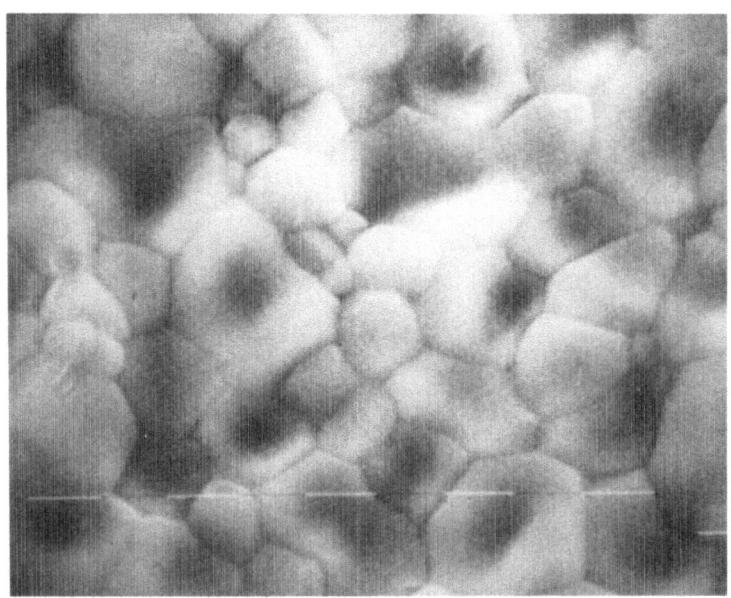

Fig. 19.   Scanning electron micrograph of a Nb$_3$Sn surface.
Magnification approximately 1000.

A depth profile of the Nb$_3$Sn layer of a niobium sample which was treated by the vapor diffusion process together with a five-cell cavity is shown in Fig. 21. This measurement was carried out using energy dispersive X-ray analysis in a scanning electron microscopic of 0.2 μm resolution at CERN. The tin content near the surface slightly exceeds that of stoichiometric Nb$_3$Sn but is still below the upper limit of the stable Nb$_3$Sn phase[61]. It is observed that removing the first 0.5 μm of the Nb$_3$Sn surface by oxipolishing[62] significantly reduces the residual resistance of a Nb$_3$Sn layer. Therefore, all cavities are oxipolished by this amount and rinsed with demineralized, filtered water and dust-free methanol before they are mounted in the test system.

To learn more about the seemingly high residual resistance of Nb$_3$Sn and its significant field dependence, and about field limitations specific to Nb$_3$Sn, the temperature mapping technique was applied to single and multicell cavities. One component of the residual resistance was found to be dependent on the cool down cycle. The Q$_o$ versus E$_a$ (Fig. 22) clearly shows the significant difference between the residual losses after a fast and a slow cool down of the cavity. A careful study in both cases indicates that even the residual losses after a slow cool down are, at least in part, caused by the same mechanism. The origin of these losses is unclear.

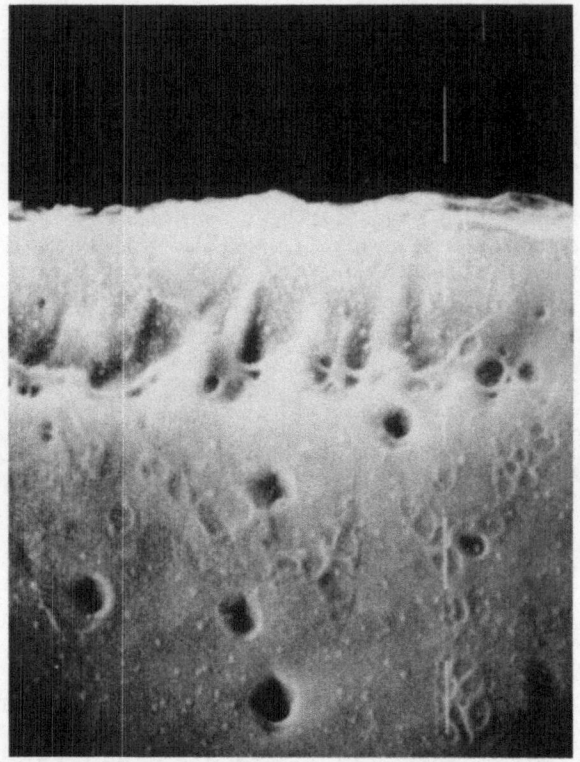

Fig. 20. Cross sectional view of a Nb₃Sn layer on top of a niobium
substrate, taken with a scanning electron microscope.
Magnification approximately 2500.

At present it is assumed that frozen-in magnetic flux produced by thermo-
electric currents and excited at the Nb3Sn interface is responsible.

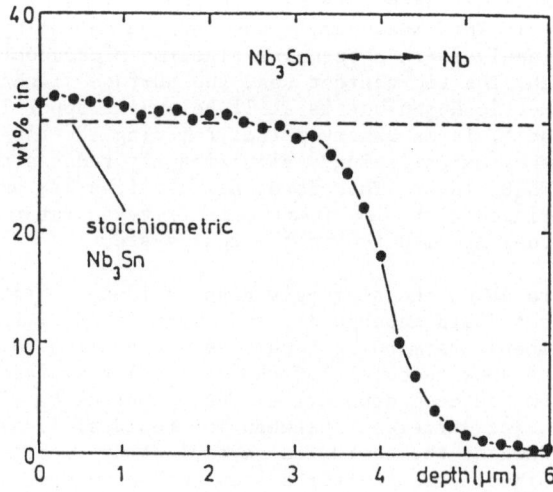

Fig. 21. Depth profile of the Nb3Sn layer on a niobium sample.

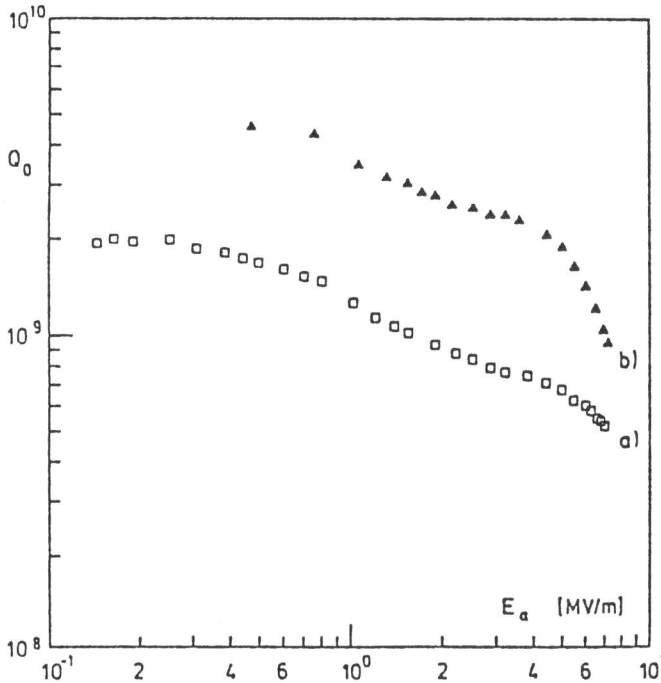

Fig. 22. Dependence of the cavity Q_0 on the accelerating field and
on the cool down procedure.
a) after fast cool down
b) after slow cool down.

Measurements at 20 GHz, 8 GHz and 3 GHz performed at Wuppertal show
that the minimum residual resistance increases with frequency. The lowest
residual resistance found so far was measured in a five-cell 3 GHz structure
and was 27 nΩ. Scaling this with $\sqrt{\omega}$ to 1.5 GHz would result in a cavity Q_0
of about $1.5 \cdot 10^{10}$ at 4.2 K. In fact 1.5 GHz is the design frequency of the
superconducting 4 GeV electron accelerator presently considered for construction at CEBAF in Newport News, Virginia. This accelerator will be
composed of niobium cavities of the Cornell design[63] and has to be operated
at 2 K. A later conversion to Nb_3Sn covered resonators appears feasible
and makes a further investigation of Nb_3Sn cavities worthwhile. At 4.2 K,
a theoretical Q_0 of about $9 \cdot 10^{10}$ is expected for Nb_3Sn accelerating resonators at this frequency.

The accelerating fields obtained in Nb_3Sn cavities are comparable to
results from cavities fabricated from low purity niobium. Temperature maps
taken on a five-cell Nb_3Sn 3 GHz cavity at different field levels, shown in
Fig. 23, show the existence of microscopic regions of weak superconductivity.
Already at low surface fields (10 mT) these regions switch to a high loss
state and lead to thermal instabilities. At present one can only speculate
about the nature of these switching defects. Impurity inclusions in the
niobium base material which disturb the uniform Nb_3Sn layer and which become
weak superconductors by the proximity effect are one explanation. The use
of the new high purity niobium for the production of Nb_3Sn cavities is
therefore a next experimental step.

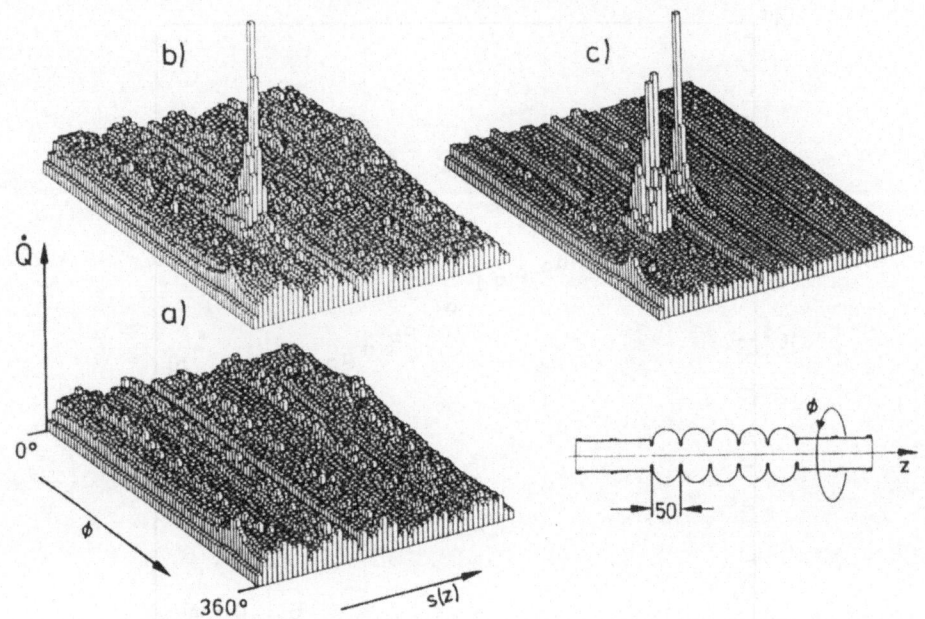

Fig. 23. Spatial distribution of the rf losses in a Nb_3Sn coated five cell cavity taken at 2.2 K in subcooled He at E_a = 2.55 MV/m (a), 2.6 MV/m (b) and 3.9 MV/m (c). With increasing field a few presumably microscopic regions switch into high loss areas (Q in arbitrary units).

5.2. Microwave properties of the new superconductors

The most spectacular experimental event [2,3] in superconductivity since the Meißner Effect is the discovery of the new superconducting perovskites of the composition $Y_1Ba_2Cu_3O_{7\pm\delta}$ (where δ is small and yttrium can be replaced by other elements of the rare earth family). These ceramic materials show at room temperature a metallic conductivity with a specific resistivity of about 1 mΩ cm resembling the electrical properties of graphite. They become superconducting between 90 and 120 K and show diamagnetic behaviour typical for superconductors in the Meißner state. Cylindrical pellets of typically 1 cm diameter and a few mm thickness are prepared quite easily by sintering stoichiometric powder mixtures of Y_2O_3, $BaCO_3$ and CuO in an O_2 atmosphere slightly below 1000°C with repeated regrindings[64]. Very recently[65] the successful deposition of thin films by a coevaporation technique has been announced. The photograph in Fig. 24, which shows a little permanent magnet floating over a 1.3 cm diameter pellet of $Y_1Ba_2Cu_3O_{7\pm\delta}$, has become a hallmark for the new superconductors.

Because of the importance of superconducting cavities to particle accelerators, the microwave properties of the superconducting perovskites are of great interest. Also, for the basic understadig of the specific nature of the superconducting state in the materials their investigation in RF fields should be persued.

Heinz London[20] was the first to suggest the investigation of the conduction mechanism in superconductors by exposing them to a microwave field. As already discussed in section 3.2.1., the nonvanishing Joule losses in rf superconductivity are attributed to those conduction electrons which are not condensed into Cooper pairs. The rf magnetic field penetrates into the sample by the London penetration depth, which in the new superconductors is about 2000 Å [66], and induces a time varying electric field.

Fig. 24. Permanent magnet (diameter 0.3 cm) floating over a
$Y_1Ba_2Cu_3O_7$ pellet (diameter 1.3 cm) which is in contact
to a liquid N_2 bath.

Because of the inertia of the Cooper pairs, this electric field is not
shunted to zero. The unpaired conduction electrons will gain energy from
this rf field, transfer it to the lattice ions, and thereby produce Joule
heating.

If a sample is placed in the field of microwave cavity, its Joule
losses change, and the cavity Q and its current distribution alters the
cavity frequency. A niobium cavity, which itself becomes superconducting
at 9.2 K and has an intrinsic Q or more than 10^8 at 4.2 K, is a useful
"microwave laboratory" to investigate the charge transport properties of
superconducting samples. Such an experiment has been performed two months
ago[67] and is reported here:

Two samples of the composition $Y_1Ba_2Cu_3O_{7\pm\delta}$ (δ was not determined),
one prepared at Wuppertal and one at Los Alamos, 13 mm in diameter and
1.5 mm (3.8 mm repectively) in thickness have been sintered in an O_2-atmos-
phere at about 1000°C from stoichiometric powder mixtures of Y_2O_3,
$BaCO_3$ and CuO with repeated regrindings.

The rf properties of both pellets have been studied separately by
placing them in a microwave cavity fabricated of reactor grade niobium and
operated at about 2.9 GHz in the TM_{010} fundamental mode (Fig. 25). The
evacuated cavity has been tested in a temperature range from 4.2 to
300 K by immersing it first in liquid helium and then warming it up with
several heating resistors mounted at the bottom of the helium cryostat.
The temperature of the cavity wall is controlled over the full range by
a combination of germanium and platinum resistors two of each. Stretching
the total warm-up time to about 6 hours has resulted in temperature
gradients of less than 2 K and provided sufficient heat exchange between
the samples and the cavity wall. The latter was verified at low rf power
level by comparison of the Joule losses between 77 and 100 K before and
after filling of the cavity with helium gas (1 atm at room temperature).

The cavity shown in Fig. 25 was designed to investigate the performance
of superconducting niobium surfaces at very high electric and magnetic rf
fields[45]. Because of its appropriate size and very low rf losses for T≤4.2 K,

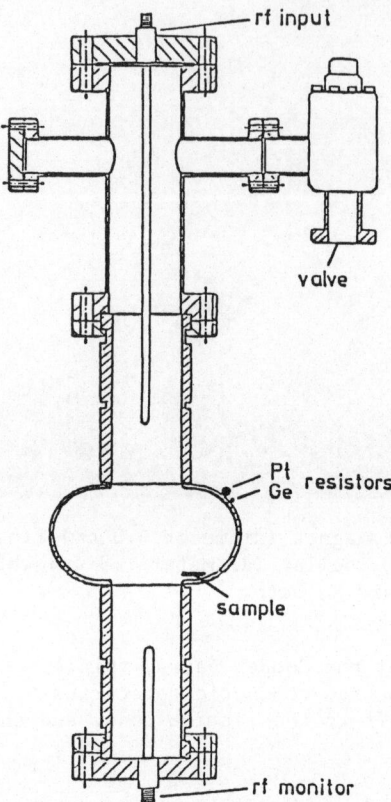

Fig. 25. 3 GHz niobium cavity with $Y_1Ba_2Cu_3O_7$ sample.

it is well suited for an exploratory experiment on the rf properties of the new superconductors. The samples were just laid on the flat part of the cavity surface. This establishes an electrical and thermal contact between the sample and the niobium wall. The Joule losses P_s of the sample have two components. One is due to the rf losses of the surface magnetic field, and the other is related to contact currents flowing between the niobium surface and the sample. The latter are undefined and therefore constitute an unknown contribution to P_s.

The loaded quality factor Q_L of the cavity is determined by the sum of all rf losses, i.e. the Joule losses in the cavity wall P_w, in the sample P_s, and in addition the power P_r radiated by the coupling antennas

$$Q_L^{-1} = (\omega_o W)^{-1} (P_w + P_r + P_s) \tag{24}$$

where ω_o is the angular frequency and W the stored energy of the TM_{010} mode. We have measured Q_L and P_r by using standard microwave techniques, resulting in an error of ± 3% for the remaining unloaded quality factor Q_o.

$$Q_o^{-1} = (\omega_o W)^{-1} (P_w + P_s) = Q_w^{-1} + Q_s^{-1} \tag{25}$$

Q_s can be separated easily by two consecutive measurements with and without the sample. The magnetic surface field at the location of the sample can be determined from W, with the assumption that it is only slightly influenced by the presence of the sample. The magnetic surface field H_s at the sample is equal to the local surface current I_s.

Figure 26 shows the d.c. resistance of one of the samples and its temperature dependence. The onset of superconductivity starts at about 92 K and at 82 K full superconductivity appears to be reached. The rf properties of a comparable sample prepared in the same way can be seen in Fig. 27, where Q_0^{-1} is plotted for the niobium cavity with and without the sample. As the difference between both of these measurements in Fig. 28, Q_S^{-1} is obtained, which is directly proportional to the Joule losses in the sample. At 4.2 K the host cavity is superconducting, and its Q_0 without the sample is $1.2 \cdot 10^8$. At this temperature the losses of the cavity with the sample are therefore dominantly determined by the sample which lowers the cavity Q_0 to about 10^6. The inverse of this residual Q_0 measured at very low rf power level gives $Q_S^{-1}(4.2 \text{ K})$ and is shown as the dashed horizontal line in Fig. 28. The same measurements were carried out with the second sample.

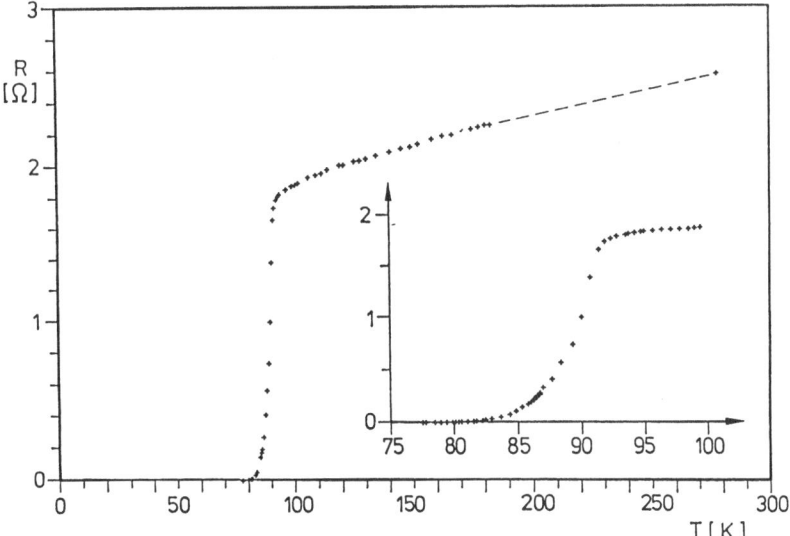

Fig. 26. Temperature dependence of the dc resistance of the $Y_1Ba_2Cu_3O_{9-\delta}$ sample 1

Both samples show a very similar behaviour if one compares their respective Q_S^{-1} values. First Q_S^{-1} reduces gradually as the temperature is lowered from 300 K to slightly above 100 K. This reduction of the losses is due to the normal increase in the mean free path of the conduction electrons with falling temperatures. At a temperature of 92 K (for sample 1) and 98 K (for sample 2) the Joule losses reduce as sharply as the corresponding d.c. resistance indicating a transition to superconductivity. Table 3 combines the most interesting data which can be obtained from the measurements for both samples.

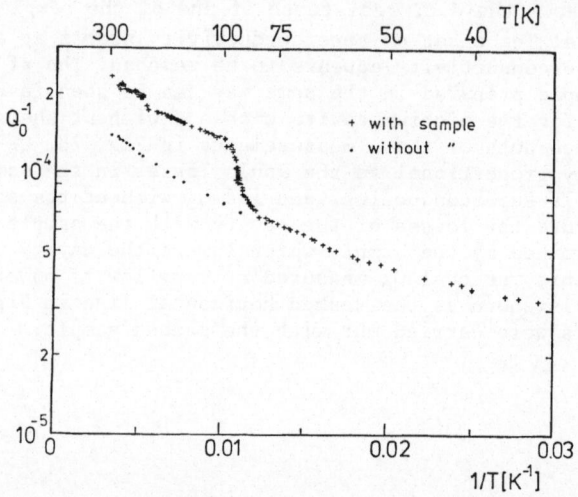

Fig. 27.   Temperature dependence of the cavity (Q_0^{-1}) with and without sample 1.

Table 3

T_C, microwave losses at 100 and 4.2 K, given by Q_S^{-1} and the derived surface current density j_s of two $Y_1Ba_2Cu_3O_{7\pm\delta}$ samples

	T_c [K]	Q_s^{-1} (100 K)	Q_s^{-1} (4.2 K)	$\eta = \dfrac{Q_s(4.2K)}{Q_s(100K)}$	j_s [A/cm^2]
sample 1	92	$0.6 \cdot 10^{-4}$	$6.7 \cdot 10^{-7}$	90	$5 \cdot 10^4$
sample 2	98	$1.5 \; 10^{-4}$	$1.1 \cdot 10^{-6}$	136	$3 \cdot 10^4$

The improvement factor $\eta = Q_s(4.2 \text{ K})/Q_s(100 \text{ K})$ is a true figure of merit for a given sample. In a perfectly superconducting sample, the residual losses at 4.2 K should be close to zero and η therefore very large. For superconducting niobium, for example, a residual surface resistance of 30 nΩ is obtained, typically with a 3 GHz cavity at temperatures of about 1.4 K. Such low residual losses would correspond to an improvement factor of about 10^7 at 15 K for the new superconductors.

174

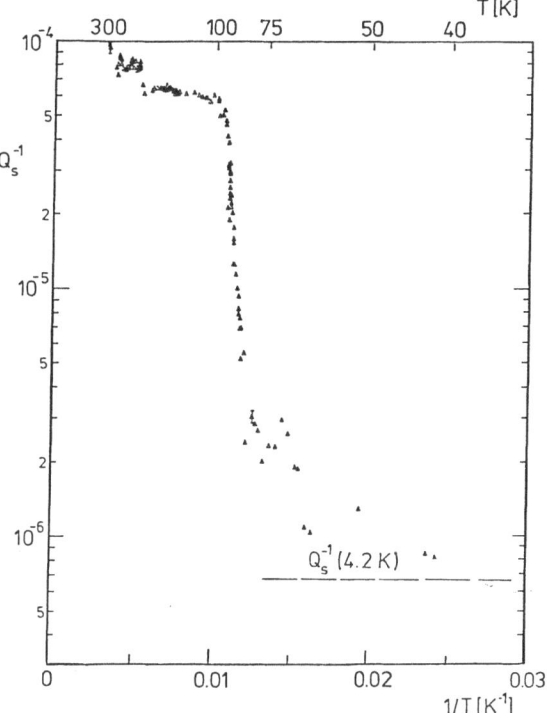

Fig. 28. Temperature dependence of the microwave losses (Q_s^{-1}) of the $Y_1Ba_2Cu_3O_{7\pm\delta}$ sample 1.

Although the η values found in these first experiments with samples of the new superconductors are by five orders of magnitude smaller, they provide evidence that only a fraction of conduction electrons smaller than 1% remain unpaired at very low temperatures. Because losses from non-superconducting impurities below the one percent level should already contribute significantly to Q_s^{-1} (4.2 K), the above results may well indicate that in a stoichiometrically pure sample of $Y_1Ba_2Cu_3O_{7\pm\delta}$ all conduction electrons are subject to the pairing interaction. That impurities were present in the samples tested in this experiment is shown in the micrograph of Fig. 29. It shows a scanning electron image of the surface of a $Y_1Ba_2Cu_3O_{7\pm\delta}$ sample prepared together with sample 1. The microscopic stoichiometry of this sample was measured as described in section 5.1. It was found uniform, aside from an apparently unreacted CuO particle to the right of the center of Fig. 29. Only further experiments will show if lower residual losses, and thereby higher improvement factors, can be obtained.

Very recently the experiment described here has been improved and new data were obtained[68]. The host cavity (Fig. 25) was anodized. The excellent insulation of the Nb_2O_5 layer covering the cavity surface suppresses all contact currents and associated losses between the cavity surface and the

Fig. 29.   Scanning electron micrograph of the surface of an
$Y_1Ba_2Cu_3O_{7\pm\delta}$ sample.

perovskite sample. The surface resistance R_s of the sample is related to Q_s^{-1} by

$$R_s = G^{loc} \cdot Q_s^{-1} \qquad\qquad (26)$$

where G^{loc} is a local geometry factor which is given by

$$G^{loc} = \frac{\omega_o W}{\int H_s^2 da} \qquad . \qquad\qquad (27)$$

In (27) W is the energy stored in the cavity, A is the surface of the sample and H_s is the magnetic field on the surface of the sample. G^{loc} can be determined experimentally by placing a sample of known surface resistance (for example made from graphite or stainless steel) at the same location as the superconducting sample. It was found that for the arrangement used in the above experiment the local geometry factor is given by $G^{loc} = 6900$ (for sample 1). The data of Fig. 28 therefore result in a surface resistance of 0.4 Ω at 2.9 GHz and 300 K. They also show a residual resistance of 4.6 mΩ, which most likely is reached already at much higher temperatures than at the 4.2 K where it was determined.

From these first experiments one can conclude that the microwave losses in the new superconductors are as yet very high. At room temperature they resemble cavities made from graphite, and in the superconducting state they would perform similarly to copper cavities.

Despite the poor cooling conditions for the sample in the described experiment, its current carrying capability was tested by increasing the rf field. In Fig. 30 the cavity Q_o at 4.2 K, which is entirely defined by the Joule losses of the sample and therefore equal to Q_s, is plotted as a function of the magnetic field H_S at the location of the sample. Considering a possible field enhancement by the sample, H_S is a lower bound for the field on its surface. For both samples a decrease of Q_S, already at very low surface fields ($>$ 10^{-2} Oe), has been observed. One can attribute this behaviour to the granular structure of the superconductor. The rf field produces losses in each grain at the surface of the sample. Many of these grains will have a very poor thermal contact to the bulk of the sample. Therefore already small rf losses can increase their temperature above T_c, and a rapidly growing area of the sample will turn normalconducting as the surface magnetic field is increased. At a limiting rf field of 1.2 Oe(100A/m) nearly the whole sample becomes normal-conducting, causing the niobium cavity to become thermally unstable and therefore quenching the field. Although this explanation is qualitative, the data allow to give an estimate of the maximum surface current of 100 A/m. Assuming a penetration depth of the magnetic field of 2000 Å [66], we obtain a maximum current density J_S of $5 \cdot 10^4$ A/cm^2 at the sample's surface.

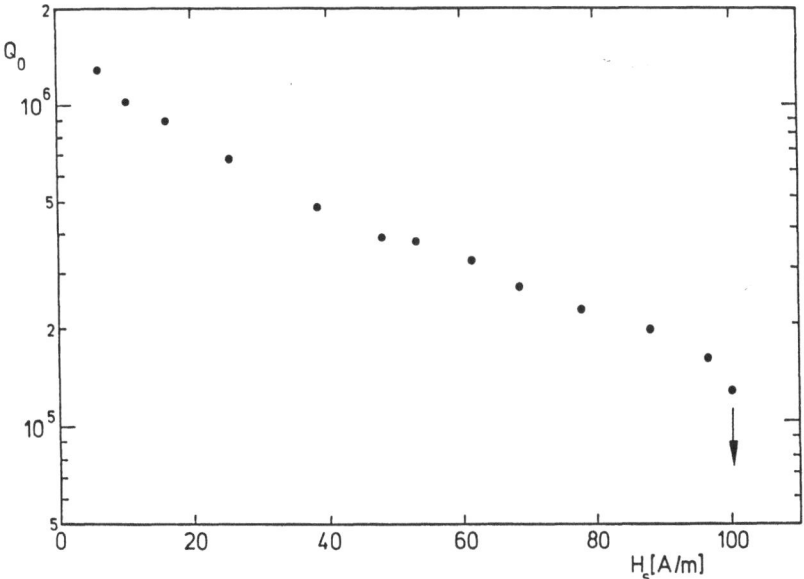

Fig. 30. Q_o of the cavity with $Y_1Ba_2Cu_3O_{9-\delta}$ sample 1 as a function of the surface magnetic field.

In very recent experiments[68], and in a pulsed operation surface fields as high as 1850 A/m, corresponding to a surface current density of about 10^6 A/cm^2 were obtained. If such results would be achieved in cavities made from this ceramic material than this could correspond to an accelerating field of about 0.5 MV/m.

The very high residual losses and the rather low surface magnetic fields obtained so far show that there is a lot of work to be done before the first "ceramic cavities" will ever accelerate a particle beam. The challenge to reach the limiting behaviour of the classical superconductors, like niobium and Nb$_3$Sn in rf cavities for particle accelerators, and the exciting perspective of the new superconductors, will certainly keep us busy in the future.

REFERENCES

1) H.K.Onnes, Comm.Phys.Lab., Univ. Leyden, Nos. 119,120,122 (1911).

2) J.G.Bednorz and K.A.Müller, Z.Phys. B64, 189 (1986).

3) M.K.Wu, J.R.Asburn, C.J.Torng and P.H.Hor, R.L.Meng, L.Gao, Z.J.Huang, and C.W.Chu, Phys.Rev.Lett. 58, 908 (1987).

4) J.M.Pierce, H.A.Schwettman, W.M.Fairbank, P.B.Wilson, Proc. of the 9th Int.Conf. on Low Temperature Physics, Part A (Plenum Press, N.Y. (1965), p.36.

5) C.M.Lyneis, M.S.McAshan, R.E.Rand, H.A.Schwettman, T.I.Smith and J.P.Turneaure, IEEE Trans.Nucl.Sci., NS-26, 3445 (1981).

6) P.Axel, L.S.Cardman, H.D.Gräf, A.O.Hanson, R.A.Hoffswell, D.Jamnik, D.C.Sutton, R.H.Taylor and L.H.Young, IEEE Trans.Nucl.Sci. NS-26, 3143 (1979).

7) A.Citron, G.Dammertz, M.Grunder, L.Husson, R.Lehm and H.Lengeler, Nucl.Instr.Meth. 164, 31 (1979).

8) R.M.Sundelin, J.Kirchgessner, H.Padamsee, H.L.Phillips, D.Rice, M.Tigner, E.von Borstel, Proc. of the 9th Int.Conf. on High Energy Accelerators, Stanford, 1974 (SLAC, Stanford, 1974),p.128.

9) L.M.Bollinger, IEEE Trans.Nucl.Sci. NS-30, 2065 (1983).

10) D.A.G.Deacon, L.R.Elias, J.M.J.Madey, G.J.Ramian, H.A.Schwettman, T.I.Smith, Phys.Rev.Lett. 38, 892 (1977).

11) J.P.Turneaure and Ira Weissman, J.Appl.Phys.39, 4417 (1968).

12) P.Kneisel, et al., IEEE Trans.MAG-21, 1000 (1985).

13) M.Kuntze, editor, Proc. of the Workshop on RF Superconductivity, Kernforschungszentrum Karlsruhe, KfK 3018 (1980).

14) H.Piel, Proc. of the CERN Accelerator School (1985), Oxford, England.

15) H.Lengeler, editor, Proc. of the Second Workshop on RF-Superconductivity CERN, Geneva, 1984, (CERN, Geneva, 1984).

16) M.Tigner and H.Padamsee, AIP Conf.Proc. on the SLAC Summer Acc. School, 1982, Melvin Month, editor (AIP, New York, 1983), p.801 and Cornell CNLS 82/553.

17) G.E.H.Reuter and E.H.Sonderheimer, Proc. of the Royal Soc. A195, 336 (1984).

18) H.Kammerlingh Onnes, Comm.Phys.Lab., Univ.Leyden (1908).

19) J.Bardeen, L.N.Cooper and J.R.Schrieffer, Phys.Rev. 108, 1175 (1957).

20) H.London, Nature 133, 497 (1934).

21) The Feynman Lectures on Physics Vol.III, Addison Wesley Publ.Comp. (1965) p.21-1ff.

22) W.Meißner and R.Ochsenfeld, Naturwiss. 21, 787 (1933).

23) D.C.Mattis and J.Bardeen, Phys.Rev. 111, 412 (1958).

24) A.A.Abrikosov, L.P.Gorkov and I.M.Khalatnikov, Sov.Phys. JETP, 8, 182 (1959).

25) J.Halbritter, Z.Phys. 238, 466 (1970).

26) J.P.Turneaure, Ph.D. Thesis, Stanford University (1967).

27) R.Blaschke, Recent Developments in Concensed Matter Physics, Vol.4,425 J.T.Devreese et al., editor (Plenum Press, N.Y. 1981).

28) U.Klein, Dissertation, University of Wuppertal WUB-DI 81-2, (1981).

29) G.Müller, Dissertation, University of Wuppertal, WUB-DI 83-1 (1983).

30) J.Matricon and D.St.James, Phys.Lett. 24A, 241 (1967).

31) V.L.Ginzburg and L.D.Landau, Zh.Eskperim, i.Theor.Fig. 20, 1064 (1950).

32) T.Yogi, G.J.Dick and J.E.Mercereau, Phys.Rev.Lett. 39, 826 (1977).

33) H.Pfister, Cryogenics 16, 17 (1976).

34) K.Schnitzke, A.Martens, B.Hillenbrand, H.Diepers, Phys.Lett. 45A, 241 (1973).

35) H.Piel and R.Romijn, CERN/EF/RF 80-3 (1980).

36) H.Piel, ibid. ref. 13), p.85.

37) C.M.Lyneis, M.McAshan and Nguyen Tuong Viet, Proc. of the 1972 Proton Linear Acc.Conf., Los Alamos, 1972 (LASL, Los Alamos, 1972), p.98.

38) Ph.Bernard, G.Cavallari, E.Chiaveri, E.Haebel, H.Heinrichs, H.Lengeler, E.Picasso, V.Picciarelli and H.Piel, Proc. of the 11th Int.Conf. on High Energy Acc., Geneva, 1980 (Birkhäuser, Basel, 1980), p.878.

39) R.Romijn, W.Weingarten, H.Piel, IEEE Trans.MAG-19, 1318 (1938).

40) D.Huppelsberg, Diplomarbeit, University of Wuppertal, WUD-85-3 (1985).

41) H.Padamsee, J.Tückmantel and W.Weingarten, IEEE Trans. MAG-19, 1308 (1983).

42) Ph.Bernard, G.Cavallari, E.Chiaveri, E.Haebel, H.Lengeler, H.Padamsee, V.Picciarelli, D.Proch, A.Schwettman, J.Tückmantel, W.Weingarten and H.Piel, Nucl.Instr.Meth. 206, 47 (1983).

43) H.Padamsee, ibid. ref. 13), p.145.

44) K.K.Schulze, Journal of Metals 33, 33 (1981).

45) H.Lengeler, W.Weingarten, G.Müller and H.Piel, IEEE Trans.MAG-21, 1014 (1985).

46) H.Padamsee, ibid. ref. 45, p.149.

47) H.Padamsee, ibid. ref. 45, p.1007.

48) P.Kneisel, Cornell University, SRF 840702 (1984).

49) Y.Kojima, KEK (1985) private communication.

50) G.Meuser, Diplomarbeit, University of Wuppertal, WUD-85-19 (1985).

51) H.Elias, University of Wuppertal (1985), private communication.

52) C.M.Lyneis, H.A.Schwettman and J.P.Turneaure, Apppl.Phys. Lett. 31, 541 (1977).

53) V.Lagomarsino, G.Manuzio, R.Parodi, R.Vaccarone, IEEE Trans.MAG-15, 25 (1979).

54) U.Klein and D.Proch, Proc. of the Conf. on Future Possibilities for Electron Acc., Charlottesville, 1979, J.S.McCarthy and R.R.Whitney, editors, (Univ. of Virginia, Charlottesville, 1979), p.N1-17.

55) P.Kneisel, R.Vincon and J.Halbritter, Nucl.Instr.Meth. 188, 669 (1981).

56) R.H.Fowler, L.Nordheim, Proc.Roy.Soc.Lond. A-119, 173 (1928).

57) Ph.Niedermann, N.Sankarraman, R.J.Noer, Ø.Fischer, ibid. ref. 19,p.583.

58) Ø.Fischer, University of Geneva, to be published in J.Appl.Phys.59, (1986).

59) G.Arnolds-Mayer, ibid. ref.15, p.643.

60) M.Peiniger and H.Piel, IEEE Trans.NS-32, 3612 (1985).

61) J.P.Charlesworth, I.Macphail, P.E.Madsen, J.Met.Science 5, 580 (1970).

62) B.Hillenbrand, H.Martens, H.Pfister, K.Schnitzke, G.Ziegler, IEEE Trans.MAG-11, 420 (1975).

63) P.Kneisel, J.Amato, J.Kirchgessner, K.Nakajima, H.Padamsee, H.L.Phillips, C. Reece, R.Sundelin and M.Tigner, ibid ref. 45, p.1000.

64) R.J.Cava et al., submitted to Phys.Rev.Lett. (1987).

65) IBM press release, New York Times, 11.5.87.

66) W.J.Kossler et al., submitted to Phys.Rev.Lett. (1987).

67) M.Hagen, M.Hein, N.Klein, A.Michalke, F.M.Mueller, G.Müller, H.Piel, R.W.Röth, H.Sheinberg and J.-L.Smith, to be published in Journ. of Magnetism and Magn. Mat. (1987).

68) M.Hein et al., to be published.

ONE DIMENSIONAL MODEL FOR THE INTERACTION OF A HIGH CHARGED BUNCHED BEAM

WITH A SUPERCONDUCTING CAVITY

R. Bonifacio, L. Ferrucci, C. Pagani and L. Serafini

University of Milan and INFN
Via Celoria 16
Milan (Italy)

INTRODUCTION

We present a first approach to the longitudinal dynamics of a low energy high charged bunched beam interacting with a multi-cell superconducting cavity. This study assumes as typical the working condition where the energy subctracted by the bunches from the resonating field is a significant amount of the energy stored in the cavity, i.e. in a high beam loading regime.

The model presented here is a one dimensional model. It is able to treat the field depletion effect, due to the bunch passage throughout the cavity, and to give a description of that effect in terms of a proper set of equations wich contain only one dimensionless parameter. This allows a clear exploitation of the scaling laws, and a more reliable control of the numerical integration. The effects due to the space charge forces and to the wake fields excitation inside the cavity, which in some cases can be dominant, are, in this model, neglected.

Some numerical results, related to the longitudinal phase space tracking of a low energy beam injected into the 4-cell superconducting LEP-cavity, are reported: from these results we derived, within the limits of the model, some interesting properties of this type of cavities, designed for ultrarelativistic beams, concerning the bunching and the energy spread control that can be applied to a low energy beam.

LONGITUDINAL DYNAMICS IN A SUPERCONDUCTING CAVITY

We examine here some of the problems related to the longitudinal dynamics of an electron bunch being accelerated through a superconducting cavity, paying attention mainly to the case when the charge content in the bunch is quite high.

One possible example of that situation is the acceleration of a 50÷100 nC bunch, emerging from a suitable gun at an energy variable in the 0.5÷1 Mev range, up to an energy of 10÷12 MeV by means of a four cells module of the superconducting cavities being developed at CERN for LEP. This could be, in the context of the ELFA project[1], the general scheme adopted in order to produce high charged electron-bunches, at the proper energy, to be injected into a single-passage high-gain FEL. These cavities have a resonant frequency of 352 MHz in the $\text{TM}_{010-\pi}$ resonating mode and they are designed for a synchronism condition tailored for $\beta=1$ particles (the length of each cell is $\lambda/2$). Up to now they have been succesfully operated at a level of about 6÷8 MV/m of average accelerating gradient, with Q-values in the range $1÷3\cdot10^9$. The energy stored in the cavity is, in this case, in the range 100÷130 J, a value quite larger than the typical stored energy inside a normal conducting cavity.

Since the travel time of an electron-bunch through the length of the cavity ($\simeq 2$ m) is of the order of two RF periods, i.e. 6 nsec, which is a very small time interval with respect to the operating charge-discharge time of the cavity, even in a strong overcoupled operation, the amount of energy exchanged between the cavity and the coupling system can be considered negligible with respect to the energy stored in the cavity.

Therefore, the system electron bunch + e.m. field resonating inside the cavity can be considered, as far as time-intervals of a few RF periods are concerned, an isolated system. Now, if the coupling of the electrons current to the higher order modes of the cavity is neglected (i.e. if one does not take into account the wake field effects), the electric field on axis can be written with the usual standing wave expression of the fundamental $\text{TM}_{010-\pi}$ mode:

1) $\quad E_z = E_o(z)\sin(kz)\sin(\omega t + \varphi) \qquad$ with $\quad$ k = ω/c $\quad$ and $\quad$ ω = 2π·ν_{RF}

where the amplitude E_o of the standing wave is left free to vary along z in order to take into account the field depletion effect, i.e. the continous decrease of the field amplitude, corresponding to the decrease of the stored energy, produced by the energy increase of the bunch being accelerated on the cavity axis.

Using z as the independent variable, the time t_j at which the j-th particle of the bunch reaches the position z is:

$$t_j = \int_0^z \frac{dz'}{v_j(z')} + t_{0j}$$

where v_j is the velocity of the particle and t_{oj} is its injection time. Here the assumption is made that the bunch is given by a number of particles travelling on the axis with pure axial velocity.

The field distribution is also assumed to be the one of the fundamental accelerating mode over all the acceleration process: the dependence of the field amplitude from the position z represents the renormalization produced on the field when all the particles have reached the position z and have subctracted a certain amount of energy from the e.m. field stored energy.

Simulating the bunch with N particles carrying the total charge Q_b, the equation of motions become:

2) $\quad \dfrac{d\gamma_j}{dz} = \dfrac{e}{m_e}E_o(z)\sin(kz)\sin\theta_j$

2') $\quad \dfrac{d\theta_j}{dz} = \dfrac{k}{\sqrt{1 - \frac{1}{\gamma_j^2}}} \qquad\qquad j = 1....N \qquad\qquad Q_b = N_e e = N q$

being $\quad \theta_j = \omega \int_0^z \dfrac{dz'}{v_j(z')} + \varphi_j \qquad$ and $\qquad \dfrac{d\theta_j(z)}{dz} = \dfrac{\omega}{v_j(z)} = \dfrac{k}{\beta_j(z)}$

where θ_j gives the phase of the j-th particle, at each z, with respect to the field wave, and ϕ_j is the initial value for the phase. The system of the 2N first order equations 2) and 2') must be integrated versus the independent variable z in the unknown functions γ_j and θ_j, j=1,N.

The total energy conservation, applied to the bunch + e.m. field system, states that:

$$\frac{d}{dt}\left(\sum_j \gamma_j m_e c^2\right) = -\frac{dW^{em}}{dt} \qquad \text{where} \quad W^{em} = \epsilon_o F V E_o^2(z) \quad \text{and} \quad F = \frac{\int_V \epsilon_o E_c^2(\vec{r})dv}{2\epsilon_o V E_o^2(0)}$$

being E_c the electric field of the empty cavity before the bunch is injected into the cavity. The dimensionless form factor F (which depends only on the geometry of the cavity) gives one half of the ratio between the mean energy density, stored inside the cavity before the bunch injection, and the maximum energy density available on axis: $E_o(0)$ is the peak electric field on axis when the bunch is not yet entered into the cavity (i.e. with an empty cavity). In the case of the LEP S.C. cavities the form factor F comes out to be such that $\epsilon_o FV = 7.16 \cdot 10^{-13}$ $[C^2 m^2/J]$.

After the injection of the bunch, the stored energy can be modified by the field-particle interaction: this can be taken into account by means of a renormalization of the maximum value of the electric field on the axis. As the particles of the bunches travel up to the position z (in different times t_j), the sum of their total energy change must be equal to the corresponding variation of e.m. field energy, that can be described (if the field distribution remains unchanged) by means of a renormalization of the electric field amplitude $E_o(z)$. Switching to the derivative versus z instead of t, one obtains:

$$m_e c^2 \sum_j \frac{d\gamma_j}{dz} = -2\epsilon_o FV E_o(z)\frac{dE_o}{dz}$$

so that the field equation becomes:

3) $\qquad \dfrac{dE_o(z)}{dz} = \dfrac{-e\rho_o}{2\epsilon_o F} \sin(kz)\langle \sin\theta \rangle$ $\qquad\qquad$ where $\qquad \rho_o = N/V$

and $\qquad \langle \sin\theta \rangle = \dfrac{1}{N}\sum_j \sin\theta_j$ $\qquad$ is the "bunching factor", which is

similar to the corresponding term in the FEL radiation equation[3].

The three previous set of equations, 2), 2') and 3) can be normalized, in order to reduce all the parameters to only one dimensionless quantity. This one turns out to be:

4) $\rho = \left(\dfrac{\Omega_p^2}{F\omega^2}\right)^{\frac{1}{3}} = \left(\dfrac{e}{\epsilon_o m_e}\dfrac{Q_b}{FV\omega^2}\right)^{\frac{1}{3}}$ $\qquad$ where $\quad \Omega_p = e\sqrt{\dfrac{\rho_o}{\epsilon_o m_e}}$ $\qquad$ is the plasma frequency.

Introducing the dimensionless variables:

$$A(\bar{z}) = \frac{eE_o(\bar{z})}{m_e c\omega\rho^2} \qquad\qquad \bar{\gamma}_j = \frac{\gamma_j}{\rho} \qquad\qquad \bar{z} = \rho k z$$

we get the three normalized equations:

$$\frac{d\overline{\gamma}_j}{d\overline{z}} = A(\overline{z}) \sin \frac{z}{\rho} \sin \theta_j$$

5) $$\frac{d\theta_j}{d\overline{z}} = \frac{1}{\sqrt{\rho^2 - \frac{1}{\overline{\gamma}_j{}^2}}} \qquad j = 1....N$$

$$\frac{dA(\overline{z})}{d\overline{z}} = -\frac{1}{2} \sin \frac{z}{\rho} \langle \sin \theta \rangle$$

A constant of the motion can be derived. Its expression, with the already defined dimensionless variables, is:

6) $A^2(\overline{z}) + \langle \overline{\gamma} \rangle = C_o$ where $C_o = \frac{1}{\rho}\left[\frac{W_{cav}}{W_b} + \langle \gamma_o \rangle\right]$ and $\langle \overline{\gamma} \rangle = \frac{1}{N}\sum_j \overline{\gamma}_j$

in which W_{cav} is the energy stored in the empty cavity, W_b is the total rest mass energy of the bunch, and $\langle \gamma_o \rangle$ is the average initial γ of the bunch.

Finally the set of equations reduces to:

7) $$\frac{d\overline{\gamma}_j}{d\overline{z}} = \sqrt{C_o - \langle \overline{\gamma} \rangle} \sin \frac{z}{\rho} \sin \theta_j$$

$$\frac{d\theta_j}{d\overline{z}} = \frac{1}{\sqrt{\rho^2 - \frac{1}{\overline{\gamma}_j{}^2}}} \qquad j = 1....N$$

NUMERICAL RESULTS

The above set of equations cannot be analytically solved unless the starting values of the γ_j are much greater than 1, because the phase equations, for the case of low energy electrons injected into the cavity, cannot be linearized (expanding the right hand side term of the equation to the first order in $1/(\rho\gamma_j)^2$).

Since we are interested in the longitudinal dynamics of electron bunches at an injection energy of 0.5÷1 MeV, we wrote a computer code able to integrate the above set of equations for a number of particles uniformly distributed in the longitudinal phase-space. This program allows to study the behaviour of a number of parameters which are of relevance for the requirements of the FEL radiation production: the bunch length, which determines the peak current in the bunch (and the gain of the FEL), the energy spread of the bunch at the exit of the cavity, which must be controlled to avoid a serious deterioration of the gain, and the final energy of the bunch, which determines the resonance condition of the FEL.

A typical example is shown in Fig.1), where the average energy and phase of a 1 kA bunch, extending at injection over 10° RF, are plotted, as functions of kz = $(2\pi\nu_{RF})z/c$ (expressed in degrees), along the acceleration through the four cells (whose boundary sketch is superimposed). Both these quantities are averaged over the energies and phases of the 1000 particles used to simulate the bunch, which has a charge content of about 80 nC and a length of 80 psec. The average phase exhibits a slippage from the injection value at -30° up to -15° at the exit from the cavity, due to the fact that the velocity of the bunch at

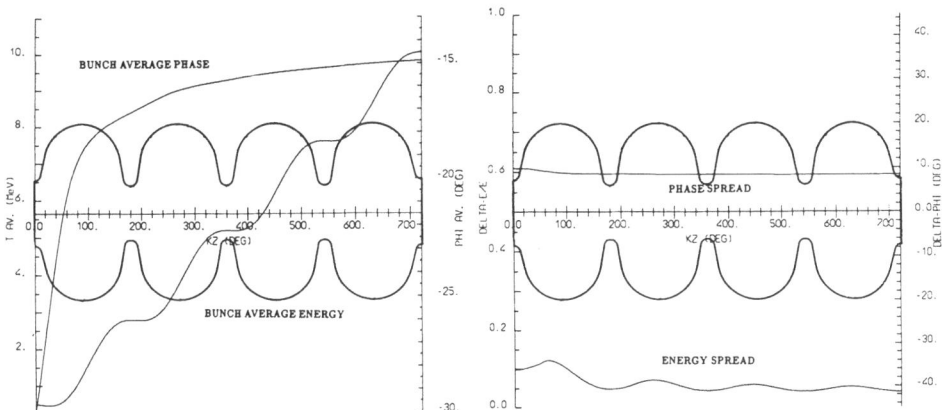

Fig. 1 - Average energy (left sc.)
         and phase (right scale)
         of the bunch (see text).

Fig. 2 - Phase spread (right sc.)
         and relative energy spread
         (left sc.) of the bunch.

injection is 0.86c (corresponding to 0.5 MeV), which is significantly lower than the sinchronous velocity c.

The length of the bunch is fairly reduced during the acceleration, as can be seen from Fig.2: this length corresponds to the maximum phase spread as deduced from the phase space at each z. A slight bunching effect is achieved, which reduces the bunch length from 10° RF at injection to about 8° RF at the exit of the cavity (this increases the bunch current up to 1250 A). The full energy spread of the bunch is also plotted in the same figure, expressed as the ratio of the maximum energy spread (full) as deduced from the phase space over the average energy of the bunch. The injection value of 10% is reduced to about 4% at the cavity exit.

The longitudinal phase spaces are plotted in Fig.3a to Fig.3d at some intermediate positions in the first cell and at the exit of the cavity (kz=720° RF) : it should be noted that the apparent phase space area reduction is due to the fact that the variable used are not the canonical variables (mainly the energy spread, which is not the absolute spread but the relative one). The non-linear effects are quite visible in the distorsion of the phase space ellipse during the acceleration process.

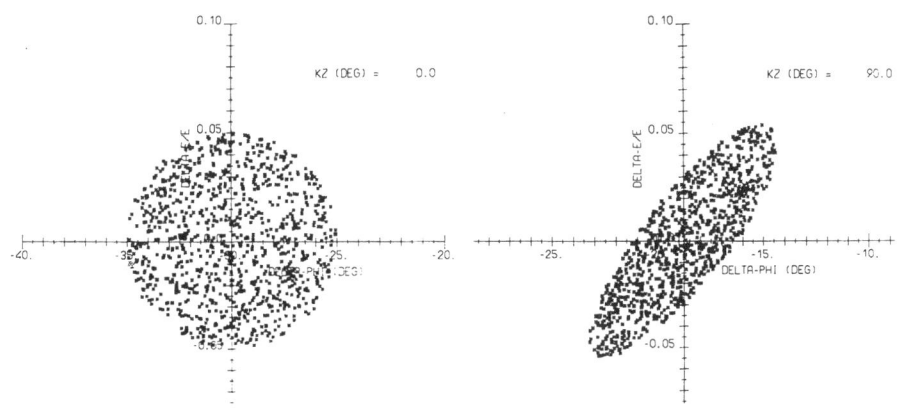

Fig. 3a - Longitudinal phase space
          at injection (1000
          particles).

Fig. 3b - Longitudinal phase space
          in the middle of the first
          cell (kz = 90° RF).

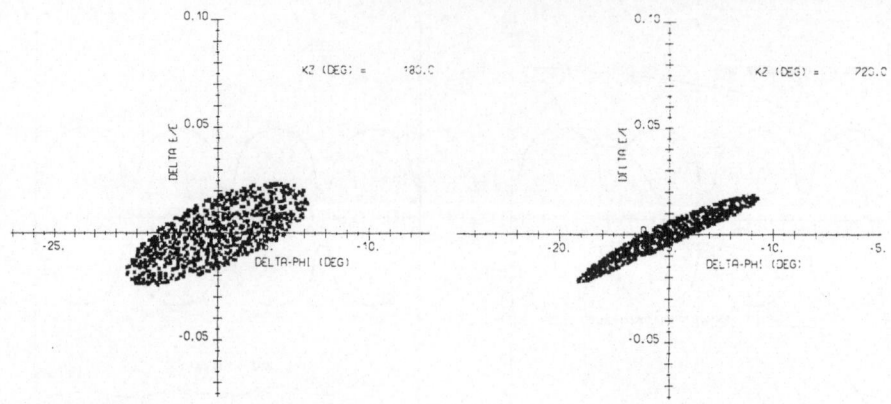

Fig. 3c – Longitudinal phase space at the end of the first cell (kz = 180° RF)

Fig. 3d – Longitudinal phase space at the exit from the cavity (kz = 720° RF)

FIELD DEPLETION SCALING LAW

From the constant of the motion of eq. 6), one can derive the behaviour of the field amplitude during the acceleration as a function of the average energy gain of the bunch. The dimensionless field amplitude can be written as:

$$A(\bar{z}) = \frac{1}{\sqrt{\rho}}\sqrt{\frac{e}{m_e c^2}\frac{W_{cav}}{Q_b} - \Delta\langle\gamma\rangle(z)} \qquad \text{where} \qquad \Delta\langle\gamma\rangle(z) = \langle\gamma\rangle(z) - \langle\gamma_o\rangle$$

represents the change in the average γ of the bunch. Substituting the expression 4) for the dimensionless parameter, the field amplitude becomes:

$$E_o(z) = E_o(0)\sqrt{1 - \frac{\Delta\langle\gamma\rangle(z)Q_b}{W_{cav}}\frac{m_e c^2}{e}}$$

from which the relative change of the field amplitude at the exit from the cavity can be deduced. That gives a figure of the field depletion effect, i.e. of the total decrease of the field amplitude: it comes out that the field depletion effect is sensitive to the ratio of the total energy gain of the bunch (which is given by $Q_b\Delta\gamma m_e c^2/e$, where $\Delta\gamma$ is the final $\Delta\langle\gamma\rangle$ at the exit from the cavity) over the energy initially stored in the cavity (W_{cav}). If this ratio is a small number, one can expand the right hand side of eq. 8) to the first order, obtaining:

$$\frac{\Delta E}{E_o} = -2.56\cdot 10^{-4}\frac{\Delta\gamma Q_b[nC]}{W_{cav}[J]}$$

This rules gives the relative decrease of the field amplitude at the end of the bunch acceleration through the cavity. In the case of a multi-bunch operation, since each bunch of the train finds an initial field amplitude $E_o(0)$ which is lower than the one of the previous bunch just of a quantity ΔE, $\Delta E/E_o$ will be equal to the relative difference in the final average energy between a bunch and the following one. If the

186

bunch-train contains N_b bunches, a first order evaluation of the energy spread of the train of bunches (i.e. the relative difference of final average energy between the first and the last bunch) is given by $N_b \Delta E / E_o$. In the previous example, where a bunch of 80 nC was accelerated from 0.5 MeV up to 9.75 MeV, with 100 J of stored energy, the relative field depletion comes out to be about 0.4%, which can produce, in a train of 4÷5 bunches, a not negligible energy spread of the train of bunches.

Keeping constant the initial field amplitude $E_o(0)$ (then the average accelerating gradient), it is well known that the energy stored per unit lenght scales like ν_{RF}^{-2}, and the change of the average γ per unit length scales unchanged. That implies that in a multicell structure the field depletion per unit length scales like the bunch charge times the square of the frequency:

$$\frac{\Delta E}{E_o} \div Q_b \nu_{RF}^2$$

On the basis of this scaling law it becomes evident that the relative low frequency S.C. cavities (able to support a quite large amount of energy stored) seem well suited to accelerate high charged bunches in a moderate field depletion regime, fact that assures a better control on the behaviour of the critical longitudinal phase space parameters like the energy spread and the length of the bunch.

CONCLUSION

The model presented here gives a simplified description of the bunch-cavity interaction, which neglects the higher order modes excitation, but gives a quick view of the energetic exchange between field and particles and a general rule for the field depletion effect.

The derived set of equations is easy and quick to be numerically integrated, allowing a simple longitudinal phase space tracking by which one can monitor the behaviour of the energy spread and of the bunch length in the presence of the field depletion effect. A simple scaling law related to this effect can be moreover extracted from the given set of equations.

Since we know that the space charge effects are quite serious at the high charge levels which we are looking at, with respect mainly to the deterioration of the energy spread, we consider the results presented in this paper as fairly indicative, and we put our effort in developing a computer program able to calculate both the space charge forces and the wake fields excitation. Presently this work is in progress and we are confident that the first results will be available at the time of the next first EPAC conference[4].

REFERENCES

1) - R.Bonifacio et al., ELFA PROJECT: Electron Laser Facility for Accelerator, to be presented at the first EPAC conf. (June '88).
2) - Ph.Bernard et al., Results with a LEP prototype superconducting cavity, CERN/EF/RF 85-6.
3) - R.Bonifacio, C.Pellegrini and L.Narducci - Opt. Comm. 50 (1984),373.
4) - L.Serafini et al., Numerical integration, in time domain, of field + particle equations in a cylindrical symmetrical cavity, to be presented at the first EPAC conference (June '88).

RF-STRUCTURES: A SURVEY

D.T. Tran

CGR MeV, BP 34
78530 BUC (France)

Abstract

This survey is divided into three parts.

In part 1, general properties of mono-periodic structures are reminded. They serve as introduction to the multi-periodic case. The coalescence phenomenon is examined in view to explain behaviors of certain wide band RF-structures and to give a basis for the study of certain types of parasitic oscillations discussed in part 3.

Part 2 is devoted to the problem of computer aided design of RF-structures used in particle accelerators or in microwave tubes. Emphasis will be made on the state of the art, in particular, on the promising finite element method. An accelerating cavity, coupled to a parasitic mode damping cavity, is given as an illustrative example of the possibility of this method.

In part 3, are examined some practical aspects considered as the most important: the beam loading mechanism and the compensation methods, oscillations and parasitic modes damping, field break-down and finally a description of a new recirculating backward wave structure with automatic beam loading compensation.

PART 1 - GENERAL PROPERTIES OF PERIODIC SLOW-WAVE STRUCTURE

SCOPE

General properties of mono-periodic structures are reminded. They serve as introduction to the multiperiodic-case. The coalescence phenomenon is examined in view to explain behaviors of certain wide band structure and to give a basis to the study of oscillations in part 3.

1.1 - The mono-periodic or single-band system

Given a geometrical periodicity, the problem is to derive the properties of the waves propagating along the structure. Let $\psi(r,z;\omega)$ be the wavefunction at a point z for a given oscillation mode of pulsation ω, the Floquet's theorem states that at a point z+L, a geometrical period L down stream, the wave function is multiplied by factor λ:

$$\psi(r,z+L;\omega) = \lambda\psi(r,z;\omega). \tag{1.1}$$

For a lossless structure λ can be written as $\exp(-j\beta L)$ where βL is a real number representing the phase-shift of the wave. The propagation constant β can be written as:

$$\beta_n = \beta_0 + \frac{2\pi n}{L} \tag{1.2}$$

where n is a positive or negative integer.

As a result, the wave-function can be considered as the sum of an infinite number of components, known as space harmonics, characterized each by an integer n:

$$\psi(r,z;\omega) = e^{j\omega t} \sum_{n=-\infty}^{\infty} a_n(r) \exp\left(-j\beta_n z\right) . \tag{1.3}$$

A particle moving down the structure will see a wave with constant amplitude $a_n(r)$ if its velocity is equal to $v_{pn}=\omega/\beta_n$ y, the phase velocity of the n-th space harmonic.

The group velocity is defined as $d\omega/d\beta$. Its meaning can be perceived by considering a wave bucket centered at ω in an interval $\pm d\omega$ small enough so that all the waves of the bucket have the same amplitude $a(\omega)$. By combining two by two waves such are $a(\omega \pm d\omega) \exp j[(\omega \pm d\omega)t - (\beta \pm d\beta)z]$, one obtains a resulting one:

$$a(\omega)\cos(d\omega t - d\beta z)\exp j(\omega t - \beta z) \tag{1.4}$$

which has a constant amplitude in a frame moving with the group velocity $v_g= d\omega/d\beta$. This frame is the same for all the space harmonics. The group velocity can be then interpreted as the propagation velocity of a perturbation which is also the velocity of energy in a lossless circuit.

At the same ω, another wave solution is obtained by changing z in its opposite:

$$\psi'(r,z;\omega) = e^{j\omega t} \sum_{-\infty}^{\infty} a'_n(r) \exp\left(j\beta_n z\right) . \tag{1.5}$$

An identical result is obtained by changing β in its opposite. As a consequence, the dispersion function $\omega(\beta)$ is a periodic even function as shown in Fig.1.

190

In an infinitely long structure, the two waves, a forward-wave ($v_g > 0$) and a backward wave ($v_g < 0$), can be excited independently. In a real structure, necessarily limited, the reflexions on the terminations give rise to a standing wave.

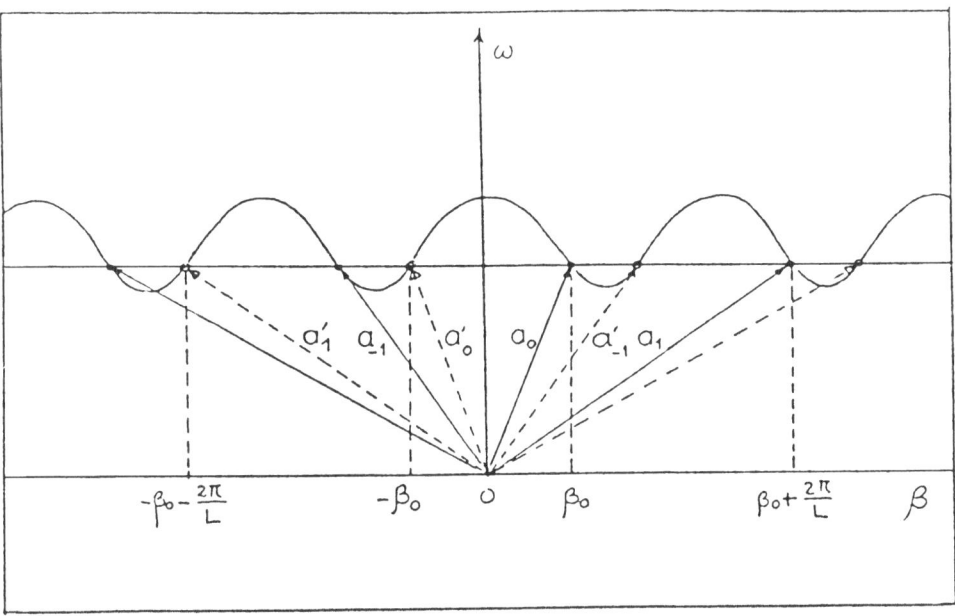

Fig. 1 - A dispersion curve $\omega(\beta)$ of a periodic slow wave structure.

In order to conserve all the properties of a periodic structure, the boundaries must be set such that the symmetries are respected. For example, a short-circuit must be placed where the tangent electric field are zero and an open-circuit, where the tangent magnetic field is zero. In these cases, one of the conditions $a'_n = \pm a_n$ holds and the resulting standing wave will have one of the following forms:

$$S\,(r,z;\omega) = e^{j\omega t} \sum_{n=0}^{\infty} a_n(r) \begin{Bmatrix} \cos \\ \sin \end{Bmatrix} (\beta_n z) \ . \tag{1.6}$$

which correspond by convention to an anti-symmetric (even) or a symmetric mode (odd).

For more precision, let's consider the genesis of the dispersion curve of the iris-loaded structure obtained by deformation from a perfectly cylindrical waveguide, shown in Fig.2. Figures (1),(2) and (3) show the TM_{01} field patterns in a cylindrical waveguides at zero-mode, π-mode and 2π-mode for a given geometrical period L. If there is no ambiguity as for the first mode, there are two possible determinations for the others. The introduction of the iris (supposedly infinitely thin) let unperturbed the zero-mode (Fig.(a)), the odd (or symmetric) determination of the π-mode (Fig.(c)) and the even (or anti-symmetric) determination of the 2π-mode (Fig.(e)). In contrary, the iris modifies drastically the even and the odd determination of the π- and the 2π-mode respectively, lowering their frequencies towards the zero- and π-mode of the cylindrical waveguide. These situations are shown in Fig.3. The successive bands are characterized by their parity.

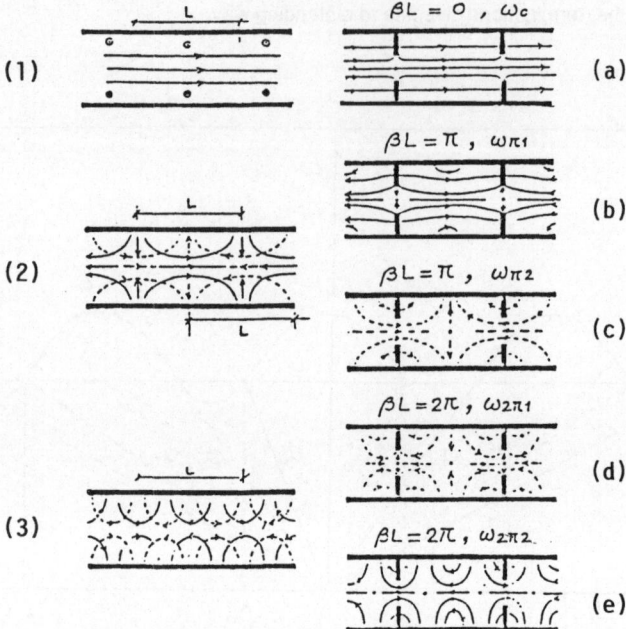

Fig. 2 - Genesis of a dispersion curve obtained by deformation from a perfect cylinder.

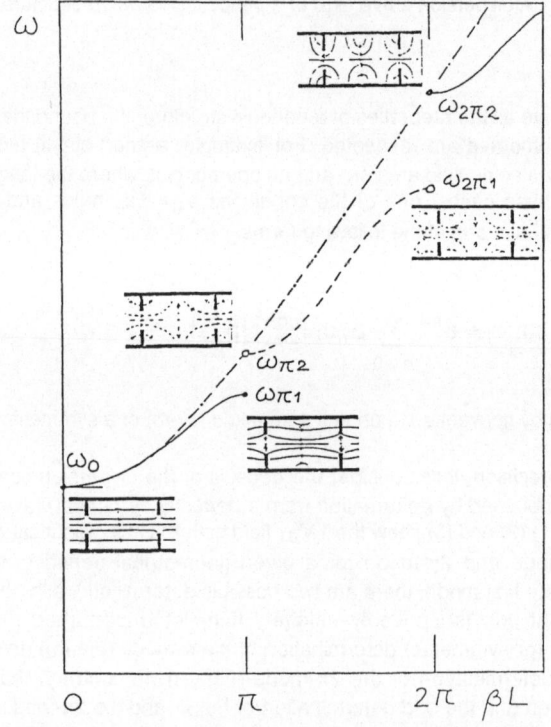

Fig. 3 - The dispersion curve of a iris-loaded waveguide.

Fig.4 shows the example of a disk-loaded structure which can be considered as a complementary of the iris-loaded structure. Similar conclusions can be drawn in particular about the parity of the different modes.

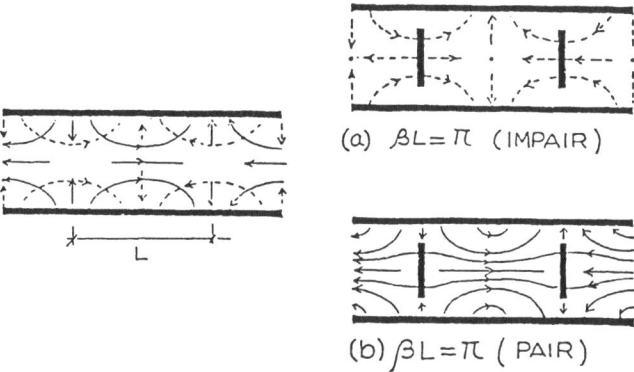

(a) $\beta L = \pi$ (IMPAIR)

(b) $\beta L = \pi$ (PAIR)

Fig.4 - The disk-loaded structure which can be seen as a complementary of the iris-loaded structure.

1.2. - The multi-periodic or multi-band system[1-3]

The two mono-periodic structures considered above can be used to build up a bi-periodic structure shown in Fig.5 which one can recognize as the disk and washer (DAW) structure. This is one of the numerous examples which could not be treated without considering at least two coupled modes. The purpose of this chapter is to examine such cases.

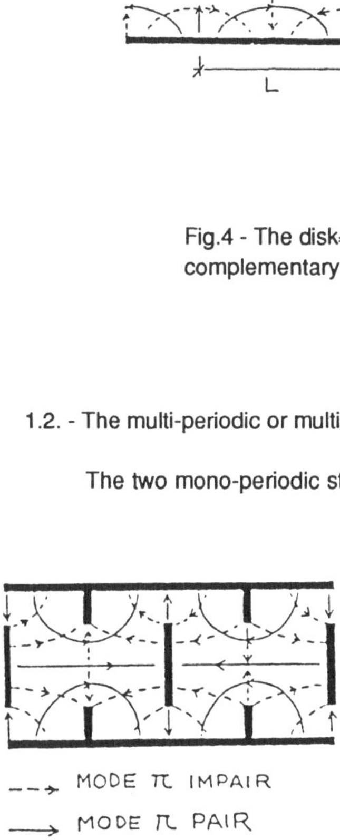

Consider a system of two interlaced periodic structures and let Ψ_1 Ψ_2 be the waves propagating along these structures. The propagation equations of the system can be written as:

$- - \rightarrow$ MODE π IMPAIR

$\longrightarrow$ MODE π PAIR

Fig. 5 - The disk and washer (DAW) structure: an example of a biperiodic structure.

$$\Delta \Psi_1 + k^2_1 \Psi_1 + D_{12} \Psi_2 = 0$$

$$\Delta \Psi_2 + k^2_2 \Psi_2 + D_{21} \Psi_1 = 0 \qquad (1.7)$$

The scalar k^2_1 and k^2_2 are the eigen values of the uncoupled system and D_{12} and D_{21} the operators representing the coupling between the elementary structures. Suppose each structure is made of resonating cells coupled through discrete coupling elements (which supposes stepwise phase-shift between cells), then the coupled system can be represented by two coupled chains shown in Fig.6.

The different kinds of graph are used to characterize the parities,

193

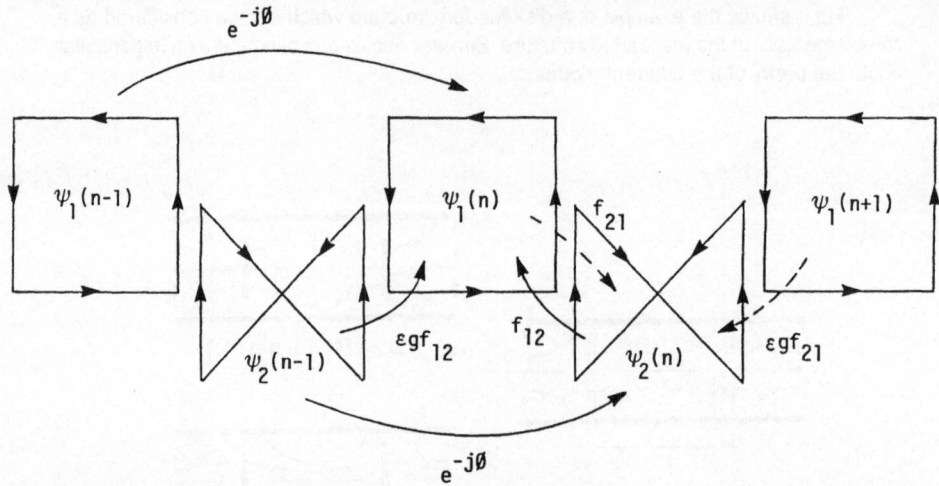

Fig. 6 - Coupled chains.

which are different in the represented case. The coefficient ε accounts for the sign of the coupling ($\varepsilon = -1$ for coupling cells of different graph and $\varepsilon = +1$ for the contrary case). The coefficient g accounts for the difference of coupling from structure 1 towards structure 2 and from structure 2 towards structure 1. The coupling coefficients f_{12} and f_{21} are not necessarily equal or of the same sign. Then the different operators can be written as, n standing for cell number:

$$\Delta\Psi(n) = {}^{\text{TM}}[\Psi(n-1) + \Psi(n+1) - 2\Psi(n)]$$

$$D_{12}\,\Psi_2(n) = f_{12}[\varepsilon\,g\,\psi_2(n-1) + \psi_2(n)] \tag{1.8}$$

$$D_{21}\,\Psi_1(n) = f_{21}[\Psi_1(n) + \varepsilon\,g\,\Psi_2(n+1)]$$

The operators D are reduced to a scalar if instead of being interlaced, the two structures are coupled parallely. Consider first the uncoupled system, i.e. for $f_{12} = f_{21} = 0$. Applying the Floquet's theorem by writing:

$$\psi(n\pm1) = \psi(n)\exp(\mp j\beta L)$$

the equations (7) give two solutions β_1 and β_2 for which the phase-shift per cell βL are given by:

$$\cos\beta_1(\omega)L = 1-k^2{}_1(\omega) \equiv F_1(\omega)$$
$$\cos\beta_2(\omega)L = 1-k^2{}_2(\omega) \equiv F_2(\omega) \tag{1.9}$$

where $F_1(\omega)$ and $F_2(\omega)$ are defined as the dispersion functions of the uncoupled system. They are supposed known in the following. If $F(\omega)$ is defined as the dispersion function of the coupled system, the new wave functions are solutions of the homogeneous equations:

194

$$\begin{pmatrix} F(\omega) - F_1(\omega) & F_{12}\left[\epsilon g \ \exp\left(j\ \beta L\right) + 1\right] \\ F_{21}\left[1 + \epsilon g \ \exp\left(-j\ \beta L\right)\right] & F(\omega) - F_2(\omega) \end{pmatrix} \begin{pmatrix} \psi_1 \\ \psi_2 \end{pmatrix} = 0 \qquad (1.10)$$

By solving (1.10), one obtains two determinations $F_+(\omega)$ and $F_-(\omega)$:

$$(F_1 + F_2) - \epsilon g f \pm \sqrt{\left(\frac{F_1 - F_2}{2}\right)^2 + f\left[g^2 + \epsilon g \ (F_1 + F_2) + 1 + g^2\right]} \qquad (1.11)$$

where $f = f_{12} \ f_{21}$. From (1.11) one can deduce the behaviour of the coupled system, in particular, the composition law. Fig.7 represents $F(\omega)$ in a general case (a) with the corresponding dispersion curve (b). The two branches of $F(\omega)$ join each other with a vertical

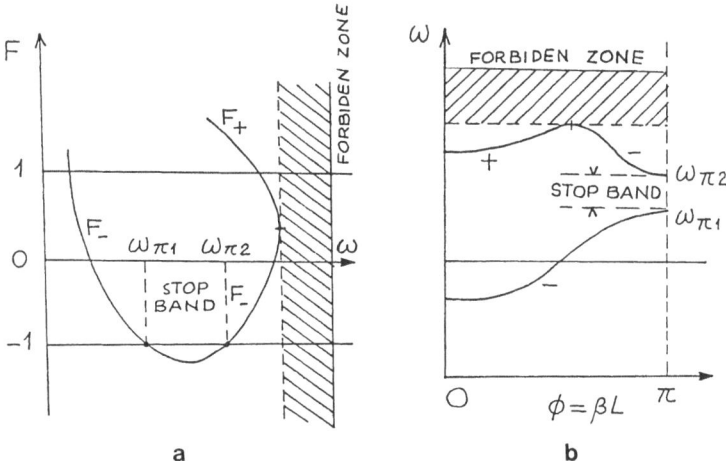

Fig. 7 - (a) Dispersion function $F(\omega)$; (b) Dispersion curve $\omega(\beta L)$.

tangent when the quantity under the square root sign cancels. According to wether this point is lying in the interval $(-1, +1)$ or not, the dispersion curve exhibits or not a horizontal slope ($v_g = 0$) in the pass-band. There exists generally two pass-bands separated by a stop-band. It is of interest the case where the stop-band disappears. It is easy to demonstrate that this situation occurs when:

$$\frac{dF}{d\omega} = 0 , \qquad \text{with} \quad F = \pm 1 \qquad (1.12)$$

$$\text{then} \quad v_g = \frac{d\omega}{d\beta} = \pm \ L \left(\left|\frac{d^2 F}{d^2 \omega}\right|\right)^{-\frac{1}{2}}$$

as shown in Figs.8 and 9. This phenomenon of mode coalescence is generally sought for in order to render the structure insensitive to mechanical errors or beam loading and,

195

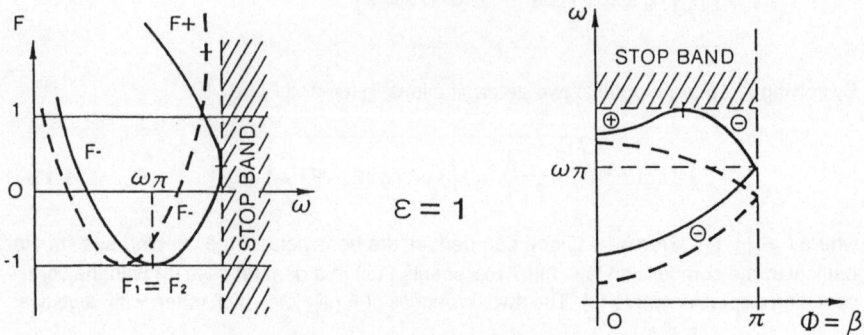

Fig. 8 - Field of same parity: coalescence at π-mode.

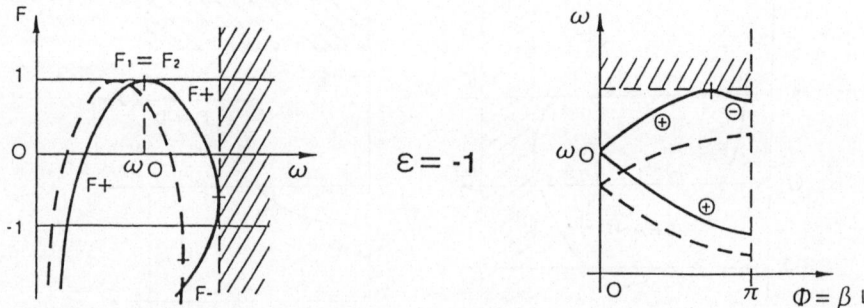

Fig. 9 - Field of different parity: coalescence at zero-mode.

in microwave electron tubes, to avoid certain parasitic oscillations. Applying (1.2) to (1.11) and studying closely its first derivative, one obtains a sufficient condition for mode coalescence:

$$g = 1$$

$$F_1 = F_2 = F = -\varepsilon$$

(1.13)

from which a composition law can be derived as follows:

> Provided the coupling is symmetric (g=1), the coalescence takes place at π-mode (F_-= -1) when the oscillation modes in the two structures are of the same parity (ε=1). It takes place at zero-mode (F_+=1) when they are of different parity (ε= -1).

These two situations are illustrated in Figs.8 and 9 respectively. Some typical examples are given in the following to illustrate this composition law.

Biperiodic cavity chains or side-coupled structure

These structures are represented in Figs.10 (a) and (b). As well known coalescence takes place at π-mode, Fig.10 (c), because the TM_{01} modes in accelerating cell and coupling cell are all even. This will occur at zero-mode however if the geometry of the coupling cell is such that an odd mode is excited instead, Fig.11 (a) and (b).

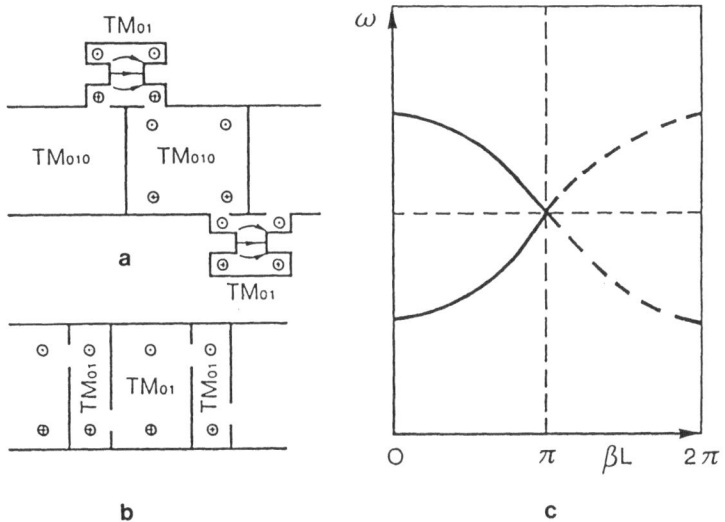

Fig. 10 - (a) Side coupled structure; (b) Biperiodic structure;
(c) Coalescence at π-mode.

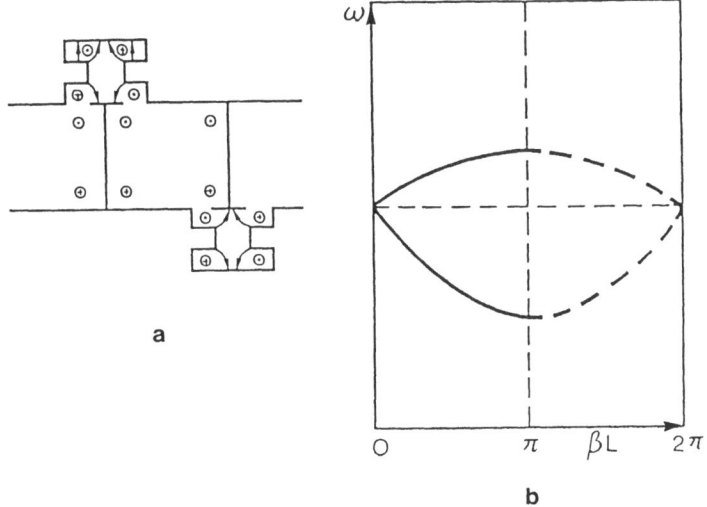

Fig. 11 - (a) Side coupled structure; (b) Coalescence at zero-mode.

Compensated Alvarez structure

It is well known also that the coalescence takes place at zero-mode, because the TM_{01} accelerating cavity mode and the TEM stem mode are of different parity, Figs.12 (a) and (b).

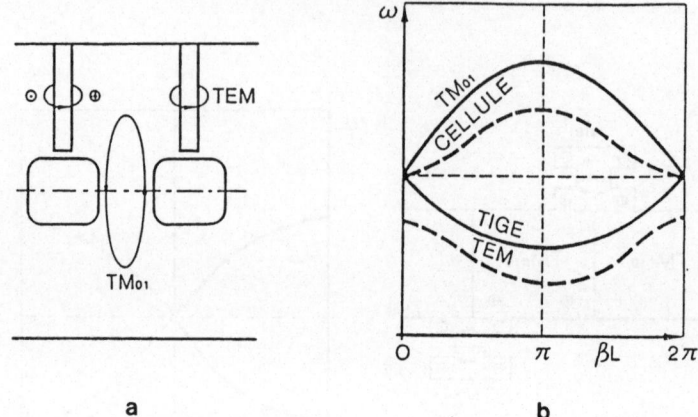

a	b

Fig. 12 - (a) Alvarez compensated structure; (b) Coalescence at zero-mode.

Monoperiodic resonant slot coupled structure

This structure is commonly used in TWT, Fig.13 (a). Cavities are coupled together through slots. When the slot is small, cavity and slot bands are far from each other. When the slot is enlarged, its resonant frequency becomes lower and the two pass-bands come close. The system behaves as a biperiodic structure and coalescence could happen. As the slot-mode is dominantly of TE-type, it has a parity opposite to that of the TM_{01} accelerating mode.

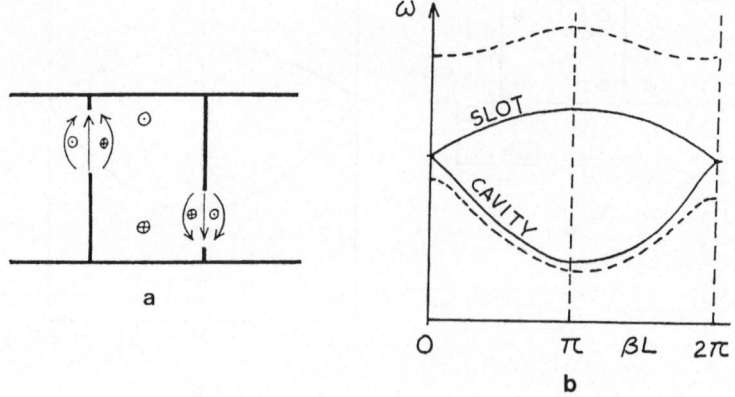

Fig. 13 - (a) Resonant slot coupled structure; (b) Coalescence at zero-mode.

Coalescence will then take place at zero-mode, Fig.13 (b). This situation is often sought for in order to broaden the band width, it is also used to avoid band edge oscillations in TWT, in this case coalescence should be slightly surpassed. This phenomenon will be examined in part 3.

Disk and washer (DAW) structure[4]

In the DAW-structure the coalescence occurs at π-mode because the TM-modes in a cell comprised between two successive washers and in a cell comprised between two successive disks are of the same parity. If the period is long enough so that the frequencies of the two TM_{01} π-modes are much lower than the frequency of the TM_{02}-mode, the coalescence will take place between these two π-modes as shown in Fig.14 (a) which corresponds to a structure for high energy particles. If in contrary the period is short enough so that the odd TM_{01} π-mode is thrown away, the coalescence will take place between the TM_{01} even π-mode and the TM_{02} odd π-mode as shown in Fig.14 (b) which corresponds to a structure for low energy particles.

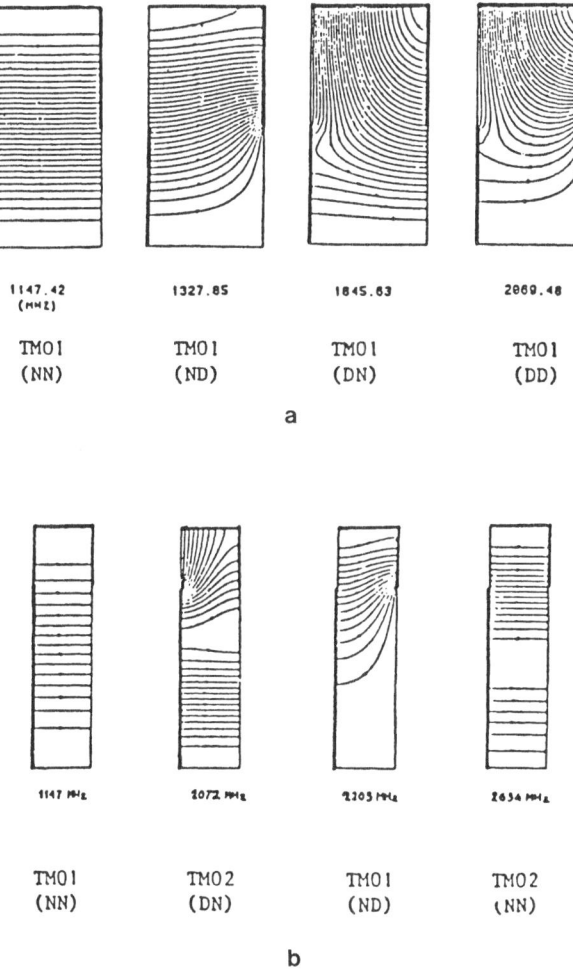

| 1147.42 (MHZ) | 1327.85 | 1845.63 | 2069.48 |
| TM01 (NN) | TM01 (ND) | TM01 (DN) | TM01 (DD) |

a

| 1147 MHz | 2072 MHz | 2205 MHz | 2634 MHz |
| TM01 (NN) | TM02 (DN) | TM01 (ND) | TM02 (NN) |

b

Fig. 14 - Field patterns DAW half cell.
(a) Coalescence TM_{01}/TM_{01}: for high velocity particle;
(b) Coalescence TM_{01}/TM_{02}: for low velocity particle.

In this case, when the structure is open enough, for example when the cavity diameter is increased in order to improve the Q value, it can happen that there is a zero-group velocity for a phase-shift per cell intermediate between 0 and π, as predicted by theory (see Fig.8). This situation should be avoided to prevent parasitic oscillations (see part 3).

PART 2 - COMPUTER AIDED DESIGN OF RF-STRUCTURE

SCOPE

Existing computer codes used to solve Maxwell's equations in RF-cavities, for particle accelerators or electron devices, are reviewed with emphasis on the last development of three-dimensional code. Illustrative results are given for URMEL 3D a finite difference code (FDM) and for HELMOT 3D a finite element code (FEM).

2.1.- Existing computer codes

Several computer codes exist and are being used currently to solve Maxwell's equations in RF-cavities for particle accelerators and electron devices. Most of them are based on finite difference method (FDM) and a few, on finite element method (FEM).

FDM-codes.

> Variational, iterative method with over-relaxation:
>
> MESSYMESH[5], 2D.
> LALA[6], 2D.
>
> Non iterative direct solution of inhomogeneous equations:
>
> SUPERFISH[7], 2D.
> ULTRAFISH[8], 2D, axis symmetrical geometry with azimuthally
>   periodic field
> URMEL[9-12], 2D and 3D.

FEM-codes

> HAX 2D[13] and 3D[14].
> HELMOT - 2D[15], 2D.
> HELMOT - 3D[15,16], 3D.

General discussion on FEM can be found in Refs. (17-22).

2.2.- Three dimensional finite difference codes

ULTRAFISH is an extension of SUPERFISH aiming to calculate the azimuthally periodic field in axisymmetrical cavities. But division 0/0 poses a numerical obstacle, since values are not exact in the finite difference technique. In fact, these cases are now calculated directly by three-dimensional codes such as URMEL-3D or HELMOT-3D.

URMEL-3D belongs to the vast family MAFIA (Maxwell's Equation by Finite Integration theory and their Applications), a joint effort between Los Alamos National Laboratory, Kernforschungsanlage Jülich, and DESY. This code computes resonant frequencies and fields in cavities of arbitrary shapes, partially filled or not with dielectric or magnetic material.

Figs.15 and 16 give an example of the possibility of this code. Fig.15 shows the meshplot representing a cavity coupled to a waveguide[12] and Fig.16 shows field patterns obtained for TM_{11}-mode.

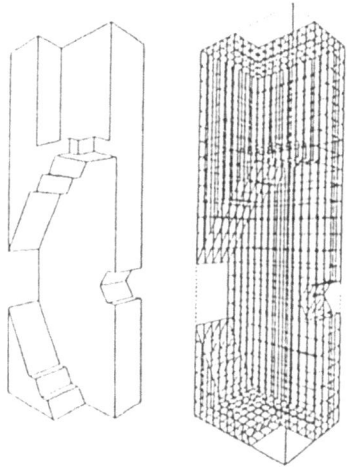

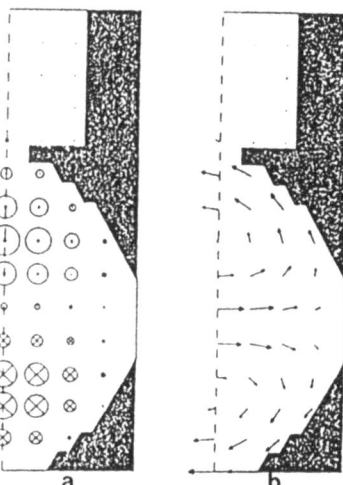

Fig. 15 - Mesh plot of a cavity coupled to a waveguide.

Fig. 16 - (a) Electric field pattern for TM_{11}-mode; (b) Magnetic field pattern for TM_{11}-mode;

2.3.- HELMOT-3D, a finite element code

A two dimensional finite element code, HELMOT-2D, was written many years ago and is currently used at Electron Tubes Division at THOMSON-CSF for the design of axisymmetrical RF-cavities.

It becomes clear that the FEM saves computer time and memory. But as long as the treated cases are limited to axisymmetrical geometries, this advantage seems to be not appreciable enough to favor this method to the FDM, given the capacity of presently available computers. In fact, the merit of the FEM becomes really appreciated for the three-dimensional problems.

The main factors which make the FEM attractive are:

(i)   the flexibility in the representation of boundaries,
(ii)   the advantage of the modularity of the FEM-codes,
(iii) the already existing libraries which make further development easy and in-
       expensive.

Unfortunately, unlike the case of scalar or two-dimensional problem, many attempts to apply the FEM to the three-dimensional Maxwellian eigen value problem have been plagued by appearance of non-physical solutions. It is not in the frame of this paper to go into the details of this problem, let's only note that we have to solve the following equations:

$$\text{rot rot } \vec{E} - k^2 \vec{E} = 0 \qquad \text{in } \Omega$$

$$\text{div } \varepsilon \vec{E} = 0 \qquad \text{in } \Omega \qquad\qquad (2.1)$$

$$\vec{E} \times \vec{n} = 0 \qquad \text{on } \Gamma$$

Ω is the volume of the cavity limited by a perfectly conducting wall Γ.

The operator rot rot is not elliptic so that the spectral theory of compact, self-adjoint operators could be applied directly. It is furthermore not suitable to the discretization by the FEM.

To render the operator elliptic in order to be treated by FEM, one has to work with the operator:

$$\Delta_s \vec{E} = s \varepsilon \mu \text{ grad div } (\varepsilon \vec{E}) - \text{rot rot } \vec{E} \qquad\qquad (2.2)$$

and consider the following eigenvalue problem:

$$\Delta_s \vec{E} + k^2 \vec{E} = 0 \qquad \text{in } \Omega$$

$$\vec{E} \times \vec{n} = 0 \qquad \text{on } \Gamma \qquad\qquad (2.3)$$

$$\text{div } \varepsilon \vec{E} = 0 \qquad \text{on } \Gamma$$

This system is reduced to the classical vectorial Helmoz's equations for s=1. Otherwise a solution of (2.3) is not necessary a solution of (2.1). Nothing ensures that this solution satisfies the original equation (2.1) with div $\varepsilon E = 0$. Such solutions are just numerical ones and have no physical meaning. Up to now (2.3) is solved with varying s and one has to discard all solutions for which the eigenvalues k^2 vary with s and to keep only that which are stationary. This is a huge task that made the FEM unpracticable. Let U = div E and take the divergence of (2.3), one has:

$$\Delta U + \frac{k^2}{s} U = 0 \qquad \text{in } \Omega$$

$$\qquad\qquad (2.4)$$

$$U = 0 \qquad \text{on } \Gamma$$

where U is a scalar. This equation can be easily solved in three dimensions and let λ_1 be its first eigenvalue (one can even avoid this task by finding a minorant $\hat{\lambda}_1$ with a simpler volume $\hat{\Omega}$ majoring Ω and for which the eigen value λ_1 is known).

If one looks for the eigen values of (2.3) in the interval of (0,k) and if one chooses the free parameter s large enough so that $k^2/s < \lambda_1$, then (2.4) can have but the trivial solution U=0, i.e. div E = 0. This procedure has been applied in HELMOT-3D and proves to be reliable. More details can be found in ref.(16).

For illustration, we consider the case of a TWT cavity coupled to a side-cavity filled with dielectric.

Consider the axisymmetrical cavity of Fig.17 (a), commonly used also in RF-linacs, coupled to a cylindrical resonator filled with dielectric of Fig.17 (b). This latter is aimed to damp a parasitic mode. Our purpose is to determine the two lowest modes and to know the influence of the coupling.

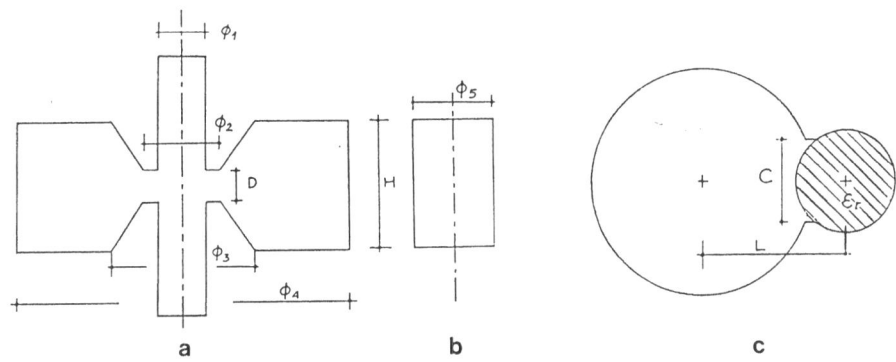

a   b   c

Fig. 17 - (a) The symmetrical accelerating cavity, with dimension in cm: φ_1= 1.5, φ_2 = 1.4, φ_3 = 1.54, φ_4 = 4.95, D = 0.5, H = 1.28.
(b) The side cavity filled with dielectric ε_r= 12, with φ_5 = 1.79, H = 1.28.
(c) Assembled system, with: L = 3.12, C = 1.5.

Applying the 3D-code to the axsymmetrical cavity, the fundamental frequency of the empty cavity is computed. With a fine mesh of 3748 modes one obtains 38.40 GHz. The fundamental frequency of the dielectric resonator, calculated analitically, is 37.04 GHz. To test the method on this resonator, the 3D-code has been also applied. When s=1, a spurious mode shows up at 29.9 GHz as predicted, as s must be chose larger than 1.7. For s=2, i.e. larger than 1.7, the spurious mode disapears and one obtains the right mode at 37.045 GHz.

Fig.17 (c) shows the parameters L and C characterizing the coupling hole. Fig.18 shows a view of mesh and domain used, defined by cutting the cavity following two symmetric plans where are imposed conditions of symmetry. The mesh comprises 2676 points and 492 elements. The boundaries are exact curves, the broken lines come from the plotting program. We know that the frequency of the coupled system is around 37 GHz then s must be larger than 1.7. To evidence the non-physical modes, s is chosen deliberately equal to 1, i.e. smaller than 1.7. The spurious mode appears effectively at 29.92 GHz as shown in Fig.19, with a magnetic field pattern very similar to that obtained with the cylinder only. With s =2, the spurious mode disappears and two right modes are

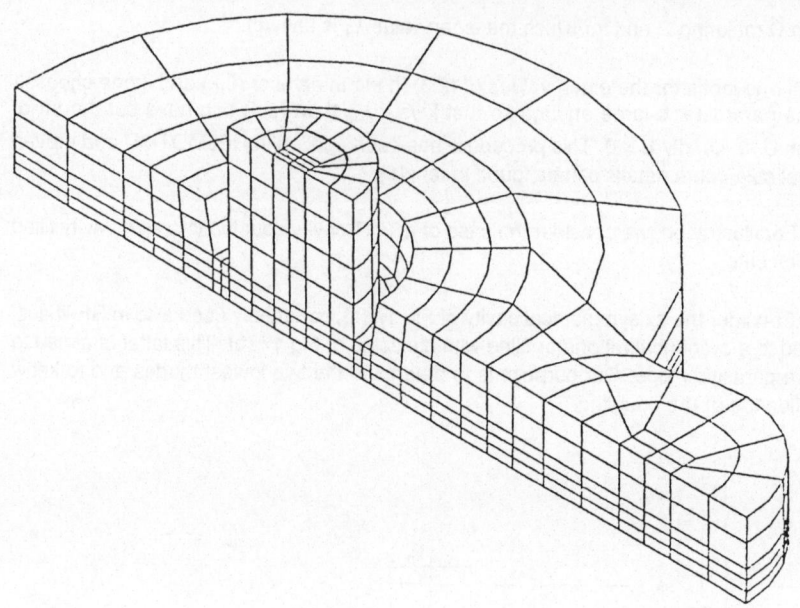

Fig. 18 - Mesh plot of a quarter of cavity: 2676 points, 492 elements.

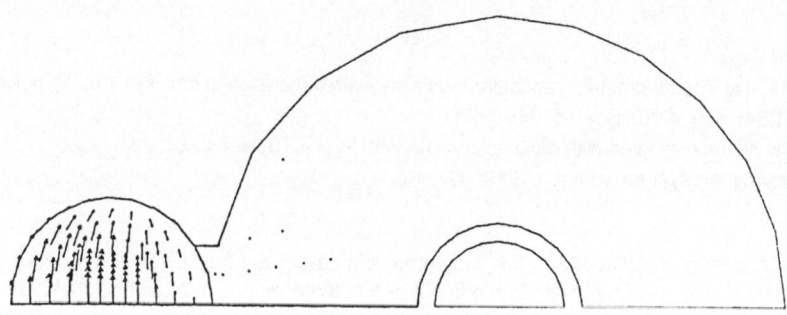

Fig. 19 - A spurious mode at 29.92 MHz for s=1.

obtained at 37.73 GHz and 39.45 GHz. The corresponding magnetic field patterns on the vertical plane are shown in Figs.20. The first mode is dominated by the side cavity TM_{01} and the second results from the coupling between the two TM_{01} modes.

<u>Remarks:</u>

By increasing s, the field patterns do not change sensibly but the frequencies increase lightly, which motivates the choice of the lowest s fo which the technique works.

Near the axis the field patterns are not very good. This is due to the fact that the elements used are rectangular parallelepipeds instead of prismas which are the best suit element for this region.

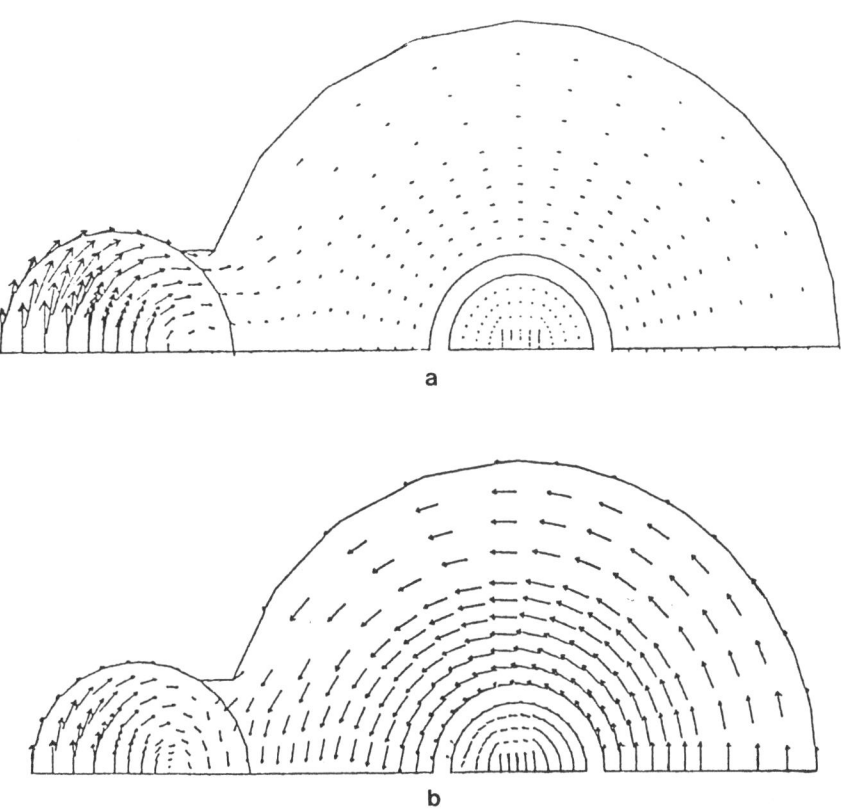

Fig. 20 - (a) Low frequency mode: f=37.73 GHz;
(b) High frequency mode: f=39.45 GHz.

CONCLUSIONS

We can say that three-dimensional codes exist now and allow to solve Maxwell's equations in cavities of arbitrary shape. The FDM has received a great attention of the accelerator community. The FEM still remain a pioneer work. We are not yet able to compare them in a systematic manner. We think however that the FEM should be less time and memory consuming and more flexible, now the we can get rid with the non-physical solution.

PART 3 - PRACTICAL ASPECTS IN RF-STRUCTURES

SCOPE

Practical criteria are given for the choice of RF-structure, followed by an examinations of some practical aspects considered as the most important such as:
- Beam loading mechanism and compensation methods,
- Oscillations and parasitic modes damping,
- RF-field break-down,

- Description of a recirculated backward wave structure with automatic beam loading compensation.

3.1. - The qualities of the RF-structure

The factors of merit of a RF-structure can be listed as follows:

Efficiency

The efficiency is measured by the shunt impedance which is considered for a long time as a major factor of merit of a RF-structure. The availability of powerfull RF-sources and a better knowledge on RF-field break-down, have attenuated now its importance. For certain applications, a too high shunt impedance even becomes harmful. Such is the case of accelerating very high charge with pulse duration much shorter than the filling time of the cavity. The stored energy is taken so fast away by the particle beam that not enough time is let to compensate the transient field drop, resulting in a large energy spread which can be written as:

$$\frac{\Delta V}{V} \sim \frac{1}{2} \frac{\Delta W}{W} = \frac{q}{E} \frac{Z_{sh}}{Q} \qquad (3.1)$$

where q is the total charge, E the accelerating field, Z_{sh} is the shunt impedance, and Q the quality factor . The structure should be therefore designed for a rather low shunt impedance but for a high quality factor, with as low surface electric field as possible for a given energy gain.

Stability

Stability should be understood as the intensitivity of the field flatness to perturbation. In single band system, the group velocity is zero at zero or π-mode. In these areas, the frequency difference between modes is so small that even a small perturbation, either due to mechanical errors or beam loading, can excite several modes at the same time, destroying the field flatness. This has an immediate consequence on beam dynamic and, in the worst case causes field breakdown. Known remedies consist in limiting the structure length, in using large coupling in order to broaden the band width or, as already pointed out, in using the coalescence technique.

Insensitivity to wake field

The wake-field excited by particle bunch of high charge could effect the beam quality and even impose a limit to the accelerated current. The shape of the cavity is of importance and, at the first place, the part of the cavity wall in the neighborhood of the particle bunch. Drift tube with small beam hole is known to improve the shunt impedance but gives rise to large wake-field[23]. A trade-off between several parameters such are bore hole, acceleration gap, cavity shape..., has to be made.

Freedom from parasitic modes

Beam breakup is known as major cause of current limitation in multisection linacs or in storage rings. This is due do the cumulative self effect of the deflecting HEM mode excited by an off-axis beam. For high current devices such are high power travelling wave tubes, auto-oscillations at the edges of the fundamental pass-band, due to the low group velocity in these areas, could also impose a limitation to the permissible output power. The case should be even worse with very high Q super-conducting cavities. The design of a RF-structure must then include parasitic modes damping devices.

"Good geometry"

Good geometry should be understood here as technological qualities such are:

- Beam focusing ability (large drift-tube in ion accelerator, small overall diameter for electron RF-structure or self-focusing capability),
- Easy cooling,
- Possibility of broad beam acceleration (positron acceleration immediately after the conversion target),
- High field capability.

Finally, the optimization of a RF-structure is becoming a very complex work. Computer aids are absolutely necessary; they are unfortunately insufficient.

3.2. - Beam loading effects

3.2.1. - Oscillating beam loading

It is usual to analyze the beam loading effects in a RF-structure by considering the particle beam as a perfect current source, i.e. with an infinite parallel impedance. This is true for a tightly bunched high energy beam. At low energy, the velocity modulation and the space variation of charged density have to be taken into account. As a result, beam dynamics are no more linear. The particle beam can still be however assimilated to a current source but with a finite parallel impedance, the amplitude and sign of which depend on beam current and energy and also on certain characteristics of the RF-structure itself.

In the case of a single gap cavity for example, a small signal theory shows that the RF-voltage V induced by the particle beam on the cavity gap is related to the beam current i by:

$$i = Y_B V, \qquad (3.2)$$

where Y_B can be assimilated to a complex admittance:

$$Y_B = G_B + jB_B,\qquad(3.3)$$

with :

$$G_B = G_0 \times \frac{1}{2}\frac{\sin D/2}{D/2}\left[\frac{\sin D/2}{D/2} - \cos D/2\right]$$

$$(3.4)$$

$$B_B = G_0 \times \frac{1}{2}\frac{\cos D/2}{D/2}\left[\frac{\sin D/2}{D/2} - \cos D/2\right]$$

where $D = \omega d/u_0$ is the transit angle (d: cavity gap, u_0 : beam average velocity) and $G_0 = I_0/V_0$ (I_0: average current, V_0: beam voltage). Figs.21 and 22 represent the reduced beam conductance $g_B = G_B/G_0$ and beam susceptance $b_B = G_B/G_0$ as functions of the transit angle $D/2$.

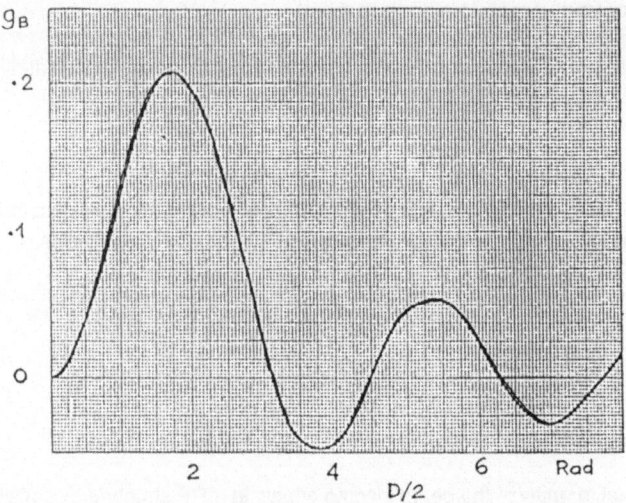

Fig. 21 - Beam conductance in a single cavity as function of the transit angle $D = \omega d/u_0$.

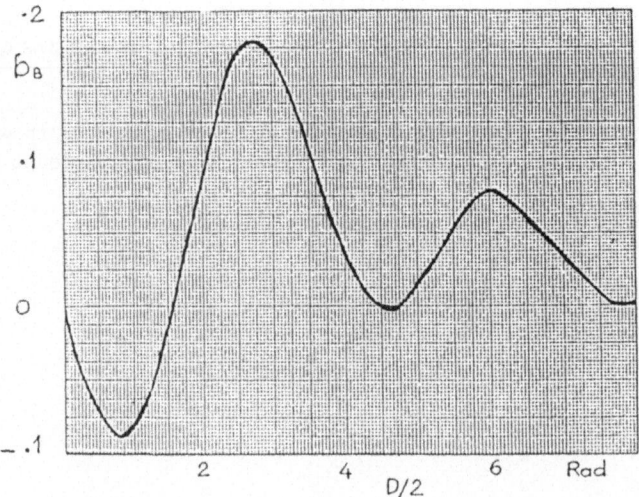

Fig. 22 - Beam susceptance in a single cavity as function of the transit angle $D = \omega d/u_0$.

It can be seen that the conductance could take a negative value. The oscillations may then appear if the cavity is not enough resistive. These oscillations are known as of the monotron-type of which reflex klystron gives a typical example.

The interaction between a particle beam and an electromagnetic wave in a multi-cell RF-structure can also give rise to a similar phenomenon. Indeed, the phase slip between the particle beam and a space harmonic travelling with a slightly different velocity could have an effect similar to that of the transit angle in the case of a single gap cavity. By a similar small signal theory [24,25], it is shown that the expressions of the beam admittance now are

$$G_B = G_0 \frac{1-\varepsilon}{\varepsilon} \frac{\sin N\pi\varepsilon}{N\pi\varepsilon} \left[\frac{\sin N\pi\varepsilon}{N\pi\varepsilon} - \cos N\pi\varepsilon \right]$$

$$\quad (3.5)$$

$$B_B = G_0 \frac{1-\varepsilon}{\varepsilon} \frac{\cos N\pi\varepsilon}{N\pi\varepsilon} \left[\frac{\sin N\pi\varepsilon}{N\pi\varepsilon} - \cos N\pi\varepsilon \right]$$

where $\varepsilon = 1 - v/u_0$, v being the phase velocity of the space harmonic, and $N = \beta L/2\pi$ is the number of wave-length on the structure length L. These expressions are but (3.4) multiplied by the factor $(1 - \varepsilon)/\varepsilon$ and where the transit angle D/2 is replaced by the phase slip $N\pi\varepsilon$. At synchronism, i.e. $v = u_0$ or $\varepsilon = 0$, the limiting values of g_B ($\varepsilon = 0$) and $b_B = (\varepsilon = 0)$ are 0 and $N\pi/6$ respectively. Fig.23 shows g_B as a function of v/u_0 with N as parameter. The extreme value of g_B increases rapidly with increasing N or length of structure. The negative extreme value occurs for a phase velocity slightly lower than the beam velocity

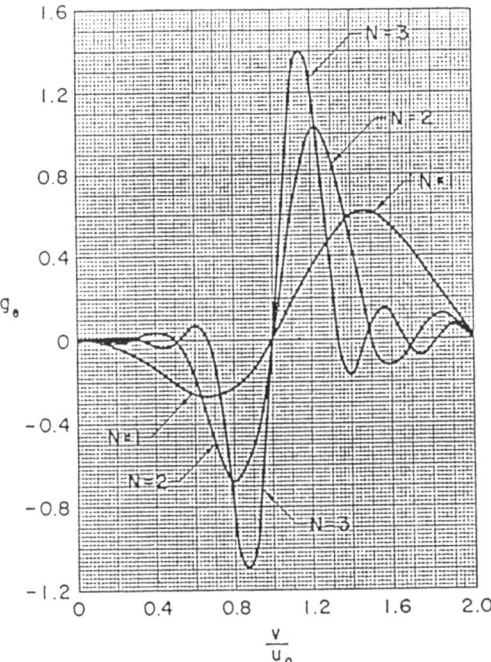

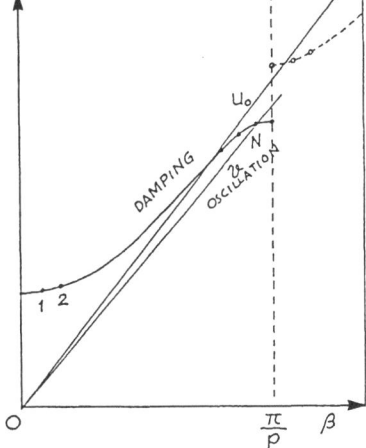

Fig. 23 - Beam conductance in a long structure as function of v / u_0. N : number on structure length.

Fig. 24 - Oscillation zone and damping zone in neighborhood of cut-off.

209

Oscillations can then start if the structure is not enough resistive and this, the more easily the more the structure is long and the more the oscillation mode is close to π-mode for which N = L/2p (p : period length). The fact that a coupling impedance is highest in the neighborhood of cut-off due to the low group velocity favourizes even more these oscillations which for this reason, are known as the band-edge oscillations. The situation is shown in Fig.24. This type of oscillation is actually sought for in certain type of oscillators such as EIOs (extended interaction oscillator) but in most of cases constitutes the major factor in the limitation of power and band width in TWTs or perhaps in the limitation of accelerated current in long superconducting structure.

3.2.2. - Beam loading compensation in RF-cavities

Let us consider first the simple case of a single gap cavity and a particle beam assimilated to a perfect current source. The system is powered by a voltage source E with an internal impedance Z_0 through a coupling loop represented by a mutual inductance M, Fig.25.

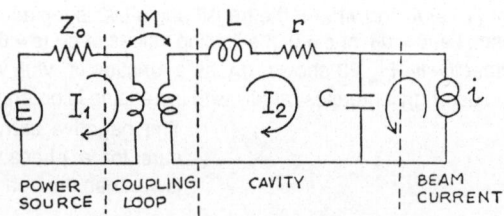

Fig. 25 - Equivalent circuit of a cavity loaded by a beam current i and powered by a source E.

Let i be the fundamental alternative component and I_0 the average component in the Fourier development of the bunched beam. I_0 is actually known as the peak current (the average current in a pulse). With a tightly bunch beam i approaches $2I_0$. Let P_{max} be the maximum power available from the power source, i.e. the power delivered to a load equal to Z_0. Then :

$$E = \sqrt{8\ Z_0\ P_{max}}$$

$$(3.6)$$

Our purpose is to find how to modify the coupling M in order to maintain the input power P_{in} always at the maximum level in spite of the beam loading.

With the notations used in Fig.25, one has :

$$\begin{pmatrix} I_1 \\ I_2 \end{pmatrix} = \frac{1}{ZZ_0 + M^2\omega^2} \begin{pmatrix} Z & -jM\omega \\ -jM\omega & Z_0 \end{pmatrix} \begin{pmatrix} E \\ -\dfrac{i}{jC\omega} \end{pmatrix}$$

$$(3.7)$$

$$V_{ac} = -\frac{I_2 + i}{jC\omega} = -\frac{i}{jC\omega} \left[\frac{-jM\omega}{ZZ_0 + M^2\omega^2} \sqrt{8 \ Z_0 P_{max}} + \left(1 - \frac{Z_0}{jC\omega \left(ZZ_0 + M^2\omega^2 \right)} i \right) \right] \tag{3.8}$$

$$P_{in} = \frac{1}{2} \ V_{in} \ I_1^* = \sqrt{2 \ Z_0 P_{max}} \ I_1^* - \frac{I_1 I_1^*}{2} \ Z_0 \tag{3.9}$$

Optimum coupling without beam

Without beam (i = 0) and at resonance ($\omega = \omega_0$, Z = r), the optimum value M_0 of the mutual inductance is such that :

$$r Z_0 = M_0 \ \omega_0^2 \qquad \text{or} \qquad Z_0 = \frac{M^2}{L^2} R_{sh} \tag{3.10}$$

where R_{sh} is the shunt impedance of the cavity (R_{sh} is half of the intrinsic shunt impedance Z_{eff} used to measure the efficiency of a RF-structure). Then the input power P_{in} is equal to P_{max} .

Optimum coupling in presence of particle beam

In presence of a particle beam, the optimum value of M is such that

$$I_1 \ (i,M) = I_1 \ (i = 0, M_0) \tag{3.11}$$

From (3.7), (3.10) and (3.11), on obtains

$$\frac{M}{M_0} = y(x) = x + \sqrt{x^2 + 1}$$

$$x = \sqrt{\frac{R_{sh} i^2}{8 \ P_{max}}} \sim \sqrt{\frac{R_{sh} i_0^2}{2 \ P_{max}}} \tag{3.12}$$

and :

$$V_{ac} = 2 \frac{\sqrt{1 + x^2}}{1 + y^2} \ V_{ac_0} = K(x) \ V_{ac_0} \tag{3.13}$$

y and K are given in Fig.26, as a function of the beam loading parameter x. These results are of course valid in the validity extent of the maximum value of M itself, which in any case cannot be larger than the self inductance of the cavity. Also the self inductance of the loop has been neglected.

It can be seen that for each beam loading parameter there is an optimum load line of which K(x) is the envelope. Fig.27 shows an example of arrangement. According to the beam current the coupling can be modified by rotating the loop or by adjusting the piston. In the latter case, a precise analysis needs an equivalent circuit more elaborated than a simple (LCr) -circuit.

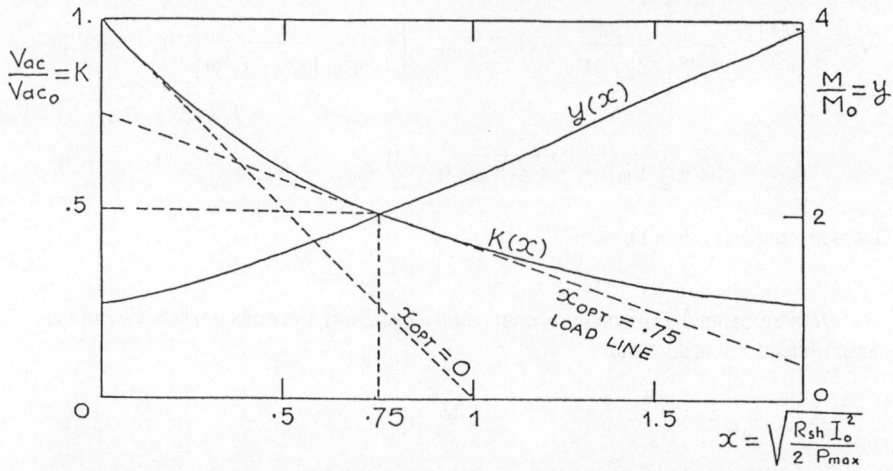

Fig. 26 - Optimum coupling y (x) and accelerating voltage K (x), as a function of beam loading parameter $\sqrt{R_{sh} I_0^2 / 2P_{max}}$.

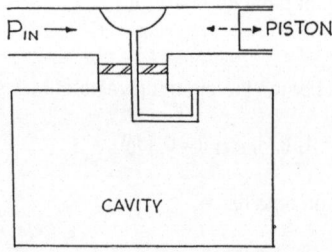

Fig. 27 - cavity with beam loading compensation system.

3.2.3. - Beam loading compensation in TW-structure

In a TW-structure, only a part of the input power serves to maintain the accelerating field and to compensate the beam loading, the other part is dissipated in a resistive load. This power varies with the beam loading. To optimize the efficiency, it could be reused by a reculating system shown in Fig.28. A part of the power coming from the RF-source goes directly to the structure and the other part to the load through a coupler.

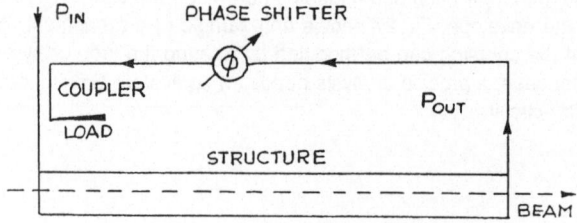

Fig. 28 - Recirculated travelling wave structure.

In the same way a part of the recirculated power goes directly to the load and a part to the structure through the coupler. For a given beam current and provided an appropriate phase shift, the coupling coefficient of the coupler can be calculated so that the two signals cancel exactly in the load, then all the available power goes to the structure. This situation is the same as in a well matched SW-structure in presence of beam. Unfortunately another value of beam current will destroy this ideal solution and a part of the recirculating power will be lost in the load.

To avoid this inconvenient one can imagine the arrangement[26] shown in Fig.29, making use of a magic - T, a 4 - port hybrid coupler and two phase shifters. The signal V_1,

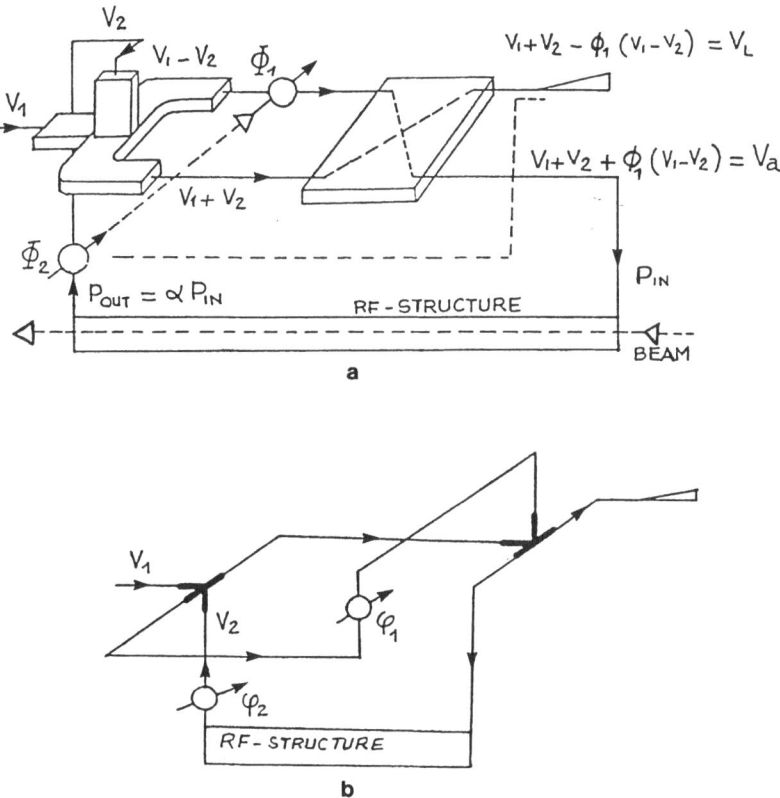

Fig. 29 - Recirculated travelling wave structure with perfect beam loading compensation.

coming from the power source is combined with the signal V_2 coming from the structure output through the magic-T to give two signals $(V_1 - V_2)/2$ and $(V_1 + V_2)/2$ which are combined again through the hybrid coupler to deliver to the load a signal V_L equal to:

$$\vec{V}_L = \frac{\vec{V}_1 + \vec{V}_2}{2} - \psi_1 \frac{\vec{V}_1 - \vec{V}_2}{2} \quad , \tag{3.14}$$

and to the accelerating structure a signal V_{in} equal to:

213

$$\vec{V}_{in} = \frac{\vec{V}_1 + \vec{V}_2}{2} + \Psi_1 \frac{\vec{V}_1 - \vec{V}_2}{2} \qquad (3.15)$$

where Ψ_1 is the rotation matrix of the first phase shift. If M is the transfer matrix representing attenuation and phase rotation due to beam loading and resistive loss on the structure wall and if Ψ_2 is the rotation metrics of the second phase shifter, one has :

$$\vec{V}_2 = \Psi_2 M \left(\frac{\vec{V}_1 + \vec{V}_2}{2} + \Psi_1 \frac{\vec{V}_1 - \vec{V}_2}{2} \right) , \qquad (3.16)$$

The first condition is:

$$V_L \equiv 0, \qquad (3.17)$$

so that no power will be wasted in the load . If V_1 and V_2 are in quadrature, it is always possible to find Ψ_1 so that (3.17) holds, because $V_1 + V_2$ and $V_1 - V_2$ have equal amplitude. The second condition is , for given M, to find Ψ_2 so that (3.16) holds.

Consider the case where M is reduced to a pure attenuation measured by a factor α (It should be noted that this is not a restriction as the rotation part of M could be taken into account in Ψ_2). Let φ_1 and φ_2 be the phase shift, then one obtains the vectorial diagram of Fig.30. (3.16) and (3.17) become respectively :

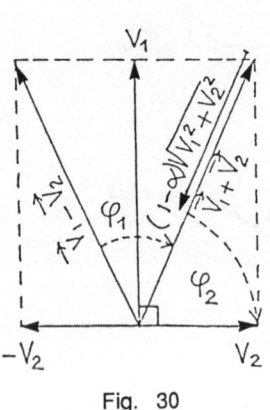

Fig. 30

$$|V_2| = \pi \sqrt{|V_1|^2 + |V_2|^2} \qquad (3.16')$$

$$\frac{1}{2} \varphi_1 + \varphi_2 = \frac{\pi}{2} + 2k \cdot \pi \qquad (3.17')$$

$$\mathrm{tg} \quad \varphi_1 / 2 = |V_2 / V_1|$$

or

$$\mathrm{tg} \quad \varphi_1 / 2 = \frac{\alpha}{\sqrt{1 - \alpha^2}} \qquad (3.18)$$

which defines completely φ_1 and φ_2 for a given α. Another possible scheme[27] is given in Fig.29, where the hybrid coupler is replaced by a second magic-T.

Relations (3.15) show the way how to compensate automatically the beam loading[27]. A possibility consists in using the signal detected on the load to drive the second phase shift Ψ_2 , while Ψ_1 is automatically adjusted so that the second relation of (3.16) is satisfied.

3.3. Oscillation and parasitic modes damping

Beam loss due to self-deflection, known as beam break-up, was evidenced at the early operation of the Stanford linear accelerator[28] . The HEM_{11}-mode is excited by an off-axis beam or simply by noise and amplified in the RF-structure.The particle beam is then deflected and, by a cumulative effect along a multi-section linac, is intercepted at last by the wall of the RF-structure. The amplification of the HEM_{11} signal is due to phase slip between the particle beam and a RF-space harmonic, following a mechanism similar to that of the band-edge type oscillations. As in the latter case, mode separation at the point the $v = c$ line cuts the HEM_{11} dispersion curve and the position of this point with respect to the π-mode, play an important role in the current limit. Beam break-up has become a major concern in the design of machines subsequent to the Stanford linac.

If beam current is high enough, an oscillation of regenerative type could take place in a single structure. Such oscillations could be of the band-edge type, backward wave type, at frequencies belonging either the HEM_{11} dispersion curve or to the accelerating TM_{01} dispersion curve. They are more frequent in TWT but could appear in circular machine or in recirculating linacs.

In certain very open structure with very wide bandwidth, such as DAW , the HEM11 passband may come to overlap the TM_{01} passband, Fig.31[29]. The deflecting modes could be excited by the power source and compete directly with the accelerating mode.

Our purpose in the following is to review the oscillation damping and parasitic modes damping techniques in use in microwave tubes or accelerating structure.

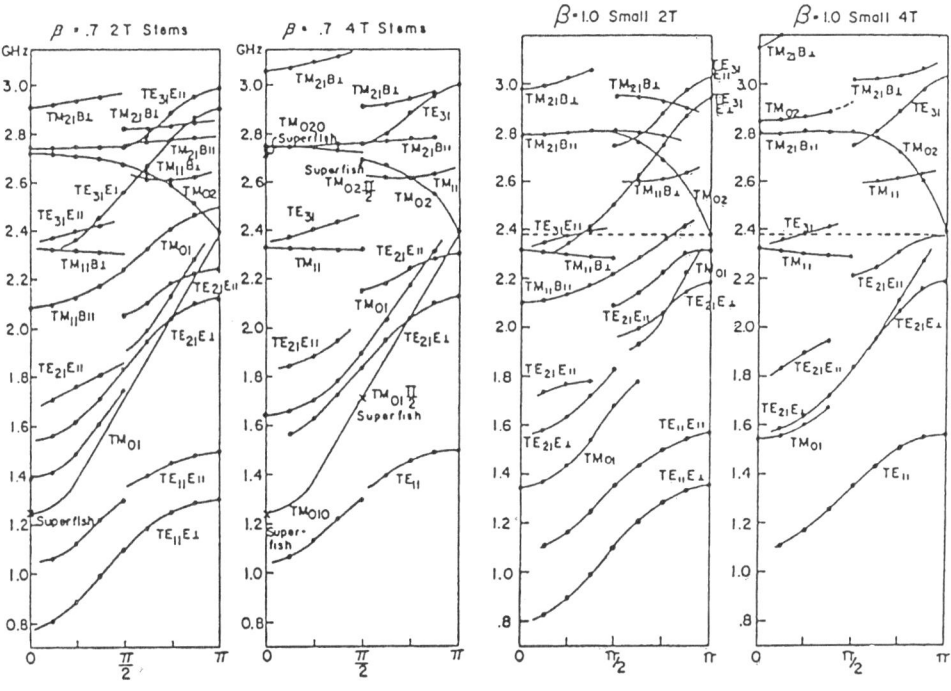

Mode spectra for ß = 0.7 PIGMI test cavity.

Fig. 31 - Accelerating modes and parasitic modes in Disk and Washer structure. (From ref.(29)).

3.3.1. Oscillation damping

As we have seen, oscillation damping is based on two principles :

(a) To have the π-mode or at least the low mode separation part of the dispersion curve in the damping zone ($v > u_0 = c$), Fig.32 (a), or if it is not possible, to have the intersection of the c-line and the dispersion curve at a high mode separation part. Fig.32 (b).
(b) To have small N or use short structure.

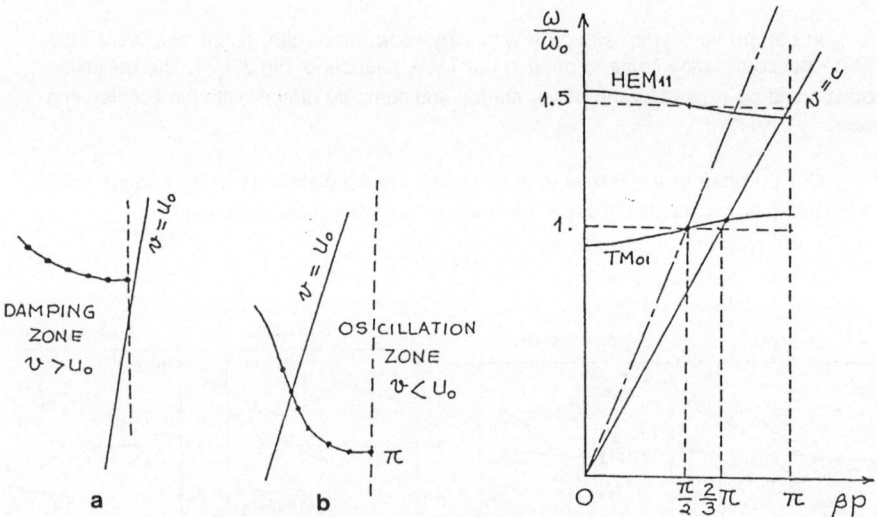

Fig. 32 - (a) π-mode and low mode separation part of the dispersion curve in the damping zone $v > u_0$; (b) $v = u_0$ line in high mode separation zone.

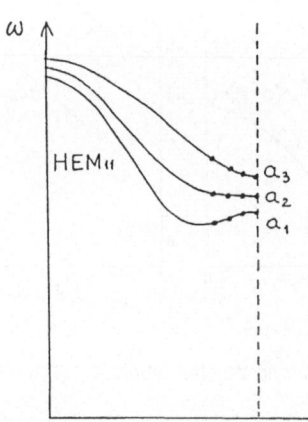

Fig.33 - Sensitivity of mode separation to bore hole radius a : $a_1 > a_2 > a_3$.

It is unfortunate that the choice of the $2\pi/3$-mode as accelerating mode, a compromise between high efficiency and high mode separation, places the intersection of the c-line and the HEM$_{11}$ dispersion curve very close to the π-mode, the ratio of the central frequencies of the deflecting and accelerating modes being close to 2/3, Fig.32 (c). Therefore the only way of improvement is to increase the mode separation in the vicinity of π-mode, which is , as shown in Fig.33, very sensitive to the iris diameter.

If the iris diameter is large, the mode separation is low. As the iris is largest at the beginning of the structure, this part must be designed carefully . From this point of view, the choice of $\pi/2$-mode would be better, Fig 32 (c).

The most commonly used linac structures are of constant gradient type. This is obtained by tappering the iris aperture. This situation is benefical for oscillation damping. Seen by the HEM_{11} , a long structure behaves then as several short structures with completely different resonant frequency, thus destroying the coherence.

For coupled cavity TWT, an artifice consists in using coalescence between cavity and slot modes as it is explained in Fig.34.

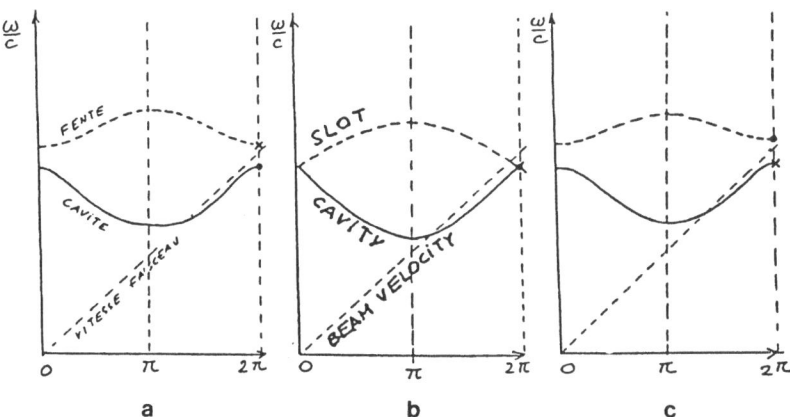

Fig. 34 - Oscillation damping with coalescence. (a) Before coalescence: favourable to oscillations; (b) Coalescence : less favourable to oscillation; (c) After coalescence : unfavourable to oscillation; cavity π-mode in damping zone.

Situation (a) (before coalescence) is less favourable to oscillation because the low mode separation zone and the π-mode are in the oscillation zone $v < u_0$. Situation (b) (at coalescence) is less favourable to oscillation, but the π-mode is still in the oscillation zone. Situation (c) (after coalescence) is the most favourable because the slot and cavity π-modes interchange. The cavity π-mode, which is the only one well coupled to the beam is now in the damping zone.

Oscillations can also be damped by introducing selective loss in the way the parasitic modes are.

3.3.2. - Parasitic modes damping

The parasitic modes can be damped inside the RF-structure or coupled outside then damped. The damping circuit could be resonant or purely resistive. Some typical devices are shown in Figs.35 to 39.

Band edge oscillation damping in helix (Figs. 35 (a) and (b))

The damping circuits are inside the structure. They are made of lossy resonant circuits (1) printed on dielectric supports or lossy resonant helix (2) based onto the envelope.

Coupled cavity (Figs.36 (a) and (b))

The parasitic mode are damped in lossy resonant dielectric buttons embedded in the cavity walls (1) or in side-cavities loaded with lossy dielectric.

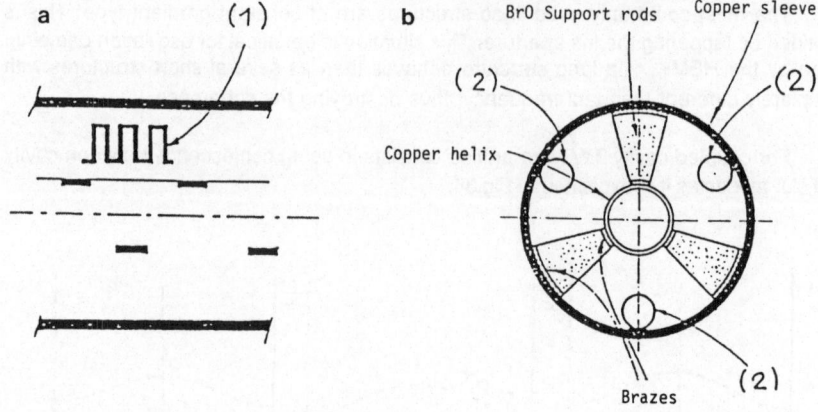

Fig. 35 - Oscillation damping with resonant lossy circuits in helix.
(a) Resonant lossy printed circuit (1); (b) Resonant lossy shorted
helix (2).

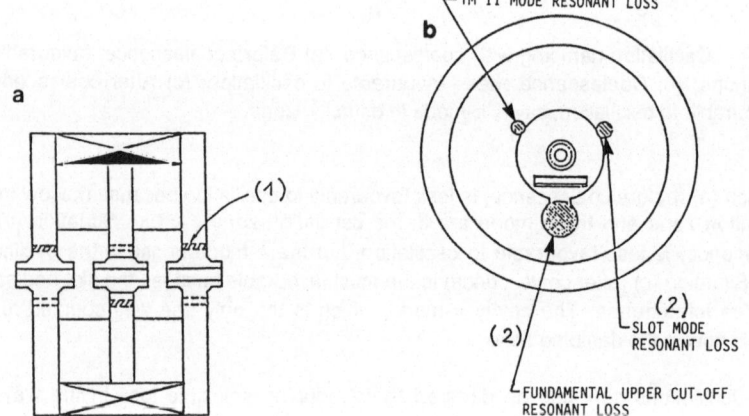

Fig. 36 - Oscillation damping with resonant lossy circuits in and outside the cavity.
(a) Resonant lossy dielectric buttons in cavity wall;
b) Resonant side-cavity filled with lossy dielectric.

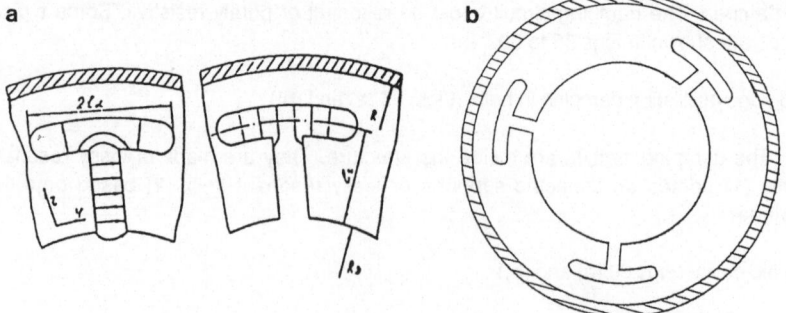

Fig. 37 - Parasitic modes damping in DAW with T-shaped circuit.

Fig.36 (b) shows three side-cavities each tuned to a different mode to be damped.

Coupled cavity or DAW (Figs.37 (a) and (b)).

The damping circuits are T-shaped slots cut in the disk wall or in the washers. The slots can resonate either in symmetric or in asymmetric mode.

Iris loaded structure (Fig.38)

The radial slots are cut at 120° in the disk in order to cut the current of the HEM_{11} mode, letting undisturbed the TM_{01} mode. To increase the effect, the disks are turned of 60° from each other.

Damping by outside circuits
(Figs.39 (a) and (b))

As the frequency of the parasitic modes are generally higher than the frequence of the accelerating mode, the unwanted modes can be coupled to overside through coupling hole at cut - off for the accelerating mode (a). To reduce the size of the coupling hole, ridged wave guide can be used (b).

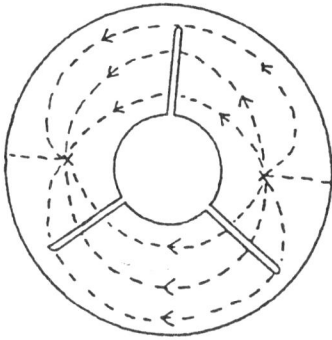

Fig.38 - HEM_{11} - mode damping in iris-loaded structure.

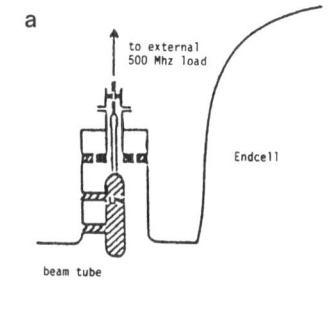

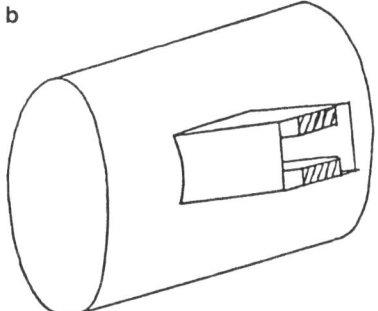

Fig. 39 - High mode coupler (a) Hera system; (b) Ridged wave guide coupler.

3.4. - RF break-down

There are several aspects in the RF break-down phenomenon.

The first question is : How does the disruptive electric field vary with frequency? The Kilkpatrick criterion gave an answer. The relation between these two factors reads :

$$f_{MHz} = 1.6 \ E^2 \ exp \ (-8.5/E) \ , \ E \ in \ MV/m$$

or more conveniently :

$$E_{MV/m} = 2.17 \, f_{MHz}^{.365}$$

in a large range of frequency. We know that this criterion is surpassed sometimes at least by a factor of two. At low frequency, for example in a 23.6 MHz resonator, Nikolaev in Soviet Union found a factor even an order of magnitude greater[30]. In this range of RF-field, Peter[31] gave another criterion:

$$f_{MHz} = (62/E) \, exp \, (17/E), \quad E \text{ in MV/m}$$

which still remain pessimistic compared to experimental data. The surface conditioning must play a great role. A coating of hydrocarbon-based should improve voltage threshold[32].

The second question is, Given a frequency; how does the threshold voltage vary with pulse length? A measurement at CERN on a deflecting structure gave a hint[33] that the law would be:

$$E \, \tau^{1/4} = const.$$

τ stands for the pulse length.

The third question is even harder to answer. In practice, a RF-structure never stand alone; there is at least a RF-window and, in the case of a microwave tube components such as electron gun, have to be included in the whole system. For the electron tube engineer, the question would be then : what is the scaling law, for a given tube, should he apply to extrapolate one operating point to another, provided the state of the art is applied to all the critical components? An attempt has been made in this sense on a high power klystron[34], the scaling law seams to be :

$$E \, \tau^{1/3} = const.$$

These kinds of criterion could be found lacking of objectivity but are useful at the end.

Another difficulty is how to translate a result obtained from a test bench to a real system. As an example it is well known that ceramic windows behave much better when tested in a resonating ring than in an actual tube, simply because the stored energy in the ring is so low that the starting spark extinguishes itself by lack of energy. In a same way, electric field threshold measured in a single cavity should not be the same as in a long structure. Beside the effect of the stored energy, the field flatness would be another factor to be taken into account.

Fig. 40 shows experimental data and proposed threshold criteria. It should be noted that the Palmer's curve refers to accelerating field E and not the surface field E_s. Provided that the ratio E_s/E is the same for all frequency, the Palmer's criterion seems to be very close to the experimental data.

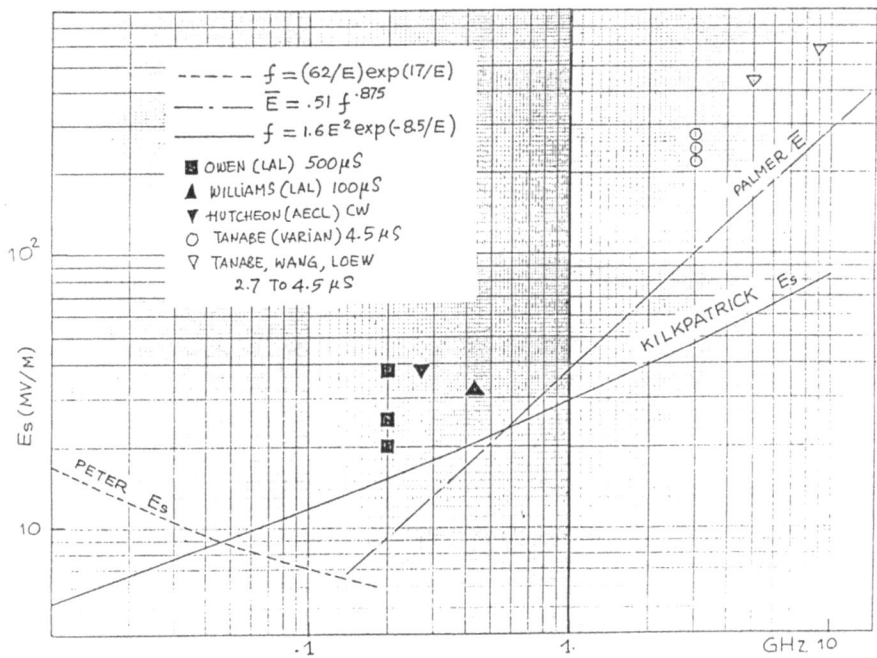

Fig. 40 - Break-down criteria.

CONCLUSION - RECIRCULATED BACKWARD SYSTEM WITH AUTOMATIC BEAM LOADING COMPENSATION

Travelling wave or standing wave?

As conclusion, let's pose this question once again. The answer is clear for slow particles such as protons or heavy ions for which low frequency is necessary. Indeed, TW-structure is to be discarded because the attenuation, which varies as $\omega^{3/2}/v_g$, is not large enough to keep the power remaining at the end of the structure at low level. For electrons, the answer has never been clear simply because each system has its advantages and inconvenients which are more or less important depending on the specific applications.

One can list the advantages and inconvenients of a TW-structure as follows:

Advantages:

1. Rather insensitive to frequency variation, TWS is well suit to multi-section linac.
2. As TWS behaves almost like a resistive load in a large frequency range, there is no need for expensive systems such as isolators or circulators to protect power tubes against reflections.
3. Filling time can be designed short so that long RF-pulse is not necessary. This allows for very high peak power capability of power tubes.

Inconvenients:

The only, but important, inconvenient of TWS is its low efficiency. There are many reasons:

1. Part of the power is wasted in the load.
2. The compromise between efficiency and high group velocity fixes the operating point far from π-mode which is the most efficient.
3. If one keeps on mind the iris-loaded waveguide which is of the forward wave type, there is a conflict between the need for group velocity suitable for large bore hole, and high efficiency, which asks for the power circulation, which asks for the contrary in order to better concentrate accelerating field on beam path.

For standing wave structures, of which the side coupled cavity would be a good representative, one can say that the advantages and inconvenients stated above exactly reverse.

The ideal is to imagine a TWS having the advantages of a SWS, i.e. an equivalent efficiency. A solution consists in separating the functions of power circulation and beam passage. This is performed by using cells with the same Ω-form as in a SWS. Beam hole is made just suitable for beam clearance and drift-tube may be used to increase the transit factor, while coupling holes are located off-axis, in the lateral wall of the cell. One obtains a backward wave system proposed recently [35].

In order to reuse the remaining power at the end of the structure, this latter is recirculated, making use of one of the recirculating system described above. Fig. 41 shows a possible lay-out. It should be noted that this system remains always optimized in spite of varying beam loading.

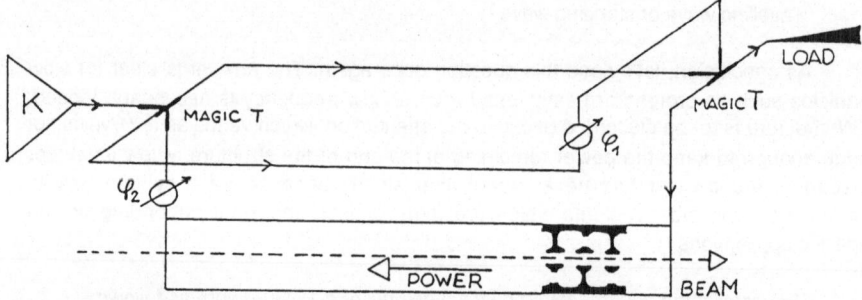

Fig. 41-Recirculated backward way system with beam loading compensation.

As conclusion, one could say that the backward wave system with automatic beam loading compensation, frees the designer from the embarassing choice of the structure length, which can be as short as allowable by the breakdown limit.

REFERENCES

1. D.T.Tran, Etudes des structures multiperiodiques au moyen des circuits à éléments localisés, Nucl. Instr. and Meth. 98:285 (1972).
2. M.Bres, A.Chabert, F.Foret, D.T.Tran and G.Voisin, The interdigital H-type structure, an accelerating structure for low energy beams, Particle Accelerators 2:17 (1971).
3. D.T.Tran, Bandwidth broadening in RF-structure, IEEE Trans. Nucl. Sci NS-32:2948 (1985).
4. J.Manca, E.A.Knapp and D.A.Swenson, High energy accelerating structures for high gradient proton linac application, IEEE Trans. Nucl. Sci. NS-24:1090 (1977).
5. T.M.Edwards, Proton linear accelerator cavity calculations, MURA reprt 622 (1961).
6. H.C.Hoyt, Designing resonant cavities with the LALA computer program, Poceedings 1966 Linear Accelerator Conf., Los Alamos, USA, Report LA 3609 (1966).
7. K.Halbach and R.F.Holsinger, SUPERFISH - A computer program for evaluation of RF cavities with cylindrical symmetry, Particle Accelerators 7:213 (1976).
8. R.L.Gluckstern,R.F.Holinger, K.Halbach and G.N.Menero, Ultrafish, Generalization of Superfish to m≥1, Poceedings of the 1981 Linear Accelerator Conf., Los Alamos, USA, Report LA 9234C (1981).
9. T.Weiland, Transverse beam cavity interaction, Part II: Long range forces, DESY report 83-005(1983); and Nucl.Instr. and Meth.216:329 (1983).
10. T.Weiland, On the unique numerical solution of Maxwellian eigevalue problems in three dimensions, Particle Accelerators 17:227 (1985).
11. T.Weiland, Computer modelling of two- and three-dimensional cavities, IEEE Trans. Nucl. Sci NS-32:2738 (1985).
12. R.Klatt et al., MAFIA-A three-dimensional electromagnetic CAD system for magnets, RF structure and transient wake-field calculations,Poceedings of the 1986 Linear Accelerator Conf., Stanford, USA, SLAC-Report 303 (1986), p. 276.
13. M.Hara, T.Wada, A.Toyama and F.Kikuki, Calculation of RF electromagnetic field by finite element method, Sci. Papers of the Inst. Phys. and Chem. Res. Japan 75:Nº 4 (1981).
14. M.Hara, T.Wada and T.Fukasawa,Three-dimensional analysis of RF electromagnetic field by finite element method, IEEE Trans. Nucl. Sci.NS-30:3639 (1983).
15. D.T.Tran, Acoupled eigenmode theory applied to RF structure, Proc. 1983 Particle Accelerator Conf.; IEEE Trans. Nucl. Sci. NS-30:3636 (1983).
16. J.M.Talbot and D.T.Tran, Three dimesional calculation on RF-cavities by finite element method, To be published.
17. P.Silvester, A general high-order finite element waveguide analysis program, IEEE Trans. Microwave Theory Thech. MTT-17:204 (1969).
18. A.Konrad, Linear accelerator cavity field calculation by finite element method, IEEE Trans. Nucl. Sci. NS-20:802 (1973).
19. J.B.Davies, F.A.Fernandez and G.Y.Philippou, Finite element analysis of all modes in cavities with circular symmetry, IEEE Trans. Microwave Theory Thech. MTT-30:1975 (1982)
20. A.Konrad, On the reduction of the number of spurious modes in the vectorial finite lement solution of the three-dimensional cavities and waveguides,IEEE Trans. Microwave Theory Thech. MTT-34:224 (1986).

21. J.P.Webb, G.L.Maile and R.L.Ferrari, Finite element solution of three-dimensional electromagnetic problems, Proc. Inst. Elec. Eng. 130:N°2 (1983).

22. B.M.Azizur Rahman and J.Brian Davies, Penalty function improvement of waveguide solution by finite elements, IEEE Trans. Microwave Theory Thech. MTT-32:922 (1984).

23. T.Weiland, RF cavity design and codes, Proceedings 1986 Linear Accelerator Conf., Stanford, USA; SLAC-Report 303 (1986), p. 243; and DESY Report M-86-07 (1986).

24. T.Wessel-Berg, A general theory of klystrons with arbitrary, extended interaction fields, Microwave Lab., Stanford Univ., Report M.L. N° 376 (1957).

25. H.Golde, Extended interaction klystrons with travelling-wave cavities, Microwave Lab., Stanford Univ., Report M.L. N° 582 (1959).

26. C.Perraudin, French patent N° 8600323 (1986).

27. D.T.Tran, French patent N° 8707416 (1987).

28. R.B.Neal and W.K.H.Panofsky, Science 152:1353 (1966).

29. Y.Iwashita, Disk-and-washer linac structure with biperiodic T-supports, IEEE Trans. Nucl. Sci. NS-30:3542 (1983).

30. Y.Nicolaev, Soviet Phys. - Tech. Phys. 8:354 (1963).

31. W.Peter et al., Criterium for vacuum breakdown in RF-cavity, IEEE Trans. Nucl. Sci. NS-30:3454 (1983).

32. W.Peter, Vacuum breakdown and surface coating of RF-cavities, J. Appl. Phys. 56:N°5 (1984).

33. Ph.Bernard et al., Experimental breakdown study of CERN deflecting structure, Report Cern/DPh II/SEP 67-6 (1967).

34. G.Faillon and D.T.Tran, RF-sources for proton linac, Proc. 1986 Linear Accelerator Conf., Stanford, USA, SLAC-Report 303 (1986),p.122.

35. D.Tronc, Electron linac optimization for short RF and beam pulse lengths, IEEE Trans. Nucl. Sci. NS-32:3243 (1985).

SUPERCONDUCTING RADIO FREQUENCY CAVITIES FOR ACCELERATORS

H. Lengeler

CERN, Geneva, Switzerland

1. INTRODUCTION

In the beginning of the 70's the first operation of the Stanford recirculating linac inaugurated the area of superconducting (s.c.) linear accelerators.

Since this time s.c. cavities have already found a number of applications for acceleration of electrons and heavy ions as well as for electron storage rings and for free electrons lasers. As the field has been reviewed recently in several papers [1-4] we will mention here only two possible future applications of s.c. cavities for accelerators.

In Italy [5] plans are advanced for a large s.c. accelerator complex for molecular, nuclear and particle physics. This complex would make extensive use of s.c. accelerator cavities for electrons and one considers:

- A small, ~ 50 MeV accelerator for a free electron laser in the infrared and millimeter range.

- A CW recirculating linac with an energy up to a few GeV.

- A $e^{\pm}$ collider for "Charm" and "Tau" physics operating in an energy range of 1.5–2.2 GeV.

- A "Beauty" factory i.e. a $e^{\pm}$ collider with an energy of ~ 10–20 GeV.

- A further extension to an energy range of 50 GeV would be designed as a Z^{o} and Toponium factory. This machine could also be considered as an injector for an even larger collider as e.g. CLIC.

The proposed operating frequency of this complex is 500 MHz.

The possible application of s.c. cavities in large linear colliders
has been studied [6-7]. These machines will ask in nearly all aspects for
a tremendous extrapolation of present day accelerator technology. Cavity
performances will have to be brought up from the present ~ 10 MV/m level in
multi-cell cavities to at least 25 MV/m if not 50 MV/m. The corresponding
Q-values will have to reach $5 \cdot 10^{10}$ or even 10^{11} in order to keep the cost
and power of cryogenic systems to a tolerable level. The requirements for
economic fabrication methods of cavities, cryostats and all auxiliary items
will ask for mass-production techniques not yet applied to accelerator
construction.

The use of s.c. cavities as "drive" linacs [8-10] for very large
colliders is also considered. The performances of low frequency cavities
(350-500 MHz) already under construction for storage rings would already be
interesting for this kind of application.

In the following a few general properties of s.c. acceleration
cavities will be reviewed and a comparison with n.c. cavities for various
applications will be made.

2. ADVANTAGES OF SC ACCELERATORS

At high frequencies and for temperatures below the critical temperature
T_c the r.f. resistance of a superconductor decreases exponentially with
temperature and its value can be made typically 10^4 to 10^6 times smaller
than for copper at room temperature [3]. The corresponding decrease of r.f.
losses in s.c. cavities has attracted accelerator constructors because much
higher acceleration efficiencies and higher CW accelerating fields than in
Cu cavities can be reached.

Due to the Meissner effect the penetration depth of r.f. fields in a
superconductor is much smaller than the normal skin depth and ranges in the
region of 50-200 nm. r.f. superconductivity is therefore a <u>surface effect</u>.
One may characterise r.f. losses of a superconductor by the surface resistance R_s (in Ohm). For cavities, losses are generally expressed by the so-
called quality factor Q. For a given cavity geometry and r.f. mode, quality
factor and surface resistance are related by the relation

$$Q = G/R_s \, ,$$
(1)

where G is a constant. A typical value for electron acceleration cavities
is G = 280 Ohm.

2.1 Accelerating fields for CW operation

The low r.f. losses eliminate to a large extent the field limitation
due to wall heating encountered in n.c. cavities for CW operation or for
long pulse durations ($\tau_p \gg \mu s$)[(*)]. For practical and economic reasons, CW
accelerating fields rarely exceed 1-2 MV/m. Typical r.f. losses are limited
to a range of 5-10 W/cm^2 and can be handled only by a powerful cooling
system. As an example we quote the r.f. losses in the 350 MHz, 5-cell
Cu-cavities for LEP [12]. At an accelerating field of 1.5 MV/m r.f. losses
would amount to 86.5 kW.

In s.c. cavities the theoretical field limit is given by the critical
(superheating) field H_{sh} which is ~ 2400 Oe for Nb and 4000 Oe for Nb_3Sn
[3]. In a typical electron acceleration cavity this would correspond to
acceleration fields of 50 and 88 MV/m respectively. Despite the much
reduced r.f. losses one has already to choose wall material of high thermal
conductivity ($\lambda \gg 10$ W/m $\cdot$ K) and of sufficiently low residual resistance
R_{res} to sustain those maximum fields in a He bath cooled cavity.

It is believed today that no fundamental effect is limiting fields
below H_{sh} but in practice and in particular for large multicell cavities
typical fields range considerably below the theoretical values. This is due
to localised regions with increased r.f. losses or to electron emittors [3].
In large multicell cavities for electrons, acceleration fields between 5 and
10 MV/m can be obtained with reasonable reliability. Progress will be bound
almost certainly to improvements in the technology of cavity fabrication,
surface treatments and assembly methods.

2.2 Acceleration efficiency

Besides higher CW accelerating fields the reduced losses of s.c.
cavities lead also to an increased acceleration efficiency which is particularly important for large systems. It is worthwhile to consider the
dependence of this efficiency in some parameters like acceleration field

---

(*) Under pulsed operation condition with pulse durations $\tau_p \leq 5$ μs
very high fields are achieved in Cu cavities. At SLAC acceleration
fields up to 130 MV/m have been reached in 3 GHz test cavities [11].

E_{acc}, beam current i_b and cryogenic conditions.

We define the acceleration efficiency η_{acc} by

$$\eta_{acc} = \frac{P_b}{P_{mains}} = \frac{P_b}{(P_b + P_c)\,\frac{1}{\eta_{r.f.}}} , \qquad (2)$$

- P_b: power given to the beam,

- P_{mains}: mains power needed for the production of P_b and P_c,

- P_c: r.f. losses at cavity walls,

- $\eta_{r.f.}$: overall efficiency of r.f. production. It depends very much on the type and output power of the r.f. generator. For large CW klystrons (350 MHz–1 GHz) one reaches an efficiency of 70%; the overall efficiency is reduced by transmission losses and by higher order mode losses in the accelerator to ~ 60%; for pulsed high power klystrons as used e.g. in n.c. electron linacs we assume $\eta_{r.f.} = 50\%$.

The r.f. power given to the beam is:

$$P_b = i_b \cdot E_{acc} \cdot \sin\phi_s , \qquad (3)$$

- i_b: mean beam current,

- ϕ_s: synchronous phase angle; a typical value for electron linacs and storage rings is $\phi_s = 0.8$–0.9

The r.f. losses per unit length of a cavity are given by

$$P_c = E_{acc}^2 / r , \qquad (4a)$$

- r: shunt impedance (linac definition!).

For a comparison with n.c. cavities it is convenient to write

$$P_c = \frac{E_{acc}^2}{(r/Q) \cdot Q_o} , \qquad (4b)$$

- Q_o: unloaded quality factor of cavity, dependent on operating temperature, wall material and possibly field level.

- r/Q: shunt impedance/quality factor[(*)]. This parameter does depend on the cavity geometry and the r.f. mode but not on the r.f. losses, it is therefore a figure of merit of a given geometry and allows a more meaningful comparison between s.c. and n.c. cavities.

For n.c. cavities one gets from (2), (3) and (4b)

$$\eta_{acc} = \frac{i_b E_{acc} \cdot \sin\phi_s}{\left[i_b E_{acc} \cdot \sin\phi_s + \frac{E_{acc}^2}{(r/Q) \cdot Q_o} \right] \frac{1}{\eta_{r.f.}}} \qquad (5)$$

For a given cavity with fixed r/Q and Q_o the efficiency depends on the gradient and on the beam current.

For illustration we give the efficiencies of the n.c. and s.c. acceleration systems for LEP in table 2 of sect. 4. Part of the low efficiency of the n.c. cavities is due to the fact, that the LEP design current is comparatively small ($i_b \leq 2 \cdot 3$ mA). Higher efficiencies are possible at TRISTAN (KEK) or HERA (DESY) where beam currents of 30 mA and 60 mA respectively are considered.

For s.c. cavities, $P_c \ll P_b$ and P_c can be neglected in the r.f. power balance. However, it cannot be neglected for the electric power production because it is dissipated at low temperatures. In addition the static cryogenic losses P_s of cryostats and He transfer lines have to be taken into account. One gets for s.c. cavities with (2), (3) and (4b)

$$\eta_{acc} = \frac{i_b E_{acc} \cdot \sin\phi_s}{\frac{i_b E_{acc} \cdot \sin\phi_s}{\eta_{r.f.}} + \left[\frac{E_{acc}^2}{(r/Q) \cdot Q_o)} + P_s \right] \frac{1}{\eta_{cry}}} \cdot \qquad (6)$$

This efficiency depends in addition on the cryogenic losses and on the (total) efficiency η_{cry} for evacuating these losses at a given operating temperature.

From a survey of η_{cry} for various large scale cryogenic installations the following typical values were derived [13]

(*) r/Q can be obtained e.g. by a computer calculation or by a perturbation measurement at room temperature. In this case the value of Q at room temperature has to be taken.

$$\eta_{cry} = \begin{cases} 0.33\% \text{ at } 4.2 \text{ K} \\ 0.1\% \text{ at } 2\text{K} \end{cases}$$

For s.c. cavities the r.f. losses are generally the dominant contributions to the losses dissipated at He temperature. One tries to keep the static losses P_s of cryostats and He transfer lines well below 10-20% of the r.f. losses.

For "pulsed" bunch operation encountered e.g. in e^- storage rings and FEL's or anticipated in linear colliders with typical bunch intervals of 10-1000 μs, the difference in efficiency can be illustrated in another way.

The Q-value of a cavity is related to the decay constant τ_d of r.f. fields by

$$\tau_d = \frac{Q_o}{2\omega_o} . \tag{7}$$

For n.c. cavities the decay constants are typically of the order of μs and do not allow to store r.f. energy inbetween bunch passages. The bunch can only remove a few percent of the stored energy in order to keep the energy spread between head and tail small; the remaining stored energy is lost leading to a much reduced acceleration efficiency. In addition peak powers to be delivered by the r.f. generator have to match the peak bunch current [8].

For s.c. cavities τ can easily reach values of many ms and the r.f. power decay remains negligable inbetween bunches. The r.f. (peak) power taken away by the bunches is supplied by the stored energy which is restored under CW conditions inbetween bunch passages. Therefore the r.f. generator has only to deliver a CW r.f. power corresponding to the mean beam current.

2.3 Shunt impedances

Another advantage of s.c. cavities is linked to the fact that the shunt impedances for the fundamental (accelerating) mode and for Higher Order Modes (hom) can be made widely different.

2.3.1 Fundamental mode

As explained before, the low r.f. losses (or high Q-values) of s.c. cavities lead to extremely large shunt impedances

$$r = (r/Q) \, Q_o .$$

One should note that the high value of r is essentially given by the value of Q_o and not so much by the geometry dependent parameter r/Q which can be varied for several reasons by hardly more than a factor 2 ÷ 3. The large value of r is of course responsible for the high accelerating efficiency of s.c. cavities.

2.3.2 Higher order modes

The advantage of storing in a nearly lossless way r.f. power in a s.c. cavities turns out to be a disadvantage for higher order modes.

Particle beams have an r.f. spectrum which can excite higher order modes in cavities. Bunched particle beams with a bunch spacing τ_b and a bunch length σ_ℓ correspond to a comb like frequency spectrum with a line distance $1/\tau_b$ and which extends up to frequencies of at least c/σ_ℓ. The excited r.f. fields increase r.f. losses to be dissipated at He temperatures, modify electric and magnetic peak fields and affect the beam stability.

The basic features of hom excitation can be best visualized in the "time domain" and are largely influenced by the relative values of decay constants for r.f. fields inside the cavity ($\tau_d = Q_o/2\omega$) and the bunch intervals τ_b (we assume that the bunch length does not exceed the filling time of the cavity).

A "point" bunch with charge q crossing a cavity excites a voltage given by

$$\Delta V_b = \frac{\omega}{2} \frac{r}{Q} \cdot q = 2 kq , \qquad (8)$$

ΔV_b has a phase such to maximally oppose the motion of the inducing bunch. It is specific for each excited resonant cavity mode with frequency ω and normalised shunt impedance r/Q and it is proportional to the loss parameter $k = \frac{\omega}{4} \frac{r}{Q}$.

For a single bunch passage, ΔV_b therefore depends on ω and, via r/Q, on the geometry of the cavity bur not on the r.f. losses or Q-factor.

For a train of equally spaced bunches the total voltage induced will depend on the fact whether the induced r.f. fields decay or not to a negligible value inbetween bunch passages.

For $\tau_d \ll \tau_b$ fields do not superpose significantly and the amount of

231

r.f. power deposited inside the cavity and in a specific mode is, with $q = i_o \tau_b$

$$P_b = \frac{q}{\tau_b} \Delta V_b = i_o^2 \tau_b \frac{\omega}{4} \frac{r}{Q} . \tag{9}$$

This power does not depend on the Q-factor and is the minimum power deposited by a bunch train (τ_b, q) in a cavity of a given geometry and a given mode. It can be reduced by lowering r/Q (e.g. by a larger iris opening).

Typical beam instabilities produced by this type of cavity excitation are the <u>single bunch</u>, <u>single passage</u> instabilities (short range wakes) where the fields induced by the head of a bunch affect the tail of the same bunch but not the following bunches.

For $\tau_d \geqq \tau_b$ fields can be built up by superposition from subsequent bunch passages.

The final voltage in the limit $t \to \infty$ is obtained by summing up the fields induced by all previous bunch passages and by taking into account the correct phase relations and the field decay. The result is (for the real part of beam loading) for a given mode with frequency ω [14]

$$V_b = \frac{\Delta V_b}{2} \cdot F(\tau) = \frac{\omega}{4} \frac{r}{Q} \cdot q \cdot F(\tau) , \tag{10}$$

with

$$F(\tau) = \frac{1 - e^{-2\tau}}{(1 - 2e^{-\tau}\cos\delta + e^{-2\tau})} , \tag{11}$$

and with

$$\tau = \frac{\tau_b}{\tau_d} = \frac{\tau_b \omega}{2Q_o} \tag{12}$$

$$\delta = (\omega - \omega_h) \tau_b, \tag{13}$$

- Q_o: quality factor of hom.

$\omega_h = h \cdot \omega_r$ (h: integer) is the frequency of a specific line of the beam spectrum i.e. a multiple of the revolution frequency ω_r (in storage rings) or the bunch repetition frequency (in linacs).

One can easily show that $\delta = 0$, or $\omega = \omega_h$ corresponds to the case where the (exciting) frequency ω_h falls on the cavity resonance frequency ω (resonance case). For $\delta = 180°$, ω lies in the middle between two adjacent beam lines (anti-resonance case).

For the <u>resonance case</u> where $\delta = 0$ a maximum built-up of excited hom fields occur.

For large decay constants τ_d (or large Q_o) where $\tau \ll 1$ one gets

$$F(\tau) = \frac{1 + e^{-\tau}}{1 - e^{-\tau}} \sim \frac{2}{\tau} = \frac{4Q_o}{\omega\tau_b} \ . \tag{14}$$

Combining (10) and (14) one gets

$$V_b = \frac{q}{\tau_b} \cdot (\frac{r}{Q}) \cdot Q_o . \qquad (\tau \ll 1) \tag{15a}$$

Similarly one obtains for the power deposited by the bunch train (q, τ_b) in an excited mode

$$P_b = i_o^2 \ (\frac{r}{Q}) \ Q_o . \qquad (\tau \ll 1) \tag{15b}$$

The field built-up and deposited power thus depends on $(r/Q) \cdot Q_o$. This shows the danger of hom field built-up in s.c. cavities where Q_o is large.

It is the domain of another class of beam instabilities based on the fact that the cavity develops a "memory" for preceeding bunches or preceeding passages of the same bunch (long range wake). Typical examples are coupled bunch instabilities in storage rings or cumulative break-up in (recirculating) linacs.

Fortunately there exist a well proven method to reduce the value of the decay constant for hom in s.c. cavities; this is by loading a specific hom by dedicated couplers (hom couplers).

In analogy to the (unloaded) quality factor of a cavity

$$Q_o = \frac{\omega W_{st}}{P_c} , \tag{16}$$

- P_c: r.f. losses inside the cavity,

- W_{st}:   stored r.f. energy,

we define for each coupler and each hom

$$Q_{ext} = \frac{\omega W}{P_{ext}} , \tag{17}$$

233

where P_{ext} is the r.f. power of the hom removed by the coupler from the cavity.

The total (loaded) Q_L of the cavity for this mode becomes

$$Q_L = \frac{Q_o\, Q_{ext}}{Q_o + Q_{ext}} \qquad (18)$$

if $Q_o \gg Q_{ext}$ one has simply $Q_L = Q_{ext}$ and $\tau = \dfrac{\tau_b \cdot \omega}{2Q_{ext}}$.

In fig. 1 the resonance factor $F(\tau)$ is shown as a function of Q_{ext} and for three different δ. For low Q_{ext}, $F(\tau)$ tends to a constant value where the excitation is independent of Q_{ext}. For high Q_{ext} the resonance (and antiresonance) effect can be seen.

The development of adequate hom couplers has been pushed to a high degree of perfection and at present hom coupler designs exist for many types of accelerators and operation conditions [15]. Q_{ext} well below the ones

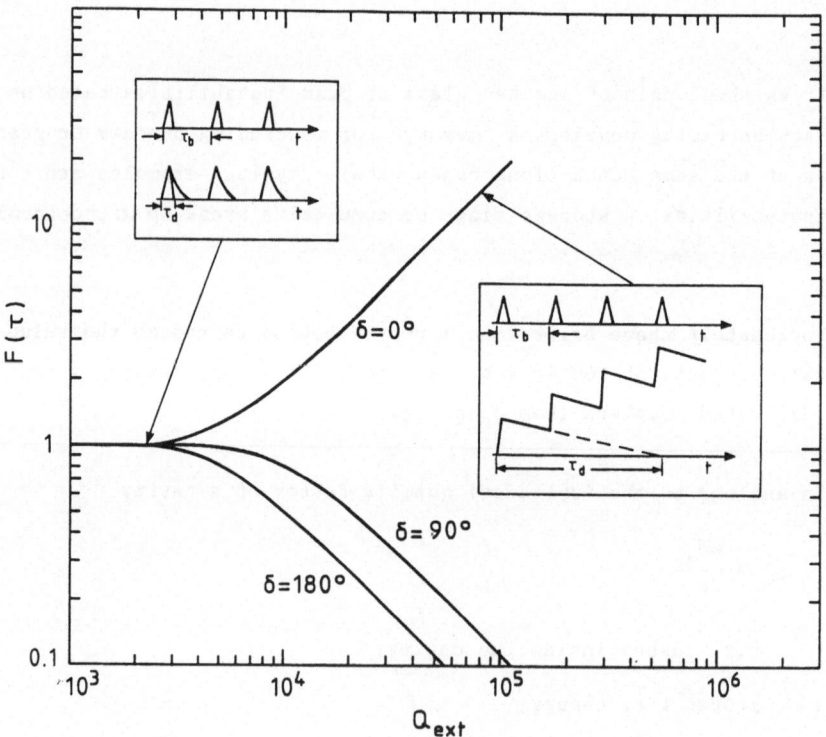

Fig. 1. Resonance factor $F(\tau)$ as a function of $Q_{ext} = \dfrac{\tau_b \cdot \omega}{2\tau}$ ($\tau_b = 22\ \mu s$, $\omega = 2\pi \cdot 6.4 \cdot 10^8$ Hz, $Q_o = 3 \cdot 10^9$). The inserts show qualitatively the behaviour of r.f. power excited by a bunch train of distance τ_b and in a given mode.

naturally existing in Cu cavities and due to the high r.f. losses
($Q_{ext} \sim 10^4$) can nowadays be achieved. In same time the loading of the
fundamental mode is kept to a negligable level by using adequate filtering
methods. Therefore the use of hom couplers allows to combine the advantages
of low losses for the fundamental mode and of strong attenuation for the
higher order modes.

Hom couplers reaching Q_{ext} well below 10^4 for the most dangerous modes
are rather sophisticated and complex devices because stringent and sometimes
contradictory requirements of r.f., tuning facilities and cooling by liquid
He have to be fulfilled. This task can be alleviated somewhat by choosing
adequate cavity geometries.

As the basic beam excitation by a single bunch is proportional to r/Q
one tries to reduce this quantity. The simplest and most usual way (in disk
loaded waveguides) is a large iris opening 2a. r/Q scales approximatly with
a^{-2} for longitudinal modes and with a^{-3} for transverse (dipole) modes. The
corresponding inevitable decrease of r/Q for the fundamental mode can
generally be tolerated and can often be compensated by a slightly enlarged
Q_o and a smaller E_p/E_{acc}. It should be stressed that this solution is
generally not acceptable for n.c. cavities because a large r/Q for the
fundamental mode is essential for an efficient use of r.f.

For high power FEL where very high bunch charges and repetition
frequencies are needed one decreases Q_{ext} by reducing the number of cells
to 1 or 2 [16]. Each cell can be equipped by hom couplers located at the
beam tubes. In this way, Q_{ext} below 10^3 can be reached [15].

3. FREQUENCY CHOICE FOR SC CAVITIES

For n.c. linacs with high gradients as needed e.g. in linear colliders
one tends to keep the mean and peak r.f. power low and to achieve a high
efficiency η for power transfer to the beam (cf below).

The scaling of these quantities with frequency is [8]

$$P \sim \omega^{-2}, \qquad \hat{P} \sim \omega^{-1/2} \qquad \text{and} \qquad \eta \sim \omega^2 . \tag{19}$$

In addition the maximum gradient which can be achieved in Cu cavities
for short pulses ($\tau \sim \mu s$) is found empirically to scale with $\sim \omega^{1/2}$ [17].

These are decisive arguments for a high operation frequency. They have however to be balanced against wake field effects which increase with frequency.

The situation is different for s.c. cavities where many arguments favour low frequencies.

A first argument for low frequencies is related to the frequency and temperature dependence of r.f. losses [3]. With

$$R_{BCS} \sim \frac{\omega^2}{T} \exp \{- \alpha T_c/T\} ,$$ (20)

and T_c: critical temperature, $\alpha = 1.76$, $Q_o = G/R_{BCS}$, $r/Q \sim \omega$, one gets

$$P_c = \frac{E^2_{acc}}{(\frac{r}{Q}) Q_o} \sim \omega \exp \{- \alpha T_c/T\} .$$ (21)

Cavity losses thus increase with frequency.

The temperature dependence tells us for a given superconductor and a given operating temperature up to which frequency we can work in order to keep r.f. losses below a given level. One tends of course to work at 4.2 K where relatively simple and much more economic refrigeration systems can be used (sect. 2.2). In Nb cavities where Q-values of a few 10^9 are needed one can work at 4.2 K up to frequencies of $\sim$ 700 MHz; for Nb_3Sn cavities frequencies well above 3 GHz still could be used [3].

A second argument is linked to longitudinal and transverse beam instabilities. Longitudinal and transverse r/Q scale like ω and ω^2 respectively for a given bunch length. This argument is of particular importance in large storage rings, FEL and linear colliders where large bunch charges are needed.

A third argument for low frequencies concerns the amount of energy removed by a bunch from the cavity. The stored energy per unit length is

$$W_{st} = \frac{E^2_{acc}}{(\frac{r}{Q} \cdot \omega)}$$ (22)

It has to be compared with the energy per unit length taken away by one bunch of charge q

$$W_b = qE_{acc}. \tag{23}$$

The ratio

$$\eta = \frac{W_b}{W_{st}} = \frac{q \ (\frac{r}{Q}) \ \omega}{E_{acc}} \tag{24}$$

should be kept small for various reasons:

the variation of acceleration field

$$\frac{\Delta E_{acc}}{E_{acc}} = \frac{1}{2} \frac{\Delta W_{st}}{W_{st}} = \frac{1}{2} \eta \ , \tag{25}$$

produced by the bunch during its passage should remain small as it leads to a dispersion in particle energy.

ΔW_{st} produces transients which change the matching of the r.f. generator to the cavity and causes additional power reflection. Another advantage of CW operation with small transients is that phase, amplitude and frequency regulation systems can be made simpler and more precise. This results also in a smaller energy dispersion of particles.

From formula (24) it appears that one should use low frequencies for keeping η low. For s.c. cavities η can be made even smaller because of the smaller r/Q (see above) and their much higher E_{acc} as compared to n.c. cavities operating under CW conditions.

A fourth argument for low frequencies are cavity, cryostat and coupler costs. These costs are strongly influenced by the number of couplers per unit length (very much the same way as the number of feedlines per unit length of s.c. magnets dominate costs).

We would like to specify this argument more in detail for electron acceleration structures. By now disk loaded cavities with rounded geometry (fig. 2) have been adopted nearly everywhere and there is a strong tendency to locate hom couplers at the beam tubes [18] and no longer at the individual cells. As a consequence the number of cells per cavity has to be kept small if efficient coupling of hom is to be achieved, and this irrespective of the frequency chosen. For high intensity and/or high peak currents this leads to typical cell numbers between 4 and 6. Low frequencies therefore keep the number of couplers per unit length low.

Fig. 2. A 350 MHz, 4 cell Nb cavity for LEP. All couplers are located at the beam tubes whose geometry has been chosen for decreasing end-cell effects of the fundamental and some higher order modes.

4. EXAMPLES FOR THE CHOICE OF SC CAVITIES

The arguments given in the previous chapter will be illustrated by a few examples where s.c. cavities are compared with n.c. cavities. It is not tried to give arguments in great detail and only the most important aspects will be discussed.

4.1 Acceleration of heavy ions [19]

The potential of s.c. cavities has already been realised at a very early stage. They are usually used as post-accelerators for Tandem electrostatic machines and have to handle only small currents (~ μA) and a comparatively small energy range (< 100 MeV). As a consequence arguments of efficiency, power consumption and higher order mode attenuation are of less concern.

Heavy ion accelerators ask for CW operation and much higher gradients as in n.c. cavities can be obtained. The much lower r.f. losses and the small currents allows one to use a r.f. power per cavity in the range of a few Watt. Therefore single cell cavities, each one equipped with a r.f. coupler become affordable. The great flexibility of a string of single cell cavities for adjusting widely different velocity profiles at different ion masses is considered today one of the greatest advantage of s.c. cavities.

There are a few specific problems connected with s.c. low velocity (low β) structures. The concentration of electric fields in the region of drift tubes leads to large E_p/E_{acc} ratios (typically ≥ 4, E_p: peak electric r.f. surface field). As a consequence field levels are nearly always limited by field emission loading originating from these regions.

A typical problem of these cavities is their low mechanical stiffness which leads to vibrations and mechanical deformations by field stresses. As the small beam loading asks for weak couplings (and small bandwidth), vibrations and the coupling between mechanical and electric forces (ponderomotive forces) can produce large and fast frequency and phase shifts. Sophisticated frequency and phase regulation systems had to be developed to master these problems.

4.2 The Continuous Electron Beam Accelerator Facility (CEBAF)

At CEBAF plans for an accelerator for nuclear physics in the energy range of a few GeV, with a current of 200 µA and a very large duty cycle have been advanced [20,1].

Original designs were based on a n.c. pulsed and recirculating linac followed by a storage ring to stretch beams to an 80% duty cycle beam. A few parameters of this version are given in table 1.

Table 1 Baseline specifications for CEBAF: pulsed n.c. linac with pulse
stretcher ring and CW s.c. linac [20]

	n.c. linac	s.c. linac
Frequency (MHz)	2856	1500
Duty factor	> 80%	100%
Emittance (πm)	$2 \cdot 10^{-7}$	$< 10^{-8}$
Energy spread ($\Delta E/E$)	$2 \cdot 10^{-3}$	$< 2 \cdot 10^{-4}$
Simultaneous CW beams at different energies	1	3
Design energy (GeV)	0.5 – 4	0.5 – 4(6)
Possible upgrade energy (GeV)	~ 6	~ 16
Current (µA)	200	200
Amount of accelerating structure (GeV)	2.0	1.0
Number of recirculations	1	3
Passes through linac	2	4

A s.c. version [20] has been finally preferred for the following main reasons (table 1): CW operation with 100% duty cycle and with an acceleration field $E_{acc} \geq 5$ MV/m was considered on overwhelming asset for high intensity nuclear physics experiments. It is combined with an improved beam quality (essentially due to the fact that bunches are distributed evenly over all r.f. cycles), savings in operation costs and the ability to deliver simultaneously CW beam at three different energies.

The choice of CORNELL type 5-cell s.c. cavities [21] with very large disk openings ($\frac{2a}{\lambda} = 0.35$) and very strong hom coupling ($Q_{ext} = 10^3 \div 10^5$) has allowed to increase the number of recirculating passes and to remain nevertheless below the most dangerous thresholds for multipass regenerative beam break up. The expected beam emittance and momentum spread are given in table 1. As a conservative gradient of 5 MV/m allows already to reach energies of 4 GeV there is room for a substantial upgrading in energy.

4.3 Superconducting cavities for LEP

LEP will be equipped initially with 128 n.c. acceleration cavities [12] but the use of s.c. cavities has always been considered for a later upgrading of energies from 55 GeV to ~ 100 GeV [22]. Therefore a comparison of both versions is of particular relevance (table 2). Similar plans exist incidently for two other large storage rings: HERA at DESY and TRISTAN at KEK [1].

For large storage rings equipped with many hundred meters of accelerating cavities, acceleration efficiencies and operating costs are of course of highest importance. For LEP the main current or luminosity limitation stems from transverse instabilities linked to the internal modes of individual bunches (short range wake fields).

These arguments have led to a frequency choice of 350 MHz for the n.c. cavities [23] where the cost and efficiency of large r.f. power sources balances favourably with cavity costs and with sufficiently low loss factors for higher order modes. For simultaneous acceleration of electrons and positrons a π-mode was chosen. An iris opening of only 100 mm ($\frac{2a}{\lambda} = 0.12$) was chosen and irises are made sligthly re-entrant for getting a high r/Q. By using slot-coupling and by choosing only 5-cells per cavity machining tolerances could be kept to an acceptable value. The economics of cooling systems limit gradients to 1.5 MV/m (CW). With this design the accelerating

efficiency was still considered too small: the large bunch spacing of LEP
(22 μs) triggered the idea of a sophisticated modulated r.f. system where
the r.f. energy is stored inbetween bunch passages in a high Q (spherical)
storage cavity. This elegant method [24] brings up the effective r/Q of the
cavities from 650 Ohm/m to 1000 Ohm/m.

The n.c. Cu cavities cause the major contribution to the total
transverse loss factor of LEP [25]: for 128 Cu cavities it is $1.1 \cdot 10^5$ V/pC
as compared to $\sim 0.4 \cdot 10^5$ V/pC for the vacuum chambers (and as compared to
$0.2 \cdot 10^5$ V/pC for 256 s.c. cavities).

For the s.c. cavities the same frequency of 350 MHz as for the n.c.
cavities has been finally adopted [22]. Once it had been shown that in
4-cell prototype Nb cavities the design gradient of 5 MV/m could be achieved
this low frequency has been favoured because very high quality factors in
excess of $3 \cdot 10^9$ can be reached already at 4.2 K.

A very advanced cavity design (fig. 2) has been worked out [18]
involving computations of higher order modes beside the fundamental modes
and for which all hom couplers are located at the beam tubes. A sufficient
(internal) coupling of hom from the middle cells to the end cells and
couplers is obtained by increasing the iris opening and by restricting the
number of cells per cavity to 4. In table 2 a few cavity parameters are
compared with the ones for the Cu cavities; the Q_{ext} compare favourably
with the ones of the n.c. cavities and are below the design values for LEP
$(5 \cdot 10^4 \div 10^5)$.

Despite the low r/Q for the fundamental mode a much higher acceleration
efficiency is obtained (table 2). A particularly striking consequence of
this is given by the fact that the total installed r.f. power of 16 MW,
which allows to reach 55 GeV with the n.c. cavities, is sufficient for an
upgrading to more than 90 GeV once (256) s.c. cavities are installed.

At a design current of 3 mA per beam and with 4 bunches per beam each
bunch in LEP has a peak current of 1330 A and a bunch charge of 67 nC. With
the formula (24) and the parameters of table 2 one calculates the percentage
of stored energy removed by one bunch. As LEP cavities are located near the
interaction regions the passages of e^+ and e^- bunches are so near in time
that the r.f. energy cannot be restored entirely inbetween. Therefore η has
to be doubled and one gets the values given in table 2 which confirm the
small amount of transients for s.c. cavities.

Table 2

A few LEP cavity parameters

	Cu cavity	s.c. cavity
Frequency f	352.209 MHz	352.209 MHz
Wavelength λ	0.851 m	0.851 m
Number of cells	5	4
Cavity active length	2.13 m	1.70 m
Iris hole diameter	100 mm	241 mm
Shunt impedance/quality factor r/Q	650 Ohm/m [a] 1000 Ohm/m [b]	276 Ohm/m
Q (Cu, 300K)	40 000	--
Q (Nb, 4.2 K)	--	$3 \cdot 10^9$
$2(f_\pi - f_o)/(f_\pi + f_o)$	1.28%	1.76%
Design accel. field E_{acc}	1.5 MV/m	5 MV/m
Accel. efficiency η_{acc} {2 · 3 mA / 2 · 6 mA}	7.5% / 13.3%	47% / 53%
Percentage of W_{st} taken by 2 bunches η_{st}	12.8%	1.6%
Total loss factor/unit length	403 $\dfrac{V}{pC \cdot m}$	46 $\dfrac{V}{pC \cdot m}$
Q_{ext}	Cu-attenuation	2 hom couplers at beam tubes
TM_{011}		$2.7 \cdot 10^4$
TM_{012}	$\sim 3 \cdot 10^4$	$7 \cdot 10^3$
TM_{110}		$4 \cdot 10^4$
TM_{111}		$1.4 \cdot 10^4$

(a) without storage cavity.
(b) with storage cavity.

5. CONCLUSIONS

The small r.f. losses of s.c. cavities allow CW operation at much higher acceleration fields than the ones in n.c. cavities. Together with the large stored energy which can be conserved with negligable losses over periods of hundreds of ms the possibility of CW operation provides great flexibility of operating conditions. In heavy ion accelerators or electron accelerators for nuclear physics particles can be distributed evenly over all r.f. periods and allow excellent detection conditions. In electron storage rings or free electron lasers large bunch charges combined with high repetition rates can be handled equally well. The use of hom couplers allows to combine small shunt impedances for the fundamental mode with Q_{ext} for higher order modes which can be made equal or even smaller than the "natural" Q_{ext} of n.c. cavities. High acceleration efficiency and large attenuation of higher order modes are therefore compatible.

For s.c. cavities many arguments favour low frequencies and large iris openings. Both choices decrease longitudinal and transverse wake fields effects. This is particularly welcome for high power, free electron lasers and for linear colliders where not only bunches with high electron populations have to be handled but where, all along the accelerator, phase space blow up and energy dispersion has to be kept small and where energy recovery is very desirable.

For linear colliders we still remain faced with the problem of achieving much higher energy gradients. As far as we can see today there exists no fundamental limit but progress will be bound to developments in the technology of s.c. cavity fabrication and surface treatments and of new s.c. materials.

REFERENCES

[1] A comprehensive review of the field will be found in Proceedings of the 3rd Workshop on RF superconductivity, ANL, Argonne (1987), Editor K. Shepard, to be published.

[2] Proceedings of the 2nd Workshop on RF superconductivity, Geneva, Switzerland (1984), Editor H. Lengeler.

[3] H. Piel, this Seminar.

[4] H. Padamsee, CLNS 87/65, Cornell University (1987) and Proceedings IEEE Particle Acc Conference, NBL (1987) to be published.

[5] See e.g. U. Amaldi, Proceedings of S. Miniato Topical Seminar on Heavy Flavours (1987) and CERN/EP 87-104 (1987).

[6] R. Sundelin, CLNS-85/709, Cornell University (1985).

[7]   U. Amaldi, H. Lengeler and H. Piel, CERN/EF 86-8 (1986).

[8]   W. Schnell, this Seminar.

[9]   W. Schnell, CERN/LEP/RF 86-06 (1986).

[10] U. Amaldi and G. Pellegrini, CLIC Note 16, CERN (1986).

[11] J.W. Wang and G.A. Loew, IEEE NS32 (1985) 2915.

[12] LEP design report, CERN/LEP 84-01 (1984).

[13] G. Passardi, private communication.

[14] P. Wilson, SLAC-PUB 2884 (1982).

[15] See e.g. G. Cavallari, E. Chiaveri, E. Haebel, P. Legendre and
      W. Weingarten, in ref. [1].

[16] H.A. Schwettman, T.I. Smith and C.E. Hess, IEEE NS32 (1985) 2927.

[17] H. Henke, CERN/LEP/RF 87-49 (1987) and Proceedings of ECFA/CAS
      Workshop, Orsay (1987).

[18] See e.g. R. Sundelin, ref. [2] p. 49;

      and E. Haebel, P. Marchand and J. Tückmantel, ibidem p. 281.

[19] See e.g. L.M. Bollinger, Ann. Rev. Nucl. Part. SC, 36 (1986) 475 and
      ref. [1].

[20] CEBAF Design Report, Newport News, Va, May 1986.

[21] R. Sundelin et al., IEEE NS32 (1985) 3570

[22] Ph. Bernard, H. Lengeler and E. Picasso, ECFA Workshop on LEP 200,
      Aachen (1986) 29.

[23] LEP design studies CERN/ISR/LEP 78-17 (1978) and 79-33 (1979).

[24] W. Schnell, CERN ISR-LTD/76-8 (1976).

[25] E. Keil, in reference [22] p. 17.

RF SYSTEM DESIGN FOR CONTROL OF HEAVY BEAM LOADING IN CIRCULAR MACHINES

G.H. Rees

Rutherford Appleton Laboratory
Chilton
Didcot
Oxon   OX11 0QX   UK

1.   INTRODUCTION

Radio frequency cavities in an accelerator or storage ring may be influenced by the loading due to the circulating beam current. When the average beam current is high, the system stability may be affected, particularly if the power imparted to the beam exceeds that dissipated in the cavities.

In this report, guidelines are given for the overall system design to compensate for the beam loading effects. The use of RF feedback and beam feed-forward techniques are considered in addition to more conventional beam control methods. The topics considered are not new but have received added attention recently in connection with the design of high current Kaon Factories.

2.   BEAM LOADING

A circulating beam excites a radio frequency field in a cavity in a manner similar to that in which an external generator excites the cavity. However, the beam may be more closely coupled to the cavity than is the generator and there is a wider spectrum of frequencies in the beam current than in the generator current.

The accelerating frequency is set at a specific harmonic of the beam revolution frequency, $h*f_{rev}$, allowing a maximum number, h, of beam bunches to be accelerated. Sometimes, two or more accelerating frequency harmonics may be used, to introduce additional non-linearities into the longitudinal focussing.

In the beam current, there are harmonics of f_{rev} present due to different numbers of particles in different bunches or due to the absence of some of the bunches and there are additional multiples of $h*f_{rev}$ present due to the bunch shape. The beam also has certain natural modes of longitudinal motion and these are characterised by a coherent synchrotron frequency, f_{s1}, and higher mode frequencies, $f_{sm} \cong m*f_{s1}$. The longitudinal motion of M bunches ($M \leqslant h$) leads to sidebands of some of the revolution frequency harmonics, $|(n \pm pM)*f_{rev} + f_{sm}|$, with $n = 1, 2 \ldots (h - 1)$ for coupled bunch motion. These frequency components may appear in the radio frequency accelerating field, directly via the beam loading or indirectly via beam control units.

The most important frequency components in the beam current and their associated beam loading effects are:

$f_{RF} = h*f_{rev}$: beam power throughout acceleration

$f_{RF} \pm f_{s1}$: uncoupled bunch, coherent dipole mode motion

$f_{RF} \pm f_{sm}$: uncoupled bunch, coherent higher mode motion

$|(n \pm pM)*f_{rev} + f_{sm}|$: coherent motion of M coupled bunches ($n \neq 0$)

When coupled bunch motion arises, it is due either to the interaction of the beam with the fundamental mode of the main cavities or with one or more parasitic higher modes. For the latter, all modes are possible in the range $n = 1, 2 \ldots (h - 1)$, while for the former, the most likely frequencies are given by:

$(h \pm 1)*f_{rev} \mp f_{sm}$; below transition energy, with $n = (h - 1)$

$(h \mp 1)*f_{rev} \mp f_{sm}$; above transition energy, with $n = 1$.

The RF cavity, amplifier and system design must ensure that, for all levels of beam loading up to the maximum design value, the beam-cavity interaction does not lead to instability of any (n, m) longitudinal mode defined by n = 0,1 ... (h - 1) and m = 1,2,3, ... The amplifiers must have the capability to provide the power dissipated in the cavities, the beam power for acceleration and the reactive power to compensate for the beam revolution frequency harmonics.

3. FEATURES OF BEAM CONTROL FOR HEAVY BEAM LOADING

Control of the beam for stable acceleration involves, for each RF cavity system, the design of a number of control loops which are coupled very strongly via the beam. The designs of the loops are linked to the basic cavity design and it is important to consider the effective impedance of the cavity at the revolution frequency harmonics near to the fundamental component in addition to that at the fundamental mode itself.

Historically, beam controlled acceleration began with four basic feedback loops for each cavity system : a cavity voltage control loop, a tuning control for the cavity resonant frequency, a beam phase control loop and a beam radial control system. In recent years, two additional controls have come into consideration : a beam feed-forward control and a short delay, RF feedback loop in the final stages of the power amplifier and cavity system. These six control loops are used to stabilise uncoupled, coherent dipole mode bunch motion. Sometimes, further loops have to be added to control coupled bunch motion, both for dipole and higher mode oscillations. Many features of such control loops are described in the report of F Pederson [1] on the TRIUMF KAON Factory RF systems.

Until the advent of beam feed-forward and RF feedback techniques, the power amplifier output stages had to be capable of providing the beam power plus at least an equal amount of power in the cavity and generator impedances to ensure the stability of the uncoupled, coherent dipole mode of oscillation (Robinson [2] - type instability). Thus for example, an instantaneous beam power of 5 MW had inferred an RF system capable of providing at least 10 MW of RF power. This restriction may now be relaxed and the ratio of beam to total cavity power, P_b/P_c, may be set at 2 or higher when using the new techniques. A ratio of 2 reduces the total RF power requirements to 7.5 MW for the example given, while a ratio of 10 requires only 5.5 MW. Such a high ratio as 10 is considered for the RF system in the Driver ring of the TRIUMF proposal.

During acceleration or storage, the beam current is not in phase with the gap voltages so that the cavities are loaded with reactive as well as resistive components of beam current. Then, for efficient operation of the power amplifiers, it is necessary to detune the cavities so as to compensate the reactive components of loading, bringing the generator currents into phase with the gap voltages. Each generator still has to supply some reactive current, however, if the beam current contains significant revolution frequency harmonics near to the fundamental. This occurs when there are some missing bunches (to accommodate kicker magnet rise and fall times, for example), and also when the beam develops some coupled bunch motion.

To obtain adequate beam control, it is necessary to use cavities with a low value for the parameter R/Q. It is also advantageous to maximise the gap voltages within the amplifiers' power capability, thus minimising the number of cavities. Both these topics will be discussed subsequently.

4. STEADY STATE FUNDAMENTAL BEAM LOADING

Basic beam and cavity parameters are defined in Appendix 1, which also includes a vector diagram of the steady state beam and generator currents in relation to an individual cavity voltage. With the aid of the diagram, it is shown that the steady state conditions are approximately:

$$I_g \cos \phi_g = I_b \sin \phi_s + (V/R)$$

$$I_g \sin \phi_g = I_b \cos \phi_s + (V/R)(2Q \, \Delta w/w_o) \qquad \ldots (1)$$

The optimum power condition for the generator is when ϕ_g is zero, which is obtained by a cavity detuning:

$$\Delta w/w_o = - (R/Q) \, I_b \cos \phi_s \, / \, 2V$$

$$\Delta f/f_{rev} = - h \, (R/Q) \, I_b \cos \phi_s \, / \, 2V \qquad \ldots (2)$$

The detuning, expressed as a fraction of the revolution frequency, may approach or exceed unity if care is not taken in the choice of all the parameters, (R/Q), h and V. If the fraction becomes unity, then one of the nearby revolution frequency harmonics coincides with the cavity resonant frequency. From stability considerations and also to reduce the demands on the automatic cavity tuning systems, it is advantageous to keep $\Delta f/f_{rev}$ less than 0.5. This limits the choice of (R/Q) and h, once V has been maximised, within the power amplifier's capabilities. Then, for example, choosing a high value for the harmonic number may lead to an impractical value for (R/Q).

The total beam loading power in the steady state is:

$$P_b \qquad = \quad I_{dc} \, V_t \, \sin \phi_s \quad = \quad I_b \, V_t \, \sin \phi_b \, / \, 2$$

For short bunches, the fundamental component of beam current is twice the dc component and ϕ_b equals ϕ_s. For long bunches, I_b is significantly less, the bunch is asymmetric and ϕ_b is greater than ϕ_s. The cavity detuning formula (2) is then incorrect to the extent that ϕ_s should be replaced by ϕ_b, with a correspondingly reduced detuning.

5.   RF FEEDBACK

By RF feedback is meant a narrow bandwidth feedback loop around the power amplifier–cavity system, centred at the cavity resonant frequency. There must be adequate phase margins at the unity gain points of the loop. In general this implies that, for a sufficient feedback gain, there must be a small signal delay, T, around the loop with the delay tracking the changing value of the frequency during the acceleration. When there is a large cavity detuning to compensate for the reactive beam loading, there has to be added correction for the loop delay.

To see the full effect of the RF feedback, it is necessary to consider:

1.   The size of the detuning for reactive beam loading compensation.
2.   The magnitudes of the open loop gain, A, and the loop delay, T.
3.   The cavity equivalent impedance at $h*f_{rev}$ and $(h \pm 1,2,..)*f_{rev}$.
4.   The modified cavity response to amplitude and phase modulations.

The required detuning is given in Appendix 1 and has already been discussed in section 4 while the magnitude of the maximum open loop feedback again, A_{max}, is derived in Appendix 2. The importance of the amplifier-cavity layout to achieve a low feedback delay, T, is evident. The derivation of A_{max} assumes a phase margin of $\pm \pi/4$ at the unity gain points and adjustment of $(f_o T)$ to equal an integer, k. Then:

$$A_{max} = Q / 4 f_o T = Q / 4 k \qquad \ldots (3)$$

The cavity equivalent impedance, in the presence of feedback, is also derived in Appendix 2:

$$Z(w) = R (1 + A e^{-j \delta w T} + j 2 Q \delta w / w_o)^{-1} \qquad \ldots (4)$$

$$\delta w(n) = w - w_o = \Delta w \pm n w_{rev}$$

At $n = 1$, that is at the nearest revolution frequency harmonic to $h*f_{rev}$, the resistive component of $Z(w)$ is approximately:

$$r(w) = R A ((1 + A)^2 + (2 Q \delta w / w_o)^2)^{-1}$$

The magnitude of $r(n = 1)$ increases as a function of the feedback gain, A, until r reaches a maximum at:

$$A \cong 2 Q \delta w(1) / w_o \qquad \text{which is:}$$

$$A = Q / h \qquad \text{at } | \Delta w / w_o | = 1 / 2 h$$

It is desirable, therefore, to have:

$$A \gg 2 Q \delta w(1) / w_o \qquad \ldots (5)$$

The modified response of the cavity to amplitude and phase modulations is obtained by solving the differential equation for the cavity voltage in terms of the excitation currents. This is given in Appendix 3 and is discussed subsequently in section 7. The feedback gain, A, is an important parameter for it modifies the form of the beam current dependence of the response and so affects the interaction between the various control loops of the system.

The complex transfer function $G(jw, ip)$, defined in Appendix 3, is:

$$G(jw, ip) = (1 + A e^{-(j \delta w + i p) T} + (2Q/w_0)(j \delta w + i p))^{-1}$$

which shows that with cavity detuning for reactive beam loading compensation, there may be a large phase shift in the response, $G(jw)$.

$$\delta w / w_0 = \Delta w / w_0 = - (R/Q) I_b \cos \phi_b / 2 V$$

This effect, in the absence of feedback, leads to the Robinson criteria, viz. instability when the beam power exceeds the cavity power. The phase shift is reduced by introducing RF feedback, especially if:

$$A \gg 2 Q (\Delta w / w_0)$$

which is the same condition as (5) for $|\Delta w / w_0| = 1 / 2 h$. Derivation of the Robinson equation for the fundamental cavity mode is given in Appendix 4, together with a modified form of the equation for the case of RF feedback. Also is included the case of coupled bunch dipole mode motion driven by the fundamental mode.

6.   BEAM FEED-FORWARD

If the beam current is sensed at a point on the machine orbit and transmitted to a cavity via a system of appropriate gain and delay, then the beam induced voltage at the cavity gap may be approximately compensated, within the system bandwidth.

A feed-forward factor, $F(w)$, may be defined:

$$F(w) = 1 - f(w) e^{j \phi_f(w)}$$

where $f(w)$ and $\phi_f(w)$ represent feed-forward gain and phase errors respectively, at the angular frequency, w. There results a reduction factor of $(1 - F(w))$ in the effective cavity longitudinal beam coupling impedance, $Z(w)$.

Operating systems have typically achieved reduction factors of 0.1 to 0.2 for $Z(w)$ at the revolution frequency harmonics near to the fundamental cavity mode, $(h \pm n)*f_{rev}$, where $n = 0,1,2 \ldots$. The systems need careful setting up, however, because of variations of amplifier gains and non-linearities in the high power stages.

The bandwidth requirements of the system are set by considerations of transient beam loading and of possible coupled bunch effects. Transient beam loading may arise during the injection of beam into a machine or due to an asymmetric arrangement of beam bunches. Feed-forward is usually more effective than RF feedback for handling the transient loading, provided there is an adequate feed-forward bandwidth. The amplifiers must be able to provide the reactive power to compensate for the transient loading.

7.   BEAM DIPOLE MODE LONGITUDINAL MOTION

Coherent dipole oscillations may be of a rigid bunch or a coupled bunch nature; the motion may be free, or driven by the main accelerating mode of the cavities, or by parasitic higher modes. The beam frequencies are those of the spectrum of the stationary bunch distribution, but with certain modulations, as may be seen in Appendix 4 in the derivation of the Robinson − type equations. For rigid bunch motion, the modulation is at the coherent synchrotron frequency, f_{s1}, whereas for coupled bunch motion, the modulation is at $f_{s1} \pm n\,f_{rev}$.

The Robinson equation describes only rigid bunch motion for no feedback, which is rarely the case in practice. In most cases, there are cavity tuning and amplitude control loops and beam radial and phase control systems. Higher order equations then result, which are further modified by the use of any RF feedback or beam feed-forward systems.

The basic interaction between the modulated beam current and the cavity is given by the complex transfer function, $G(jw, ip)$, already defined. This response is strongly influenced by the cavity tuning, which is usually adjusted for reactive beam loading compensation to optimise the generator power. An instability may result, as outlined here and in Appendix 4.

In the absence of feedback, the Robinson equation shows that instability occurs when the beam power exceeds the 'loaded' cavity power if the detuning is set to compensate for reactive beam loading. In the presence of feedback, the detuning introduces coupling between the four, basic, low frequency control loops. The system may be analysed by closing any three of the loops, in turn, and then applying the generalised form of the Nyquist criteria to the open loop transfer functions for the four cases, some of which may correspond to open loop instability. Closed loop stability may also be studied by solving for the roots of the entire closed loop equation. In general, these loops do not increase the Robinson threshold, but are required to maintain the basic acceleration.

To increase the instability threshold, it is necessary to introduce RF feedback or a beam feed-forward system. For the former, the gain has to be sufficient to over-ride the effect of the cavity detuning, as explained in the Appendices 3 and 4. For the latter, the beam feed-forward acts directly to reduce the beam loading; it is particularly effective in reducing the reactive beam loading effect at the neighbouring revolution frequency harmonics. Landau damping does not counteract this particular instability.

In the case of coupled bunch motion, dipole instabilities may be driven by fundamental or parasitic cavity modes. An equation for coupled bunch motion is derived in Appendix 4 for the case when there are no low frequency feedback loops. It has the same general form as the fourth order Robinson equation and may be extended to include the case of RF feedback. Approximate estimates for instability growth rates are also derived and these indicate enhanced growth rates for low values of RF feedback gain, as expected from the increased values of $r(w)$, noted in section 5.

Feedback control for coupled bunch motion may be applied via the main accelerating cavities or through special cavities. If a coupled bunch dipole mode is excited by a resonance near the frequency $(k h \pm n) * f_{rev}$, it is possible to apply damping via the range of frequencies near $(\ell h \pm n) * f_{rev}$, with either the + or - sign used or both. If the excitation arises from the fundamental mode ($k = 1$), the probable value of n is 1, whereas for a parasitic mode, n may have any value from 1 to ($h - 1$). Feedback via the main cavity ($\ell = 1$) is only practical if the value of n is not too large. Thus the choice of whether to use the main or special cavities for the feedback requires a knowledge of the parasitic longitudinal modes of the cavities and ring structures.

To design the feedback systems, use may be made of the formulae of Appendix 3, where transfer functions are derived for coupled bunch motion. These are given separately for upper and lower sidebands of the modulation and as a composite function for the case where both are used. For feedback via the main cavity, the transfer functions have to be modified to include the effect of any low frequency or RF feedback loops.

The feedback is complicated by the fact that the coupled bunch modes n and (h - n) appear in the beam spectrum at nearly the same frequencies. The former has frequency components at $(k\ h \pm (Q_s + n))*f_{rev}$, and the latter at $(k\ h \pm (Q_s - n))*f_{rev}$, where $Q_s *f_{rev} = f_{s1} = \Omega/2\pi$, the coherent synchrotron frequency. Applying feedback to damp the former may antidamp the latter, or vice-versa. Thus it is necessary in the feedback to transform appropriately the individual frequency components in order to damp the both modes.

It may prove possible to avoid the use of feedback and stabilise against coupled bunch instability by the use of Landau damping. At least three alternatives are available. Non-linearities may be introduced into the focussing fields, either by extending the bunch lengths or by the use of two or more harmonically related RF systems. The third possibility is to use two non-harmonically related RF systems to introduce a spread of synchrotron frequencies in the beam bunches.

8.   HIGHER MODE LONGITUDINAL BEAM MOTION

The higher modes are defined by the mode number m, with m = 2 for the quadrupole mode, m = 3 for the sextupole, m = 4 for the octupole, m = 5 for the decapole, and so on. Again, the bunch motion may be coupled, as defined by the coupled bunch number n = 0,1, ... (h - 1), with n = 0 corresponding to each bunch motion being in phase.

Excitation of the even order modes corresponds to amplitude modulation of the beam current harmonics, and that of the odd order modes, to phase modulation. Either modulation leads to both amplitude and phase modulation of the accelerating fields, but only the amplitude modulated fields can excite the even order modes and only the phase modulated fields, the odd order modes (an exception is the special case of mode m = 1). The spectrum of the possible instability excitation fields is not the same in the two cases.

A brief outline is given in Appendix 5 of the derivation of the higher mode equations of motion. These are not in agreement with other published results; the equations here are of a higher order and there is a different prediction for the frequency range of the fields that may excite the instabilities. It is of interest to note that certain beam current components may excite the modes, without leading to an instability condition.

Higher mode coupled bunch feedback systems may be designed in a manner similar to that outlined for the dipole modes. The transfer functions for beam current modulations are as in Appendix 3, but these must now be incorporated into the higher mode equations of Appendix 5. Typical modulation frequencies have the approximate values $(m\ Q_s \pm n)*f_{rev}$.

9.   DESIGN OF CAVITY – AMPLIFIER SYSTEMS FOR HIGH BEAM LOADING

The following topics are involved in the design of cavity-amplifier systems for high beam loading:

1. First there is the choice of the power source for each cavity. Factors to be taken into account are the operating frequency range and the total power requirements of the system, including that for continuous and transient beam loading. The driver stages are as important as the output amplifier in this assessment for the choice of a very high power driver may involve a large signal delay, which limits the maximum gain obtainable for any RF feedback loop. Within an amplifier's power capability, it is best to maximise the voltage at the cavity gaps, for this minimises the number of cavities and hence the total longitudinal impedance of the machine.

2. Having chosen the RF harmonic number and cavity gap voltage, use may be made of the criteria of section 4 to choose the value of R/Q for the cavity: $R/Q < V\ /\ h\ I_b\ \cos\ \phi_b$.

3. The cavity design then involves considerations of Q, R/Q,
tuning, insulators, multipactoring, spark-down, input coupling,
monitors, fabrication and the minimisation of parasitic mode
impedances.

4. Assume the use of a one-turn delay beam feed-forward and in the
design choose a bandwidth to handle the transient beam loading.

5. Assume the use of RF feedback, as outlined in section 5 and
Appendix 2, with a variable loop delay to track the changing RF
frequency:

Minimise the loop delay, T, consistent with $f_o T = n$ (=integer).
Design for a ratio of beam to cavity power, P_b/P_c, of 2 or 3.
Choose the open loop gain, A, so that $2Q \, \Delta w/w_o \ll A < Q \, / \, 4k$.

6. Next, a full scale cavity model is required to see if a
satisfactory RF feedback loop can be made to work in the presence of
simulated beam loading. If it cannot, then it is advisable to lower
the Q value of the cavity to reduce the ratio of P_b to P_c. The new
ratio depends on the characteristics of the beam loading and the
beam feed-forward system. Without a feed-forward system, the ratio
must be less than unity. If a Q-reduction is made, it may be
necessary to make a reassessment of the cavity power source.

7. The model cavity is finally used for measurements of the
longitudinal and transverse parasitic modes and development of the
systems for parasitic mode damping.

8. Having determined the cavity parameters and the parasitic modes,
estimates may be made for coupled bunch instability growth rates and
for the effects of any gaps in the beam due to missing bunches. If
coupled bunch feedback is necessary, there is the choice to be made
between feedback via the main cavities or via separate cavities.

As an example for a design, consider the following possible parameters for a Kaon Factory booster ring:

f_{rev} (MHz)	I_b (A)	P_b (MW)	V_t (MV)
$0.5 \pm 5\%$	5.0	1.8	1.0 (0.72 at $\phi_s = 0$)

Assume for the basic units, a 150 kW plate dissipation tetrode amplifier and a double gap drift tube cavity, with a gap voltage of 42 kV. Then:

Number of cavities required	12	
$h(R/Q) = V / 2 I_b \cos \phi_b$	3000 Ω	(12,000)
R/Q for 100 MHz (h = 200)	15 Ω ⟶	too low
R/Q for 50 MHz (h = 100)	30 Ω	(120)
Choose P_b / P_c	3	(3)
Cavity Power $P = P_b / 36$	50 kW	(50)
Shunt Impedance $= V^2 / 2 P$	17.4 kΩ	(70)
Cavity Q = R / (R/Q)	580	(580)
RF Feedback Delay, $T = k / f_o$	$80 \pm 5\%$ ns	(80)
$Q / h \ll A < Q / 4k$	$5.8 \ll A < 36$	
RF Feedback Factor A (k = 4)	35	(35)

The figures in brackets are for an 84 kV single gap cavity.

REFERENCES

1. F. Pederson, TRIUMF internal report, TRI – DN – 85 – 15 (1985).
2. K W Robinson, Stability of Beam in Radio Frequency System, CEAL – 1010, 27 February 1964.
3. F J Sacherer, Bunch Lengthening and Microwave Instability, CERN/PS/BR 77-6 (1977).

APPENDIX 1.   VECTOR DIAGRAM OF STEADY STATE BEAM LOADING

The beam and generator currents are assumed to drive cavities which are represented by equivalent R, L, C circuits. Parameters are defined:

I_{dc} = dc component of beam current,

I_b = fundamental component of beam current,

I_g = fundamental component of generator current,

eV = amplitude of the energy gain per cavity,

eV_t = amplitude of the energy gain per beam revolution,

P_b = beam loading power,

P_c = total power dissipation in all cavities,

P = power dissipation per cavity ($= V^2 / 2 R$),

ϕ_s = synchronous phase angle,

ϕ_b = $\pi / 2$ - (phase of I_b relative to the phase of V),

h = harmonic number of the RF system,

f_{rev} = particle revolution frequency,

f_{RF} = $h*f_{rev}$ ($= w / 2 \pi$),

f_o = cavity resonant frequency ($= w_o / 2 \pi$),

Δf = cavity detuning ($= (w - w_o) / 2 \pi = \Delta w/ 2 \pi$),

Z = cavity longitudinal coupling impedance,

R, Q = loaded values of cavity shunt impedance and Q,

L, C = equivalent cavity inductance and capacitance,

R/Q = $w_o L$ or $1 / w_o C$, a function of the cavity geometry.

The vector diagram is drawn for the condition below transition energy. Using complex number representation for sinusoidal excitations:

$$V (1 + j 2 Q \Delta w / w_o) = R (I_g e^{j \phi_g} + I_b e^{j (3 \pi/2 - \phi_b)})$$

$$V = R (I_g \cos \phi_g - I_b \sin \phi_b)$$

$$\Delta w / w_o = R (I_g \sin \phi_g - I_b \cos \phi_b) / 2 Q V$$

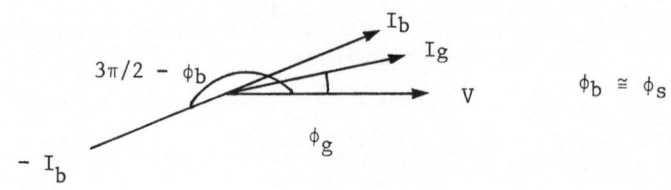

$\phi_b \cong \phi_s$

APPENDIX 2.   CAVITY - AMPLIFIER RF FEEDBACK

The cavity-amplifier feedback loop has a narrow bandwidth, centred at the cavity resonant frequency. Parameters and a block diagram are:

$Z(w)$ = equivalent output impedance,

w = a general angular frequency,

g_m = power tube mutual conductance,

$w_o T$ = $2 \pi k$ (by adjustment of the delay),

δw = $w - w_o$

A = $\beta A_1 g_m R$

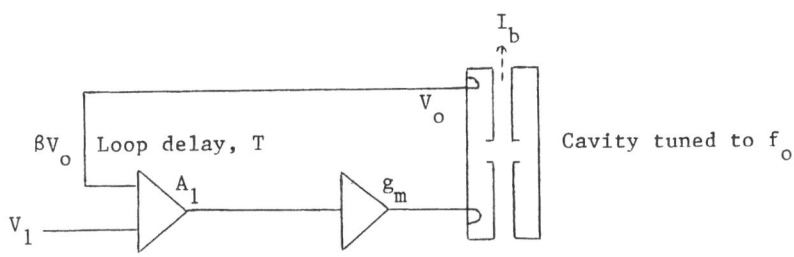

Open loop gain, $A(w)$ = $- A e^{-j wT} (1 + j 2 Q \delta w / w_o)^{-1}$

$A(w)$ = $- A e^{-j \delta wT} (1 + j 2 Q \delta w / w_o)^{-1}$

$Z(w)$ = V_b / I_b

V_b = $A(w) V_b + R I_b (1 + j 2 Q \delta w / w_o)^{-1}$

$Z(w)$ = $R (1 + j 2 Q \delta w / w_o)^{-1} / (1 - A(w))$

$Z(w)$ = $R (1 + A e^{-j \delta wT} + j 2 Q \delta w / w_o)^{-1}$

Allow phase margins of $\pm \pi / 4$ at the unity gain points of the loop.

Then, δw = $\pm \pi / 4 T$ at $|A(w)|$ = 1.

$\delta w / w_o$ = $\pm 1 / 8 k$

Max. loop gain, A_{max} $\cong$ $2 Q \delta w / w_o$ = $Q / 4 k$.

259

APPENDIX 3. RESPONSE OF A DETUNED CAVITY TO A.M. AND P.M. MODULATIONS

The analysis is simplified by the use of a dual complex number representation for modulated signals, with one complex number to represent the carrier and a second for the modulation components. Thus;

$$f(t) \quad = \quad e^{jwt} \, e^{ipt}$$

represents a carrier of angular frequency, w, with a modulation angular frequency, p. The response of an accelerating cavity, ΔV, to such a modulated current, ΔI, in given by a transfer function:

$$G(jw, ip) \quad = \Delta V \, / \, R \, \Delta I \, = R_j \, ((R_i \, (jw, ip) \, e^{ipt})) \, e^{jwt})$$

R_i is the real part of the complex argument with j assumed real and R_j is the real part of the subsequent argument with j assumed complex. R_i and R_j may be interchanged with i replacing j in the last sentence.

The differential equation for a cavity voltage in terms of its excitation currents may be written, for the equivalent R, L, C circuit:

$$\frac{d^2V}{dt^2} \; + \; \frac{w_o}{Q} \cdot \frac{dV}{dt} \; + \; w_o^2 \, V \; = \; \frac{w_o \, R}{Q} \cdot \frac{d \, (I \, - \, A \, (V \, / \, R) \, e^{-(j\delta w \, + \, ip)T})}{dt}$$

where A is the RF feedback gain, as defined in Appendix 2 and δw may be identified with Δw for the fundamental RF component or with an equivalent Δw_p ($= k \, w_{rf} \, - \, w_{op}$) for a higher harmonic component relative to a parasitic cavity resonance at $w_o = w_{op}$.

The real part of the modulation, ΔI, has the form (cos wt cos pt):

$$G(jw, ip) \, = \frac{((jw + ip) \, w_o \, / \, Q)}{((jw + ip)^2 + (jw + ip)(w_o/Q)(1 + Ae^{-(j\delta w + ip)T}) + w_o^2)}$$

$$G(jw, ip) \, = \, (\, 1 + Ae^{-(j\delta w + ip)T} + (2 \, Q \, / \, w_o)(j\delta w + ip) \,)^{-1}$$

The term involving A and T indicates the effect of the RF feedback, the terms involving δw the off-resonance effects and the factor $(2Q/w_o)$ is the cavity natural bandwidth. The detuning, Δw, is proportional to I_b for reactive compensation and the response is then strongly modified for heavy loading, unless there is an appropriate feedback gain, A. For no feedback, or for a parasitic resonance, set A = 0 in G(jw, ip).

The transfer function G(jw, ip) gives the composite response for both the upper and lower sidebands of the modulation. For the fundamental mode, the two frequency sidebands are at: $(h\ w_{rev} \pm p)$. For other revolution frequency components, consider the pair of harmonics $(k\ h \pm n)\ w_{rev}$. There are two separate pairs of sidebands:

$$(k\ h\ w_{rev} \pm p_{1,2}) \quad \text{with } p_1 = (n\ w_{rev} + p) \text{ or } p_2 = -\ (n\ w_{rev} - p)$$

The first pair are associated with the coupled bunch modes, n, where n is an integer from 0 to (h-1) and the second pair with the modes (h-n). The nearest modes to the fundamental have k = 1 and n = 1, with the coupled bunch mode 1 usually of most significance above the transition energy and the mode (h-1) below. The transfer function G(jw, ip) may be used directly with either pair of sidebands. Typically, for coupled bunch mode excitation of a parasitic resonance, the response is:

$$G(jw, ip) = (\ 1 + (2\ Q\ /\ w_{op})(j(k\ h\ w_{rev} - w_{op}) + i(p \pm n\ w_{rev}))\ )^{-1}$$

In the case of coupled bunch feedback systems, it is sometimes necessary to obtain the transfer functions for individual sidebands. A different approach is then needed. The response may be found for $(e^{j(w +p)t} + e^{j(w - p)t})\ /2$, and developed to show:

$$G\ (jw, ip)\ =\ G_u(jw, ip) + G_\ell(jw, ip)$$

$$G_u\ (jw, ip)\ =\ \tfrac{1}{2}(1 - ij)(1 + A\ e^{-(j\delta w + ip)T} + (2Q/w_o)(j\delta w + ip))^{-1}$$

$$G_\ell\ (jw, ip)\ =\ \tfrac{1}{2}(1 + ij)(1 + A\ e^{-(j\delta w + ip)T} + (2Q/w_o)(j\delta w + ip))^{-1}$$

where $p = p_1$ or p_2, G_u is the transfer function for the upper sideband $(p \pm n\ w_{rev})$ and G_ℓ that for the lower - $(p \pm n\ w_{rev})$. The terms in (ij) indicate the quadrature terms associated with single sideband modulation.

APPENDIX 4.   ROBINSON EQUATION FOR THE COHERENT DIPOLE MODE

This may be derived from 3 equations:

1. The linearised coherent phase equation for small phase $(\Delta\phi_v)$ and amplitude (v) modulations of the accelerating fields.

$$(\frac{d^2}{dt^2} + \Omega^2) \phi = \frac{d^2}{dt^2} (\Delta\phi_v) - \Omega^2 \tan \phi_s (\frac{v}{V})$$

2. The linear approximation to the Bessel function form of the dipole mode phase modulation of the fundamental beam current component, I_b.

$$\Delta I_b = I_b (\Delta\phi_v - \phi) e^{-j \phi_s}$$

3. The response of the main mode of the accelerating cavities to beam current modulations, using the complex transfer function, $G(jw, ip)$.

$$(\frac{v}{V} + j \Delta\phi_v) = R [1 + \frac{2Q}{w_o} (j \Delta w + \frac{d}{dt})]^{-1} \Delta I_b / V$$

In the first equation, Ω is the synchrotron oscillation angular frequency; in the second equation, I_b is the amplitude of the fundamental component of beam current; while in the third equation, the term (ip) of $G(jw, ip)$ has been equated with the derivative, d/dt. The 3 equations combine to give the fourth order Robinson equation.

$$((s^2 + \Omega^2) ((s + \frac{w_o}{2Q})^2 + (\Delta w)^2) + \frac{D w_o}{2Q} \frac{\Delta w \Omega^2}{\cos \phi_s}) \phi = 0$$

$$(s^4 + b_3 s^3 + b_2 s^2 + b_1 s + b_o) \phi = 0$$

$s \equiv d/dt$ or the Laplace Transform operator.

$D = R I_b / V$

$$b_o = \Omega^2 \left(\left(\frac{w_o}{2Q} \right)^2 + \Delta w^2 + \frac{D w_o}{2Q} \frac{\Delta w}{\cos \phi_s} \right)$$

$$b_1 = \Omega^2 b_3 = \Omega^2 \left(\frac{w_o}{Q} \right)$$

$$b_2 = \left(\left(\frac{w_o}{2Q} \right)^2 + \Delta w^2 + \Omega^2 \right)$$

Instability arises for 2 conditions:

1. b_o negative, which corresponds to the beam power becoming greater than the cavity power under reactive beam loading compensation.

2. $b_1 b_2 b_3 - b_1^2 - b_o b_3^2 < 0$, which corresponds to the cavity being detuned in the direction opposite to that for reactive compensation. Of importance is the resistive component of the cavity impedance at the modulation sideband frequencies.

When RF feedback is used, there is a modified form for $G(jw, ip)$ and a different fourth order equation results:

$$b_o = \Omega^2 \left(\frac{(1 + A)^2}{\left(\frac{2Q}{w_o} - A T \right)^2} + \Delta w^2 + \frac{D \Delta w}{\left(\frac{2Q}{w_o} - A T \right) \cos \phi_s} \right)$$

$$b_1 = \Omega^2 b_3 = \Omega^2 (1 + A) \, 2 \, / \, \left(\frac{2Q}{w_o} - A T \right)$$

$$b_2 = \left(\frac{(1 + A)^2}{\left(\frac{2Q}{w_o} - A T \right)^2} + \Delta w^2 + \Omega^2 \right)$$

Instability then arises for 3 conditions:

1. b_o negative, which corresponds to the following condition of beam to cavity power, $D \sin \phi_s$, for reactive compensation.

263

$$D \sin \phi_s > (1 + A) / [(1 - \frac{A w_o T}{2Q}) (1 + \frac{A w_o T}{2Q} \cot^2 \phi_s)]^{\frac{1}{2}}$$

RF feedback allows the factor to become $\gg 1$.

2. $b_1 b_2 b_3 - b_1^2 - b_o b_3^2 < 0$, which corresponds to the same condition as when $A = 0$, provided $A < 2 Q / w_o T$.

3. b_1, b_3 negative, which corresponds to $A > 2 Q / w_o T (= Q / \pi k)$. This condition sets an upper limit to A, close to the value of $A_{max} = Q / 4 k$ derived for loop stability in Appendix 2.

In addition to these equations for the rigid dipole mode, there is a similar fourth order equation for coupled bunch dipole motion if driven by the fundamental cavity mode. The nearest revolution frequency harmonics to the fundamental are at $(h \pm 1) f_{rev}$, so the cavity response is mainly that for the sidebands $h w_{rev} \pm (p \pm w_{rev})$. Then, following the transfer functions of Appendix 3, replace p by $(p \pm w_{rev})$ and the required equation is the Robinson equation, but with the term $(s + w_o / 2 Q)^2$ replaced by $(s \pm i w_{rev} + w_o / 2 Q)^2$. The positive sign is for the mode $n = 1$ and the negative for $n = (h - 1)$. The growth rate is approximately:

$$\tau^{-1} = \pm \Omega D \Delta w / 4 B w_{rev} \cos \phi_s$$

$$B = 1 + (h/4Q)^2 (1 + (2Q/w_o)^2 (\Delta w^2 - w_{rev}^2))^2$$

The coupled bunch mode n has a phase shift of $2 \pi n / h$ between individual bunches, assuming all the bunches are present. Missing bunches or unequal bunches reduce the value of I_b and also give other beam frequency components. There is still the possibility for instability to occur, however.

264

APPENDIX 5.   HIGHER MODE EQUATIONS

From the Vlasov equation, Sacherer [3] obtains:

$$(s - i\, m\, \Omega)\, R_m(r)\, e^{im\theta} = -\frac{\sin\theta}{\Omega} \cdot \frac{d\,\psi_o}{dr} \cdot (s^2 + \Omega^2)\, \phi$$

where (r,θ) are polar coordinates in $(\phi,\dot\phi)$ space,

m is the mode number of the longitudinal motion,

ψ_o is the 2-dimensional density of the stationary bunch,

$R_m(r)$ is the radial function of ψ, the perturbation in density,

Ω ($= w_s$) is the synchrotron oscillation angular frequency and

s ($= ip \equiv d/dt$), with p ($= w$), the complex coherent frequency.

Now let (r,θ) relate to $(\phi,\dot\phi/\Omega)$ space and solve for $(s^2 + \Omega^2)\,\phi$. The perturbation of the line charge density is not periodic in ϕ and so cannot be directly represented by a Fourier series. Instead, represent the fields by $e^{jk\phi}\, e^{ipt}$, as in Appendix 3; the carrier has the angular frequency $(k\, h\, w_{rev})$ and the modulations are at p_1 or p_2. The amplitude modulations (a.m.) correspond to the $\sin(k\, r\, \cos\theta)$ terms and the phase modulations (p.m.) to the $\cos(k\, r\, \cos\theta)$ terms.

The $(s^2 + \Omega^2)\,\phi$ expression is developed as in Appendix 4. Use is made of the beam current modulations and the general form of $G(jw, ip)$.

$$\Delta I_k \text{ (p.m.)} = 2\, j\, I_k\, e^{j\, k(3\pi/2 - \phi_b)}\, J_1(k\,(\Delta\phi_v - \phi))$$

$$\Delta I_k \text{ (a.m.)} = (\partial I_k / \partial\phi_b)\, e^{j\, k(3\pi/2 - \phi_b)}\, \Delta\phi_b$$

where ϕ is the coherent phase motion of the bunch centre (m odd),

$\Delta\phi_b$ is the coherent modulation of the bunch phase extent (m even),

J_1 is the Bessel function order 1, argument the amplitude of $k\,(\Delta\phi_v - \phi)$,

I_k is the kth Fourier harmonic of the unperturbed beam current and

$(\partial I_k/\partial\phi_b)$ is the derivative of I_k w r t the bunch phase extent.

Since the odd order modes appear as modulation of the I_k harmonics while the even order modes depend on $(\partial I_k/\partial\phi_b)$, the range of possible frequencies in their excitation spectra is not the same. The upper frequency limit for the odd and even modes is the same, at approximately

the inverse of the bunch time duration, but the lower frequency limits
are different. These conclusions are not in agreement with other
published results. The phase $(\Delta\phi_v)$ and amplitude (v/V) field modulations
have next to be obtained. Also required are the Bessel function
expansions:

$$\sin(k\ r\ \cos\theta) = 2\sum_{1}^{\infty}(-1)^{n+1}J_{2n-1}(k\ r)\cos(2n-1)\theta$$

$$\cos(k\ r\ \cos\theta) = 2\sum_{1}^{\infty}(-1)^{n}J_{n}(k\ r)\cos 2n\theta + J_{o}(k\ r)$$

Individual modes involve different Bessel functions and it may be
seen that the odd and even order modes are driven respectively by the
phase and amplitude modulated fields. An exception is the mode m = 1
which may be driven by either field modulation.

Mode m	Bessel Functions J(k r)	Field modulations
1	$J_o + J_2$	p.m. or a.m.
2	$J_1 + J_3$	a.m.
3	$J_2 + J_4$	p.m.
4	$J_3 + J_5$	a.m.

$$J_{m+1}(k\ r) + J_{m-1}(k\ r) = 2\ m\ J_m(k\ r)\ /\ k\ r$$

The mode equations may now be developed, using small amplitude
approximations, to the following forms:

$$(s^2 + m^2\Omega^2)\ (\ (1 + \frac{2Q}{w_o}\ i\ p_{1,2})^2 + (\frac{2Q}{w_o}\ \delta w\ )^2\ )\ X =$$

$$H\ X\ \Omega^2\ (\ (1 + \frac{2Q}{w_o}\ i\ p_{1,2})\ \cos k(3\pi/2 - \phi_b) + (\frac{2Q}{w_o}\ \delta w)\ \sin k(3\pi/2 - \phi_b)\ )$$

$$p_{1,2} = p \pm m\ w_{rev}\quad \text{and}\quad s^2 = -p^2$$

$$X = \phi,\quad H = m^2\ D\ k\ I_k\ /\ I_b\qquad\qquad \text{for } m = 3,5,7\ \ldots$$

$$X = \Delta\phi_b,\quad H = -D\ (\partial I_k\ /\ \partial\phi_b)\ /\ I_b\ \cos\phi_b\qquad \text{for } m = 2,4,6\ \ldots$$

COMPONENTS FOR HIGH-POWER RF SYSTEMS IN MODERN ACCELERATORS

Georg Schaffer

Institut für Experimentelle Kernphysik
der Universität Karlsruhe, D 7500 Karlsruhe / FRG

ABSTRACT

We discuss the most significant basic components for generation and transmission of RF power in particle accelerators and storage rings of recent design, with emphasis on high duty factor or CW operation. Examples of super power klystron and tetrode amplifiers in the UHF and VHF ranges are presented. Some information is added on waveguides, on power combining and splitting devices, ferrite isolators and circulators, and on a variety of other components used in power transmission systems and accelerating structures. An idea about the cost per MW installed RF power emerges from HERA and LEP cost estimates.

OUTLINE

1. Introduction
2. Power Transmission from Source to Cavity
3. Powerful Klystrons in the UHF Range
4. Waves and Waveguides
5. Means of Measurement
6. Coupling and Decoupling Devices
7. Dielectric Material in High RF Fieldstrength
8. Vacuum Windows
9. Ferrites and their Applications
10. Isolators and Circulators
11. Megawatt-Tetrode Amplifiers
12. Cost Estimate

1. Introduction

Before we consider how big and important RF and microwave sources have become today when we produce huge amounts of RF power by them in particle accelerators and storage rings, let us briefly recall the birth of the proper kind of power tubes some 50 years ago.

It was not primarily for the purpose of accelerating particles when at that time Arsenjewa-Heil and Heil invented and developed a new method to obtain electromagnetic waves of short wavelength and high intensity as illustrated by fig. 1. There was a natural challenge for physicists to find something new because of the facts that for very high frequencies

(1) electrons in grid controlled (=density modulated) tubes are slow, hence their transit time between control electrodes becomes comparable to, or even larger than an oscillation period, connected to the loss of efficiency,

(2) the dimensions of the tube electrodes have to be made smaller and smaller since they represent the capacitance of the resonant circuit, which is in conflict with limits to carry away the heat resulting from electron bombardment.

The solution was to use the otherwise so troublesome transit time in a constructive manner. In the <u>Heil tube</u>, this was accomplished in the following way:

(1) A well focused linear high density electron beam of adequate speed became a separate part of the new microwave source.

(2) By passing this beam through an RF control gap, the particle speed was modulated, with the result that

(3) the beam, when subsequently travelling through a field-free drift tube of properly chosen length, formed particle bunches which could deliver an essential fraction of their energy to the coupled resonant circuit.

The same basic principle is used in the famous <u>Varian invention</u> from 1937 which was published in 1939. Here again, the essential progress was the constructive use of the beam transit time. A linear continuous electron beam was emitted from an oxide coated cathode and accelerated, by some 300 to 4000 V negative voltage, towards a microwave cavity structure at ground potential. This structure consisted of a 'buncher' cavity resonator (RF input), a drift tube, and a 'catcher' resonator (RF output). Some details are illustrated by fig.2. An optimum efficiency of 58 % was calculated.

Russel and Sigurd Varian worked enthusiastically on their new device and called it a 'klystron', expressing the breaking of waves on a beach, according to the Greek verb 'klyzo'. It is worthwile reading their articles to see how much they struggled to get the proper design of cavity resonator, sufficient beam quality, etc. The team of the Stanford University working on the klystron project included also David Webster (klystron theory), William Hansen (microwave resonators) and (since 1939) Eduard Ginzton.

The way to produce klystrons with multi-megawatt output pulses capable to accelerate electrons to record energies was now open.

We sum up the <u>advantages of the klystron amplifier</u> in comparison with grid-controlled electron tubes (triodes, tetrodes wherever applicable):

Eine neue Methode zur Erzeugung kurzer, ungedämpfter, elektromagnetischer Wellen großer Intensität.

Von **A. Arsenjewa-Heil** und **O. Heil** in Bormio.

Mit 10 Abbildungen. (Eingegangen am 20. April 1935.)

Ein neues Prinzip zur Schwingungserzeugung wird beschrieben, bei dem die Schwingungsenergie aus einem Elektronenstrahl herausgekoppelt wird, ohne daß die schwingenden Elektroden von Elektronen getroffen werden. Nach der Berechnung kann sich bis zu 35% der Elektronenstrahlenergie in Schwingungsenergie verwandeln. Die Berechnung gibt außerdem Aufschluß über das Einschwingen, die Stabilität und die Modulierbarkeit des Generators.

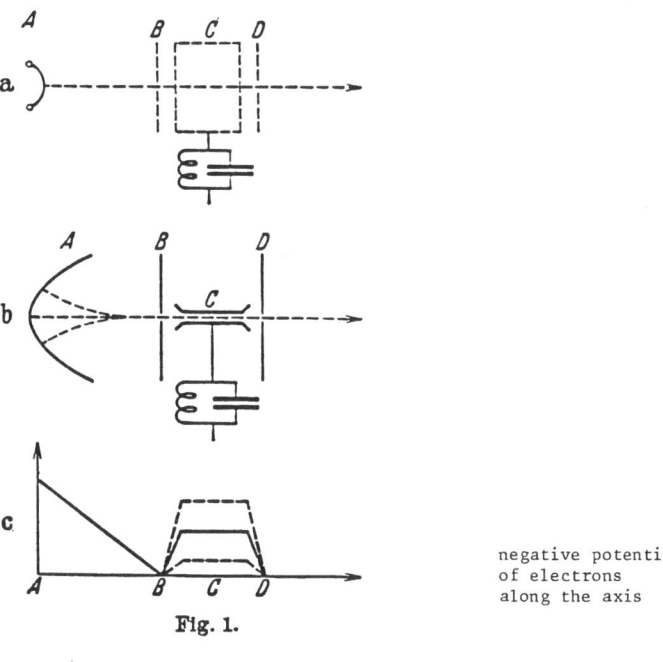

negative potential
of electrons
along the axis

Fig. 1.

Fig. 1.   Principle of the microwave generator invented by Oskar Heil. An assumed RF voltage on electrode C will modulate the speed of electrons entering the drift tube. At the gap between C and D the beam has formed bunches which supply energy to the resonator.

(reproduced from Zeitschrift für Physik 95, July 1935)

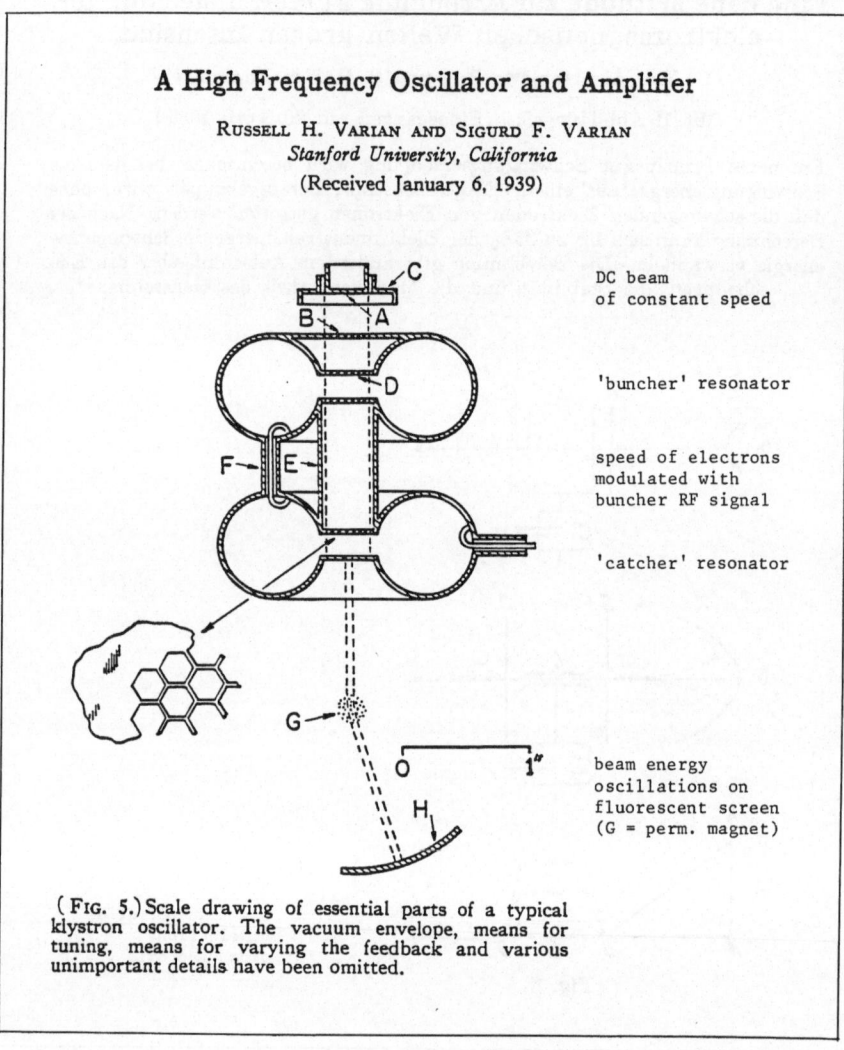

A High Frequency Oscillator and Amplifier

RUSSELL H. VARIAN AND SIGURD F. VARIAN
Stanford University, California
(Received January 6, 1939)

DC beam
of constant speed

'buncher' resonator

speed of electrons
modulated with
buncher RF signal

'catcher' resonator

beam energy
oscillations on
fluorescent screen
(G = perm. magnet)

O   1″

(FIG. 5.) Scale drawing of essential parts of a typical
klystron oscillator. The vacuum envelope, means for
tuning, means for varying the feedback and various
unimportant details have been omitted.

Fig. 2. The Varian 'klystron' as oscillator, designed for 10 cm wavelength.

(reproduced from Journal of Applied Physics 10, May 1939)

- the separation of the cathode system to obtain a beam of high intensity and energy,

$$I_B = K \cdot V^{3/2},$$

with the 'perveance' K of the electron gun.

Examples: $K = 1 \cdot 10^{-6}$ A/V$^{3/2}$ (typical value)

$V = 50$ kV (in air), or 200 kV (under oil)
$I_B = 11$ A   90 A
$V \cdot I_B = 550$ kW   18 MW = available beam power.

- the separation of the RF part (cavities and drift tubes) from cathode and anode. No internal feedback. The RF part is on ground potential.

- the separated anode ('collector') which can be built for enormous power dissipation. The collector is run at (or near) ground potential.

- the very high amplification factor which is possible if several inter-mediate cavities for enhanced bunching are used. These are driven by the beam alone. The cost of the driver stage becomes negligible.

The only extra investment is the necessary magnetic field along the klystron 'body' in order to keep the beam transversely focused.

Trend in frequency: The reduction of stored energy in accelerator structures i.e. a small capacitance per resonant cell, is in general attractive for lowering the filling time. But more important, the available shunt resistance per unit length increases with the square root of frequency f. The skin depth δ in a metal of conductivity K, $\delta = 1/\sqrt{\pi f \mu K}$, reduces the shunt resistance per cell with the square root of f, but the number of cells which can be accomodated in a given length is proportional to f.

An increase in frequency is desirable as long as the power transmitted to the beam is smaller than the ohmic losses in the structure. Therefore, the demand on microwave power is unlimited in frequency. Extreme examples are linear colliders for e^+e^-.

On the other hand, frequency limits may be set by

(1) the beam size (intensity) in relation to the transverse dimensions of the accelerating structure,

(2) transverse beam oscillations in circular accelerators,

(3) superior stability against breakdown with larger cavity size, in case of superconducting structures.

Typically, e^+e^- synchrotrons and storage rings are designed for frequencies between 300 and 800 MHz. For proton linear accelerators below 1 GeV we find frequencies between 200 and 800 MHz, for heavy ions up to 108 MHz. The lower frequencies in these cases are chosen because of the much lower charge-to-mass ratios, low beam brilliance and, in subsequent synchrotrons, also by the need to tune the accelerating cavities over a large or very large frequency range (by HF or VHF ferrites) before relativistic particle velocities are reached.

Trend in power: After the successful operation of the DESY I synchrotron with 500 MHz long-pulse klystrons, the storage rings DORIS and PETRA were designed, and equipped with larger klystron amplifiers of up to 600 kW

CW output power. As shown by fig.3, the demand in total RF power for collider rings increases drastically at the end of the present decade. HERA will need about 13 MW CW at 500 MHz for its electron ring, LEP will require about 100 MW CW at 353 MHz. The specified output power level for a LEP klystron is 1 MW.

A considerable effort has also been devoted to upgrade the output power of pulsed klystrons for electron and positron linacs operating at 2856 MHz (SLAC from 21 to 38 MW per klystron, pulse length about 5 /us, repetition rate 60 - 360 pulses per second).

Finally, we should also mention the development work on heavy ion linacs such as the Unilac of GSI. Here, the capability of tetrode amplifiers has been stabilised to supply 1.6 MW output power in long-pulse (12.5 ms) service with a 25 % duty cycle at 108 MHz.

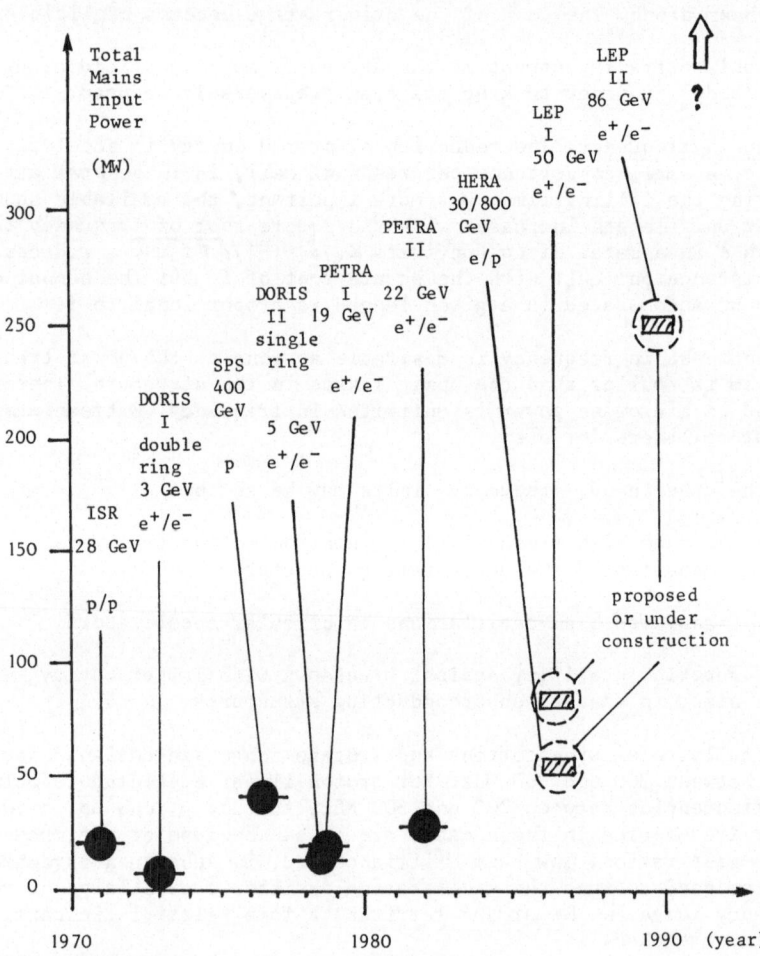

Fig. 3. European 'High-Energy' Facilities.
Total mains input power vs. year of starting operation (after DESY).

2. Power Transmission from Source to Cavity

Some operating conditions of RF amplifiers feeding an accelerating cavity merit special attention since they deviate significantly from those of transmitter stages feeding an antenna.

Common to both cases is the matching of load input and generator output impedances to the characteristic impedance of the interconnecting transmission line. But, as is illustrated by fig.4,

(1) an accelerating cavity has a narrow-band (high Q) characteristic and is easily detuned from resonance by small tuning errors, for instance by uncompensated changes of temperature,

(2) the voltage step up ratio at the cavity feeding point (window, coupling loop or else) is extremely high, typically between 100 and 1000,

(3) the presence or absence of the particle beam changes the matching conditions. The same happens if the beam changes its bunch structure, phase and intensity.

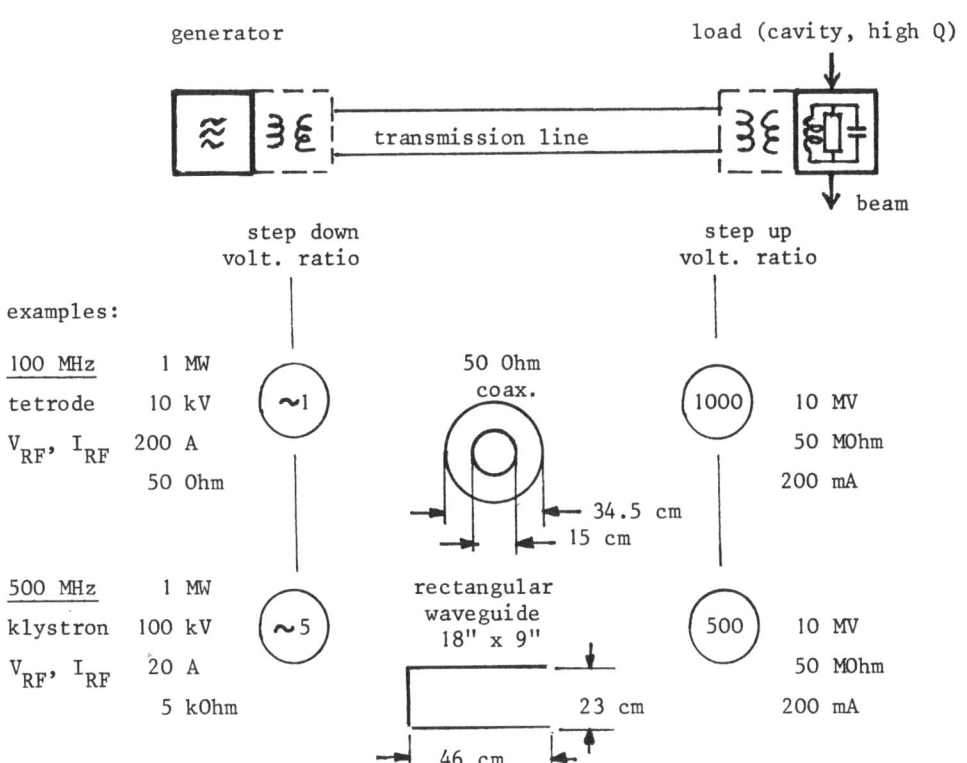

Fig. 4. Typical power transmission scheme.

273

Numerous problems and questions arise from these circumstances.

Let us begin with the <u>filling time</u> of a high Q cavity, for instance at 50 MHz, assuming Q = 50000. The time constant for filling = Q/π periods of 20 ns is about 300 μs. A significant fraction of the generator power will be reflected to the generator until the cavity has nearly reached its full voltage. How to regulate the cavity voltage and/or phase fast enough if required ?

Next, if a significant amount (or even most) of the RF power is to be transmitted to the beam, what happens if the beam intensity or its phase changes: <u>control of beam loading</u> ?

What are the <u>limits</u> of the <u>beam loading ratio</u> = beam power to total power, or beam current I_B to cavity current I_C ?

What will be the consequences for <u>superconducting</u> cavities with Q-values of 10^9 for example ?

A few answers can be given straight away:

Evidently, against routinely reflected power, the output power stage needs adequate protection. If we are forced to operate it at maximum ratings, we risk to break the tetrode ceramic or other parts, or the output window of a klystron. Ferrite isolators or circulators are available for the 1MW UHF range. Their action is to deviate the reflected wave to a high power load. Else, one may choose the electrical length of the transmission line such that the reflected wave arrives in opposite phase at the critical point.

<u>Modulation of the RF output</u>: It will be necessary to balance fast or slowly the effect on cavity voltage and phase by the beam, and to match the cavity voltage to the bucket requirement during the acceleration cycle. Fast action = 'anti-beam-wave' at injection for instance, means to cancel the otherwise unavoidable transient in cavity voltage and phase, by the superposition of the beam induced wave with a compensatory wave from the generator.

With tetrodes, the modulation of the RF output can be performed by rapid or slow modulation of the RF drive. The inherent phase shift with change of amplitude is sufficiently small. The efficiency remains good.

With klystrons, fast modulation of amplitude and phase can also be performed by modulating the drive power, provided the RF output level is not in saturation (fig.5a). Again, the inherent phase shift with change of amplitude will be small. The efficiency goes down with smaller output power because the klystron beam current stays constant.

An additional possibility on klystrons is to insert a modulating anode between cathode and body. The klystron beam current can then be changed via the voltage between modulating anode and cathode, while the total beam voltage remains constant (fig.5b). The current interception by the modulating anode is small. With modest modulator power, the changes of beam current are relatively slow, for instance with a time constant of 30 μs, since the capacitance of the modulating anode and modulator output will be of the order of a few hundred pF, and usually a swing over the full beam voltage is needed. There will be an inherent change of RF phase of about 100 - 200 degrees for the full range. The klystron efficiency remains good because lower output power means also lower beam current.

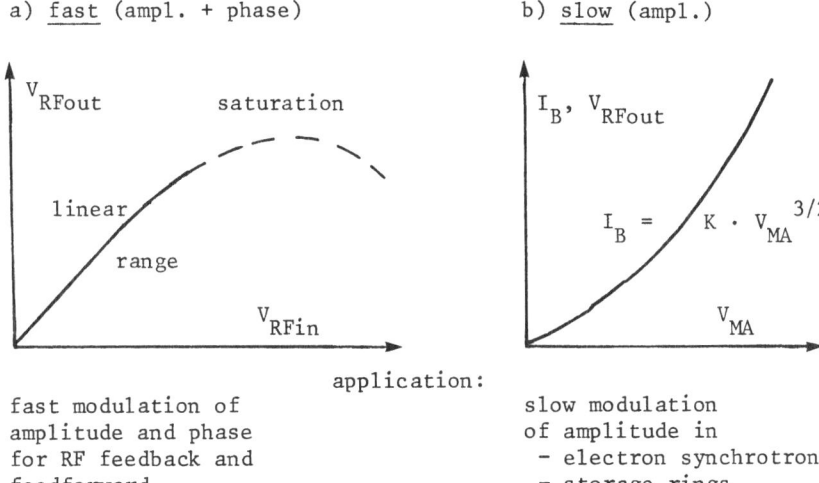

a) <u>fast</u> (ampl. + phase) b) <u>slow</u> (ampl.)

application:

fast modulation of
amplitude and phase
for RF feedback and
feedforward

slow modulation
of amplitude in
- electron synchrotrons
- storage rings

Fig. 5. Modulation of the klystron output signal,
    a) by the RF input signal, b) by the voltage of a modulating anode.

As we recognise, a klystron amplifier permits slow and fast feedback
or feedforward to stabilise the acceleration mechanism for an intense particle beam.

3. Powerful Klystrons in the UHF Range

The first UHF klystron which was selected (in 1958) to accelerate electrons to 6 GeV in a rapid cycling synchrotron (DESY) was made by Eimac at
San Carlos / USA. It was an experimental type (X 602) but from a normal
production line. It produced 200 kW pulse and 50 kW average RF output at 500
MHz with an efficiency of 42 %. Four external cavities permitted frequency
tuning from 325 to 500 MHz. Max. gain was 50 db. The klystron was equipped
with a modulating anode. Both cathode and modulating anode were air insulated, the max. beam voltage was 55 kV.

This klystron is shown in fig.6. The synchrotron RF system consisted of
16 three-cell cavities with a total shunt resistance of about 160 MOhm,
Q = 40000. In 1964, acceleration to 6 GeV was immediately achieved after
correct setting of synchrotron orbit and magnet excitation conditions.
2 klystrons working in parallel over a waveguide diplexer gave an electron
energy of 7 GeV.

It was then decided to specify klystrons for 500 kW peak and 250 kW
average RF output power, in order to extend the energy range of the synchrotron to still somewhat higher limits, and to be able to build a 3 GeV storage ring (DORIS) which required about 1MW CW power. At that stage, European
manufacturers (CSF) entered the marked for the production of this size of
klystrons.

In the meantime there is a large choice of high-power UHF klystrons for
CW or long-pulse operation on the worldwide market. The efficiency has been
improved by adding a beam-driven second harmonic cavity to the intermediate
cavities. The state of the art of 1980 (Valvo) is shown in table 1.

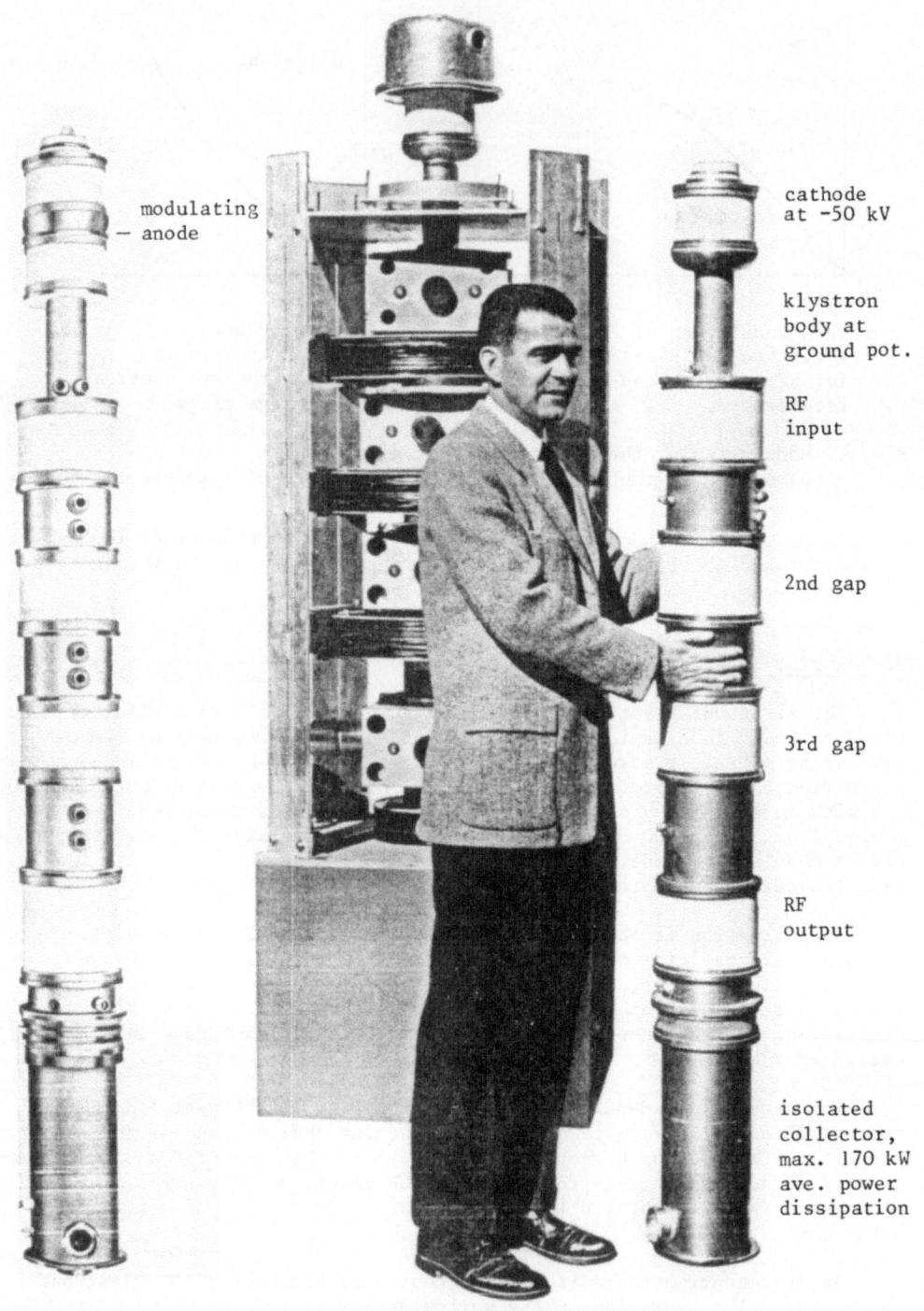

modulating
— anode

cathode
at -50 kV

klystron
body at
ground pot.

RF
input

2nd gap

3rd gap

RF
output

isolated
collector,
max. 170 kW
ave. power
dissipation

Fig. 6. Eimac 4-gap klystron amplifier X 602 K (left), tuning range 325 to
500 MHz, for max. 200 kW pulse and 50 kW average output power,
selected in 1958 for the 6 GeV electron synchrotron DESY.

Table 1. Large UHF CW klystrons from Valvo GmbH (after Demmel).

		YK 1300	YK 1300[1]	YK 1300[2]	V 107 SK	V 109 SK
frequency of operation	MHz	500	500	500	320-360 [3]	1000
bandwidth (1 db) of saturation curve	MHz	3	1	2 1	>1	4
output power	kW	600	528	800[4] 650[5]	1000	400
beam voltage	kV	60	65	75 75	85	55
beam current	A	16.7	11.6	16.4 12	15.7	12.1
beam perveance	$/uAV^{-3/2}$	1.1	0.7	0.8[6] 0.6	0.6	0.9
efficiency	%	60	70	65[6] 72[6]	75[6]	60[6]
gain	db	43	34	42 36	40	40
height incl. stand	cm	390	390	≈ 400	≈ 500	≈ 300
number of resonators (fundam. + 2nd harm.)		4+1	4+1	4+1	5+1	4+1

1) efficiency optimisation
2) improvement YK 1300
3) fixed tuning to midband frequency
4) power optimisation
5) efficiency optimisation
6) design goal

Table 1 is no longer up-to-date but it contains useful data on operating parameters, dimensions, etc. Today, the LEP klystron, 354 MHz, 1.1 MW, 3.8 m long (without horizontal support) would have to be added. There is also a 220 MHz klystron for 3 MW long-pulse service (for Eiscat / Tromsö) with a tube length of 6.0 m.

Another type is shown in fig.7, a test klystron for long-pulse operation at 324 MHz of a 'disc and washer' structure of a 1.1 GeV proton linac. The output power could be raised until 1.5 MW with 10% duty factor.

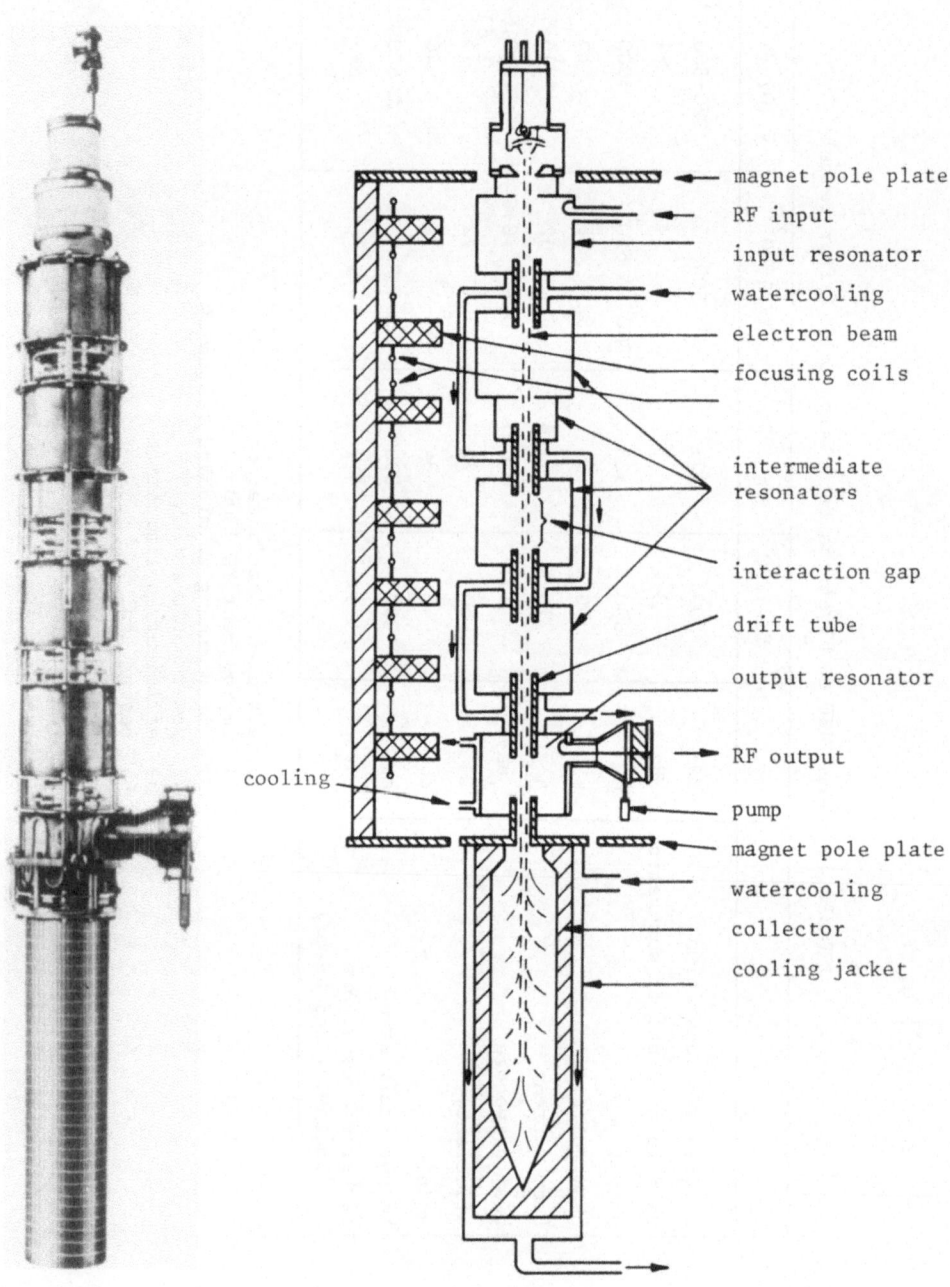

Fig. 7. 324 MHz test klystron, developed by Valvo for SNQ.

4. Waves and Waveguides

We have to say a few words on waves and waveguides with respect to our power transmission scheme.

Up to 200 MHz, the coaxial waveguide is exclusively used. A typical size of inner and outer conductors of a large rigid line has been shown in fig.4. This line can handle up to 500 kW CW power at 100 MHz and up to about 350 kW CW at 200 MHz, if reflected waves are negligible. Without forced cooling, the inner conductor will then run at about 100°C. The attenuation $R^*/2Z$ is of the order of 0.3 db and 0.4 db per 100m for the 2 frequencies. The resistance R^* per unit length can be calculated on the basis of the penetration depth δ. $Z = 60 \cdot \ln D/d$ is the characteristic impedance for the TEM mode. Other modes are not permissible. We would run into trouble with higher frequencies as soon as the wavelength approaches and falls below the circumference of the inner conductor.

The normal choice for high power transmission at higher frequencies is the rectangular waveguide in standard dimensions (EIA), with a (nominal) ratio 2 : 1 between width (a) and height (b). For efficient operation in the desired TE_{10} mode, (a) must be greater than $\lambda_o/2$ and less than λ_o, the wavelength in air. The wavelength λ inside the waveguide is greater by the factor $\lambda/\lambda_o = 1 / \sqrt{1 - (\lambda_o/2a)^2}$. Generally, one operates between 60 % and 95 % of the cut-off frequency of the next higher mode TE_{20}.

In our example for 500 MHz we take an 18" x 9" aluminium waveguide (WR 1800). We are then at the midband frequency. The attenuation per 100m is about 0.2 db. In 1 MW CW operation (eg PETRA), the normal power loss with negligible reflection is 400 W/m. PETRA waveguides are hot during operation but extra cooling is not needed. Nevertheless it may be good.

Sometimes we find series or parallel connections in waveguide systems, for instance as shown in fig.8, or transitions between different types of waveguides, usually at the klystron output, cavity input and on coaxial dummy loads. Another component of interest is a hybrid junction or 'magic-T', a superposition of parallel and series connections as illustrated by fig.9. This is a typical bridge circuit or 'diplexer', which is useful to combine the output signals of 2 power sources, without mutual interaction.

5. Means of Measurement

The status of power transmission over a transmission line is characterised by incident (forward) and reflected (reverse) waves, with voltages V_f and V_r. Of immediate interest is the ratio $V_r/V_f = p$, which gives the voltage standing wave ratio VSWR = $1+p/1-p = 1/m$.

The directional coupler, schematically shown in fig.10 for a coaxial waveguide, is the most important measuring probe for V_f and V_r. It works in a simple manner. The induced signals on A and B result from the superposition of capacitive and inductive components. With proper adjustment of the mutual inductance M and the coupling capacitance, the superposition will result in A showing no voltage for reverse wave, and B no voltage for forward wave. A- and B-signals are proportional to the respective forward (or reverse) voltages and currents

$$I_f = V_f/Z$$
$$I_r = V_r/Z,$$

with Z = characteristic impedance of the line.

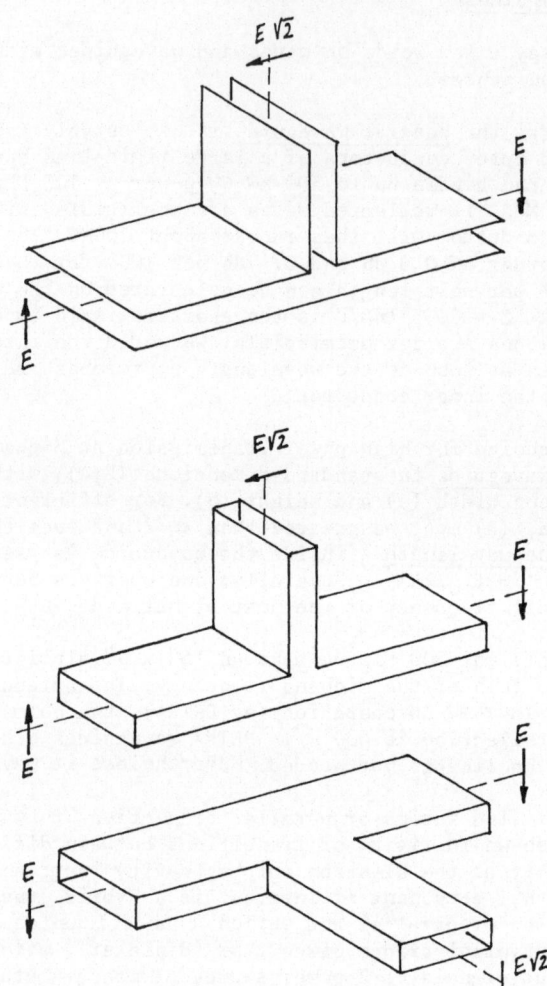

Fig. 8. Similarity of a stripline series connection (above) and a rectangular waveguide lateral series connection (middle). A parallel waveguide connection is shown below (after Grivet).
(Note that an impedance transformation will be necessary).

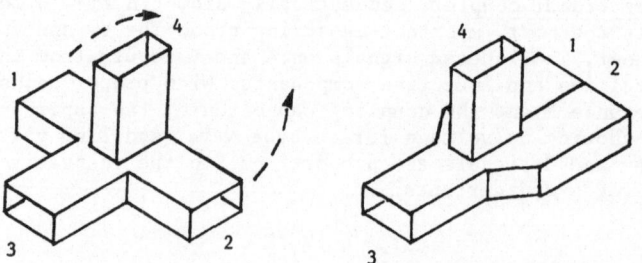

Fig. 9. Conventional magic-T (left) and H-plane folded T.

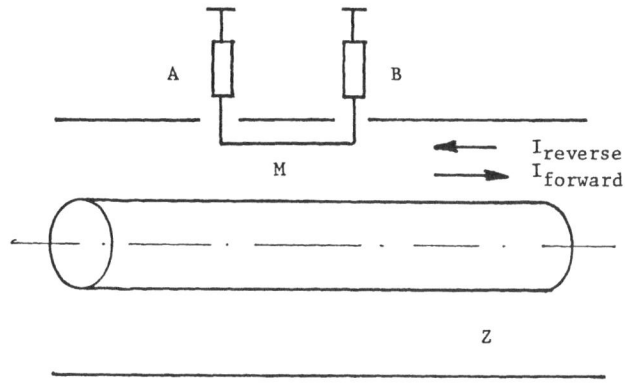

Fig. 10. Directional coupler.

The perfection of the coupler, its 'directivity' (eg 40 db) and its 'attenuation' (eg 50 db) determine the accuracy of the signals.

Amplitude and phase of the 2 signals may be used for display on the polar diagram of an oscilloscope, for impedance measurements over a range of frequencies of interest (see fig.11 and fig.12).

In practical operation, directional couplers are used for control and monitoring purposes, in particular for accurate tuning of accelerating cavities and for power level measurements.

6. Coupling and Decoupling Devices

The directional coupler of fig.10 can also be manufactured with low attenuation values, ultimately down until 3 db.

A waveguide version has been shown by fig.9, the magic-T.

As schematically shown by fig.13, such 3 db couplers are used for many different purposes. Apart from power splitting and power combination, they offer the interesting possibility to dump reflected waves from two equal loads (cavities) into a dummy load. Fig.14 illustrates how this principle can be applied to a group of 16 cavities, fed by 2 klystron amplifiers. Another possibility is shown in fig.15: A chain of directional couplers with different attenuation values is used to feed a group of 8 cavities from one generator.

7. Dielectric Material in High RF Fieldstrength

The application of dielectric material is a critical matter in most cases of very high RF fieldstrength, in particular

    (1)   for the support of the inner conductor of a coaxial waveguide,

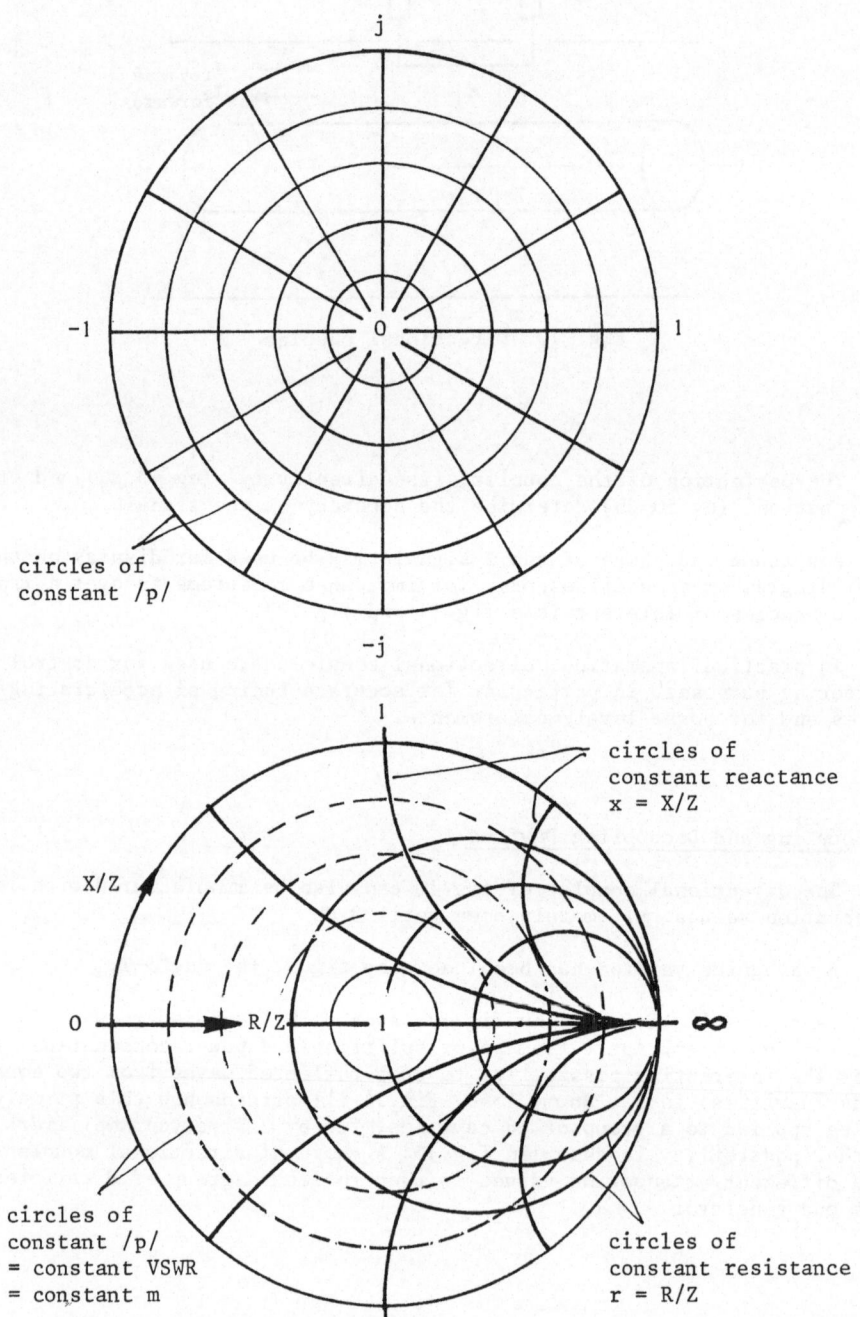

Fig. 11. Polar diagram for the complex reflection coefficient $p = /p/e^{j\varphi}$ (above), and resulting diagram for the complex load impedance $R/Z + jX/Z = r + jx$ (below).

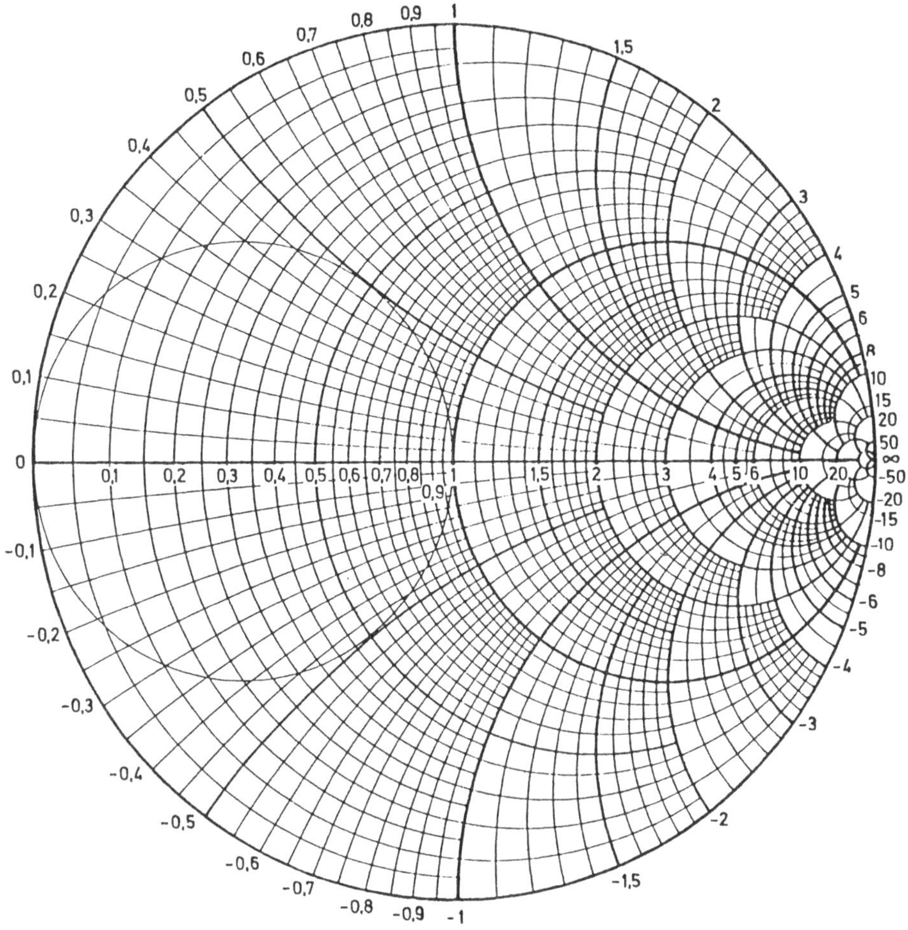

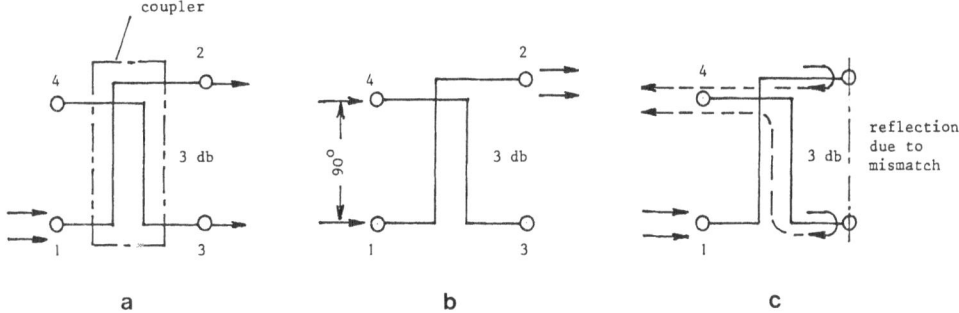

Fig. 12. Smith diagram.

Fig. 13. Application of a 3 db coupler (a) for power splitting and (b) for
power combination. Case (c) shows the flow of reflected power in
case of equal mismatch on output channels 2 and 3 (after Stadler).

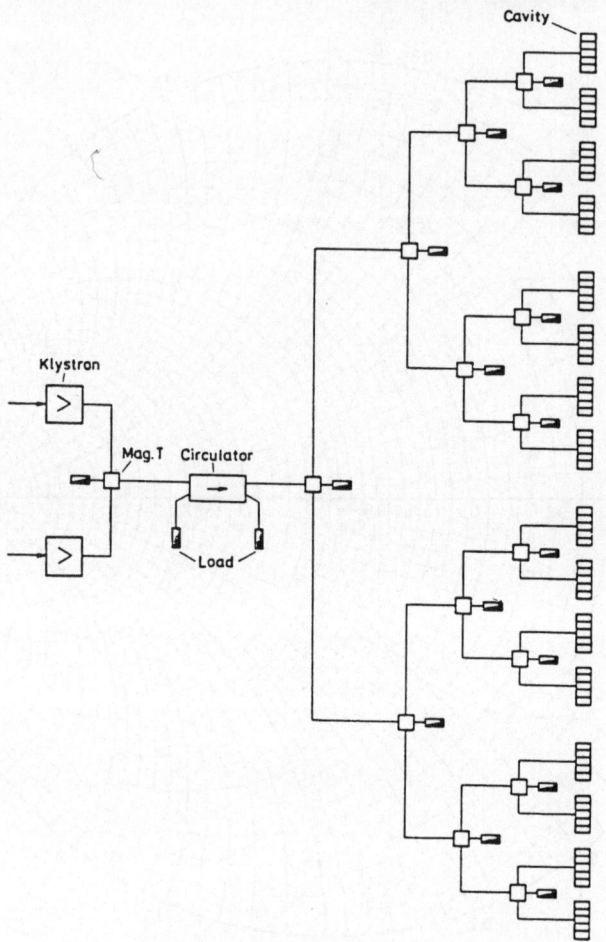

Fig. 14. Power distribution to a group of 16
cavities (DESY).

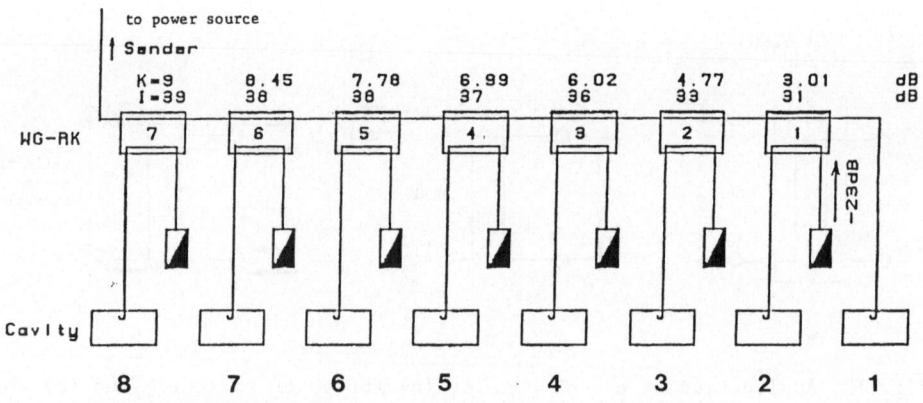

Fig. 15. Alternative method of feeding a group of cavities by directional
couplers of varying coupling strength (after Dwersteg and Musfeldt).
K = coupling attenuation, I = directivity.

(2)   for the vacuum window at the feeding point of an
      accelerating cavity, and

(3)   for the output window of a high power amplifier tube.

The question is not just simply whether the material will stand the dielectric losses, or whether the breakdown fieldstrength of the dielectric material is sufficiently high. We have more questions. A very important one is: what will happen to the <u>surrounding field</u> if a dielectric with a higher ε is introduced?

This should be illustrated by fig.16. Using higher ε insulating material means that the fieldstrength at transitions from gas or from vacuum to the dielectric may rise by a factor up to $\varepsilon_2/\varepsilon_1 = \varepsilon_r$ of the insulator. Hence, the following questions turn up:

(1)   will the gas surrounding the piece of insulator hold the
      increased fieldstrength? If yes, also at higher temperature?

(2)   will the rest gas on the surface of a vacuum window lead to
      discharges (multipactor enhancement?) on spots of increased
      fieldstrength?

We will consider a few examples.

Fig.17 illustrates the destruction of Mycalex supports in a large co-axial transmission line. There are 3 reasons for the failure, namely

(1)   the material has a low critical temperature (600°C),
      a high tgδ ($\sim 10^{-3}$), and a low thermal conductivity,

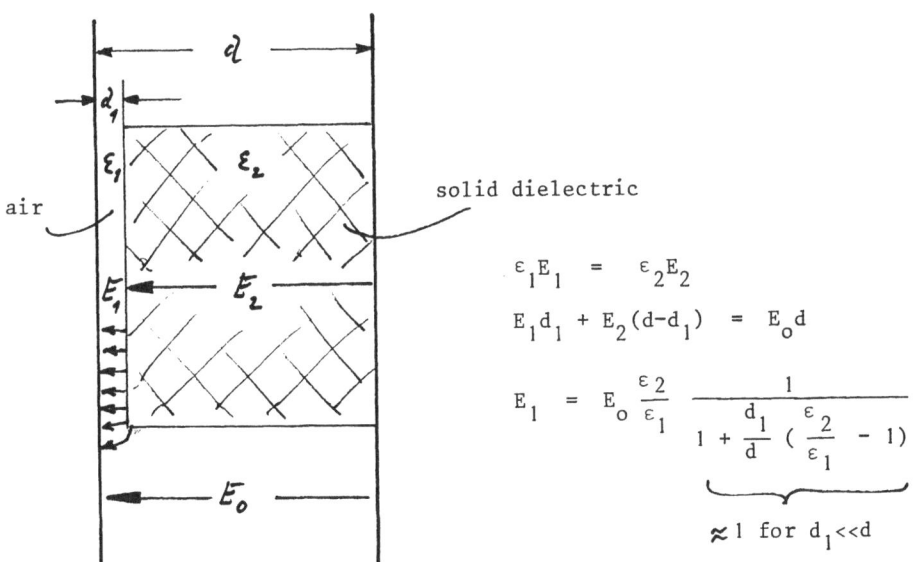

$$\varepsilon_1 E_1 = \varepsilon_2 E_2$$

$$E_1 d_1 + E_2 (d-d_1) = E_o d$$

$$E_1 = E_o \frac{\varepsilon_2}{\varepsilon_1} \underbrace{\frac{1}{1 + \frac{d_1}{d}(\frac{\varepsilon_2}{\varepsilon_1} - 1)}}_{\approx 1 \text{ for } d_1 << d}$$

Fig. 16. Increase of fieldstrength in air gaps by a high ε dielectric.

285

Fig. 17. Result of using Mycalex supports for a 2 MW coaxial transmission
line, inner conductor 14 cm diameter (photo GSI).
Frequency of operation 108 MHz, duty factor 25 %.

(2)   the supports have been fixed to the inner conductor
      by metallic screws which penetrate into the field area,
      the discharge mechanism starts from there,

(3)   the cross-section of the supports diminishes towards the
      inner conductor.

While reason (1) and (2) may appear rather obvious, (3) may not. We
shall come back to it.

Let us first consider what other choices we have if we want to select
material of higher electric and mechanical stability. Some data are shown
in table 2.

The highest stability offers alumina ceramic which has a very good
thermal conductivity. But on the other hand, the high dielectric constant
of $\varepsilon_r = 9.2$ will cause considerable field distortions on transitions. If we
take a dielectric material with lower ε_r like Teflon or Polypenco, the advantage is low loss factor $\varepsilon_r tg\delta$ and easy machining but, as the last column
of table 2 shows, there is almost no stability against destruction in case
of an arc.

Concerning the insulating supports of the above mentioned 2 MW/ 500 kW
coaxial transmission line (which is connected to the Alvarez tanks of the
Unilac) a considerable improvement has been possible by the use of steatite
rods, and by an improved mechanical fixing.

Table 2. Comparative data of a few dielectric materials.

	critical temperature ϑ_u (°C)	dielectric constant ε_r	$tg\delta$* 10^{-4}	loss factor* $\varepsilon_r tg\delta$ 10^{-4}	thermal conductivity* λ_M (mW/cm°C)	stability against destruction** $\lambda_M(\vartheta_u - 20)$ (W/cm)
alumina 99.5 % (Al_2O_3)	2030	9.2	~2	~18	210	~420
steatite KER 221 DIN 40685 eg Frequenta, Stettalit	1340	~6	~4	~24	23	~30
quarz eg Rotosil, Ultrasil	1700	4.2	~10	~42	~15	~25
mica products eg Micaver, Mycalex, Supramica	~600	~7	~10	~70	~7	~4
polytetrafluorethylene eg Teflon	~270	2.1	~3	~6.3	~2.4	~0.6
polydichlorostyrene, cross linked eg Polypenco Q 200.5	~225	2.5	~1	~2.5	~2.4	~0.5

* at 20°C
** without heat radiation

The performance limitations of supports of different material can easily be tested in a quarter wavelength resonator of simple construction like the one sketched in fig.18. The Q factor of this resonator is 3500 at 108 MHz. An amplifier with about 10 kW output power is sufficient to produce voltage pulses up to 40 kV peak at the open end of the inner conductor. By this way one gets useful information on temperature increase of the supporting rods, on their flashover voltage and the dependance of flashover voltage on temperature.

In fig.19, a picture of a flashover test is shown.

Isodynamic support: We refer to argument (3) on fig.17. The strength of the electric field $E(\varrho)$ in a coaxial waveguide is inversely proportional to the radius ϱ, namely $E(\varrho) = V/\varrho/\ln(D/d)$. A gain in flashover voltage, and a more equal distribution of dielectric losses can be obtained if there is a way to correct the field distortion. To a certain degree, this can be done by proper shaping of the insulating supports. An analogous problem is the feedthrough on a HV transformer from tank to air. In this case, field equalisation is achieved by a concentric stack of capacitive layers such that their length decreases with radius ϱ. Similarly, for a coaxial waveguide, supports should be thicker on the inside than on the outside. Some approximations are shown in fig.20.

8. Vacuum Windows

The critical condition, life or destruction, of vacuum windows on klystron output and cavity input circuits should be illustrated by fig.21. Like so many others, this picture has been taken during the development phase of 20 MW S-band klystrons at Stanford. This was in the 1960s. Many cures were developed to improve the situation. Nevertheless, the risk of window failure stays with us and has to be reconsidered whenever higher pulse or average RF power is planned for an accelerator.

High-purity alumina as material is the normal choice, some experience has also been made with beryllia. Metallisation of the surface to reduce multipacting is widely applied. Clean vacuum conditions are crucial.

Disc windows as shown in fig.21 are used in S-band klystrons (short sections of cylindrical waveguide) and for inductive coupling holes in cavity walls (eg DESY I). Other types are dome windows for cavity coupling loops (eg Unilac), cylindrical windows (eg tetrode anode circuits) and coaxial windows. The coaxial type is used in large UHF klystrons and in accelerating cavities.

Let us look in some detail at the PETRA cavity which is shown in fig. 22. The resonant cell in the center carries the flange for mounting the input coupler with its double disc window. The cone of the coupling loop is watercooled, and there is some aircooling between the 2 discs. Under regular operating conditions, 60 kW CW per cavity means 2.5 kV peak voltage at the coaxial feedthrough. Runs were also made with up to 250 kW CW per cavity. The situation at the klystron output window is, on the one side, more severe (9 kV for 800 kW forward power). But, on the other side, the vacuum conditions inside an accelerator are subject to higher risks.

Work on cavity development and accelerator commissioning for DESY I gave the following lessons: Since the restgas pressure of the synchrotron vacuum chamber was initially very poor, worse than 10^{-5} torr which meant

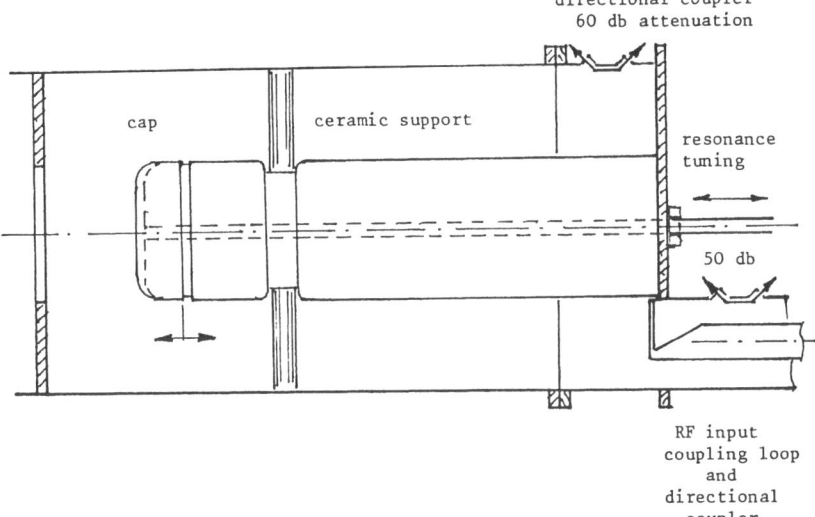

directional coupler
60 db attenuation

cap

ceramic support

resonance
tuning

50 db

RF input
coupling loop
and
directional
coupler

Fig. 18. Quarter wavelength test resonator for flashover and lifetime tests of ceramic supports of a 40 kV coaxial transmission line for 108 MHz. Two directional couplers permit the measurement of forward and reflected power (a) inside the resonator ($P_f = P_r$) and (b) at the feeding line in front of the adjustable coupling loop.

Fig. 19. Flashover test with ceramic supports in the resonator of fig. 18.

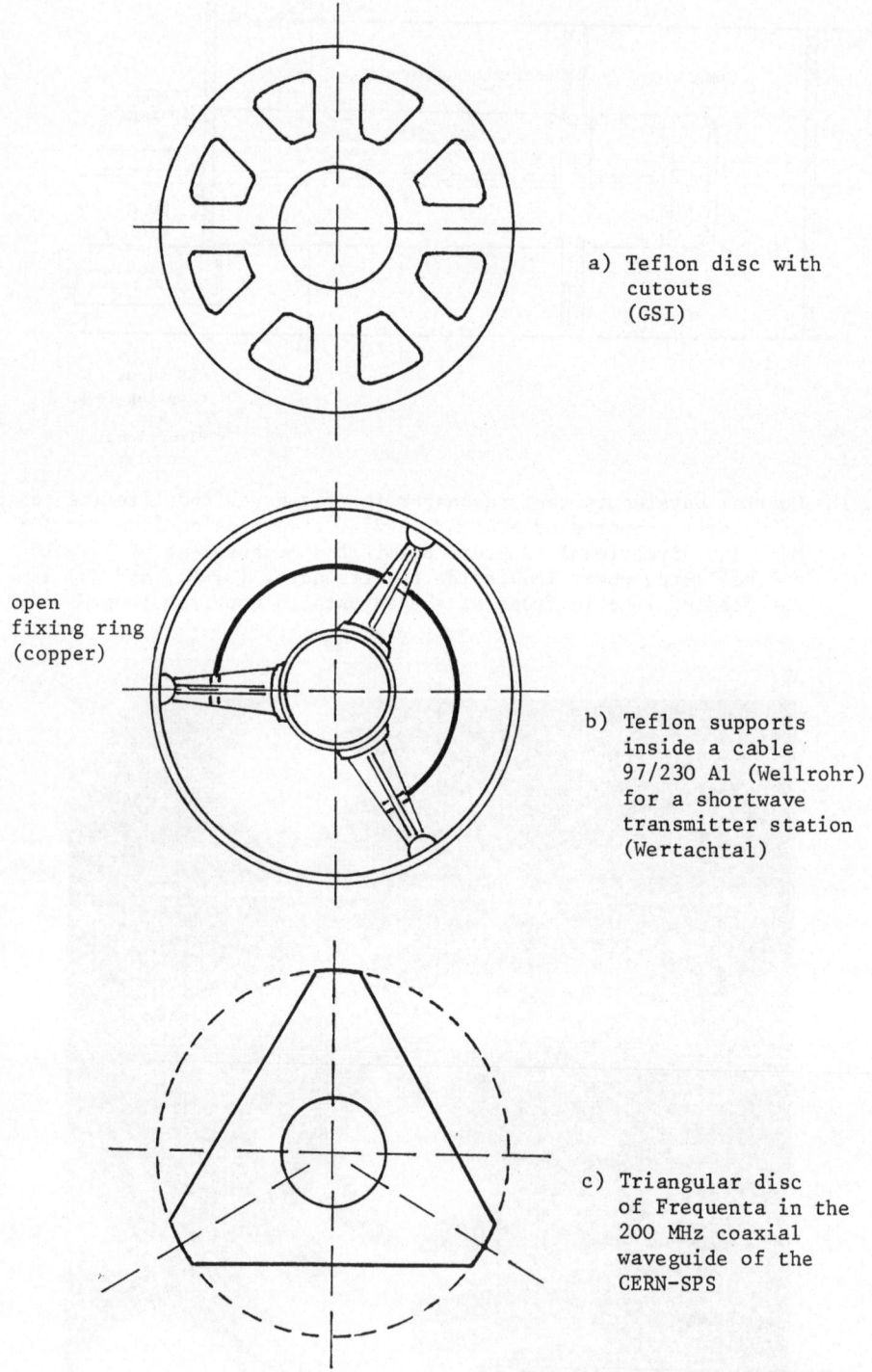

a) Teflon disc with
   cutouts
   (GSI)

open
fixing ring
(copper)

b) Teflon supports
   inside a cable
   97/230 Al (Wellrohr)
   for a shortwave
   transmitter station
   (Wertachtal)

c) Triangular disc
   of Frequenta in the
   200 MHz coaxial
   waveguide of the
   CERN-SPS

Fig. 20. Approximations to an isodynamic support of the inner conductor
in coaxial waveguides.

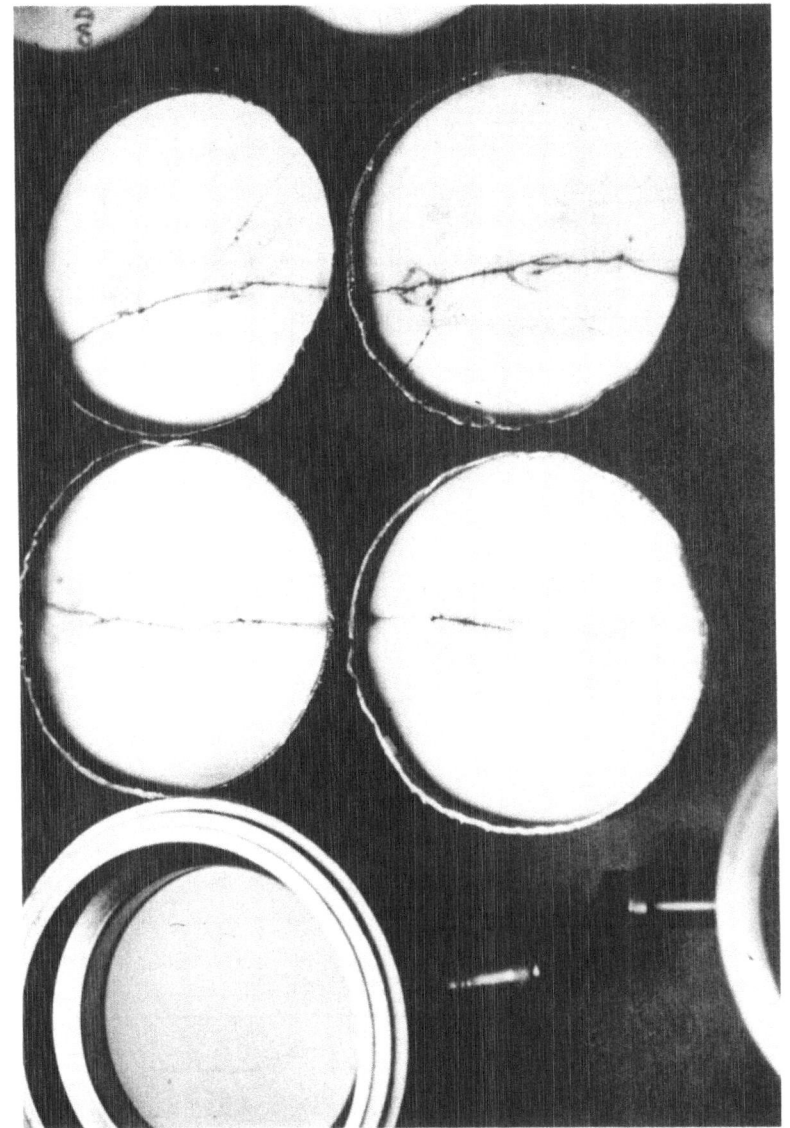

Fig. 21. A few alumina disc windows after failure (SLAC), and a beryllia window (left).

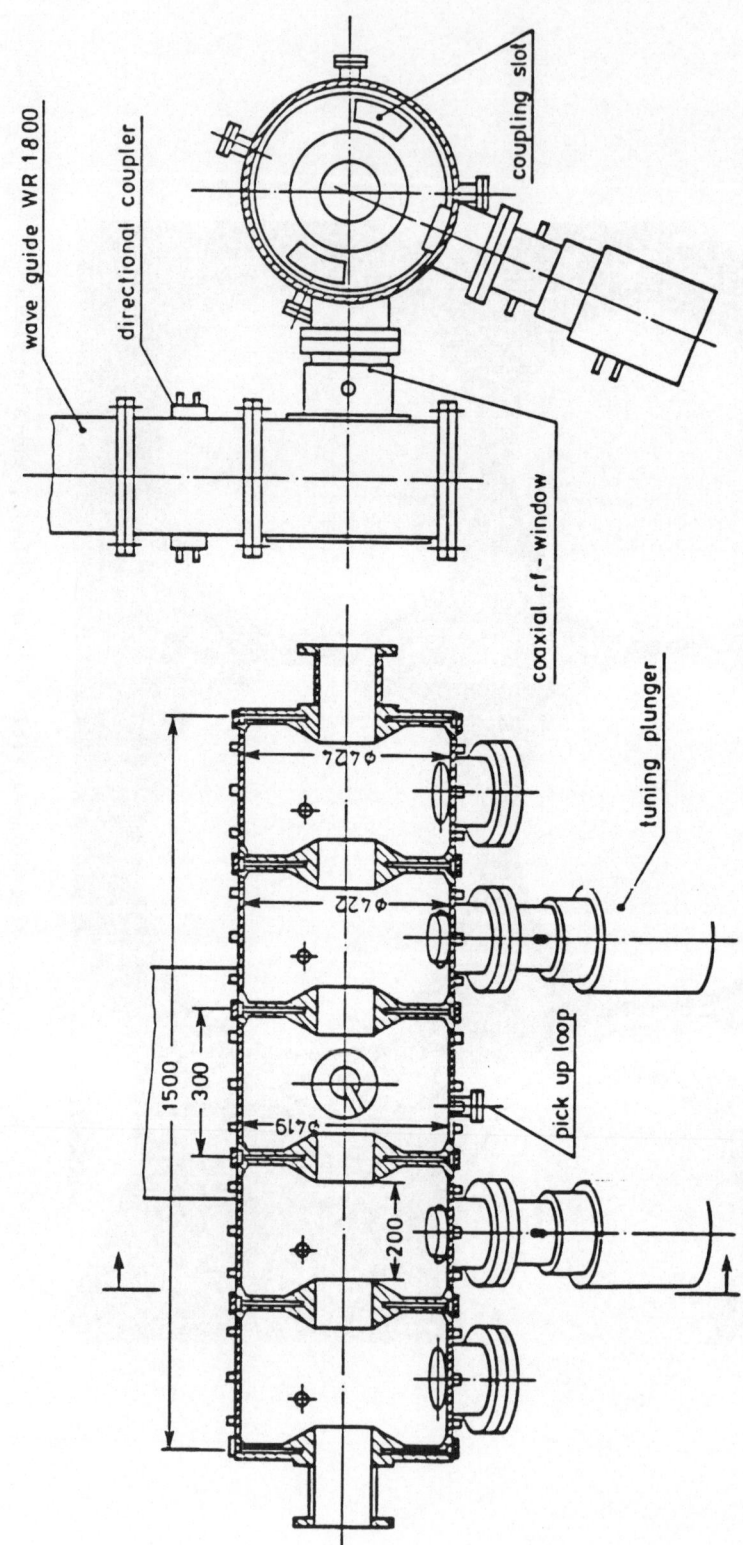

Fig. 22. PETRA accelerating structure and feeding. The coaxial RF window is placed between the RF input coupling loop and a waveguide-to-coax transition. The directional coupler monitors incident and reflected waves.

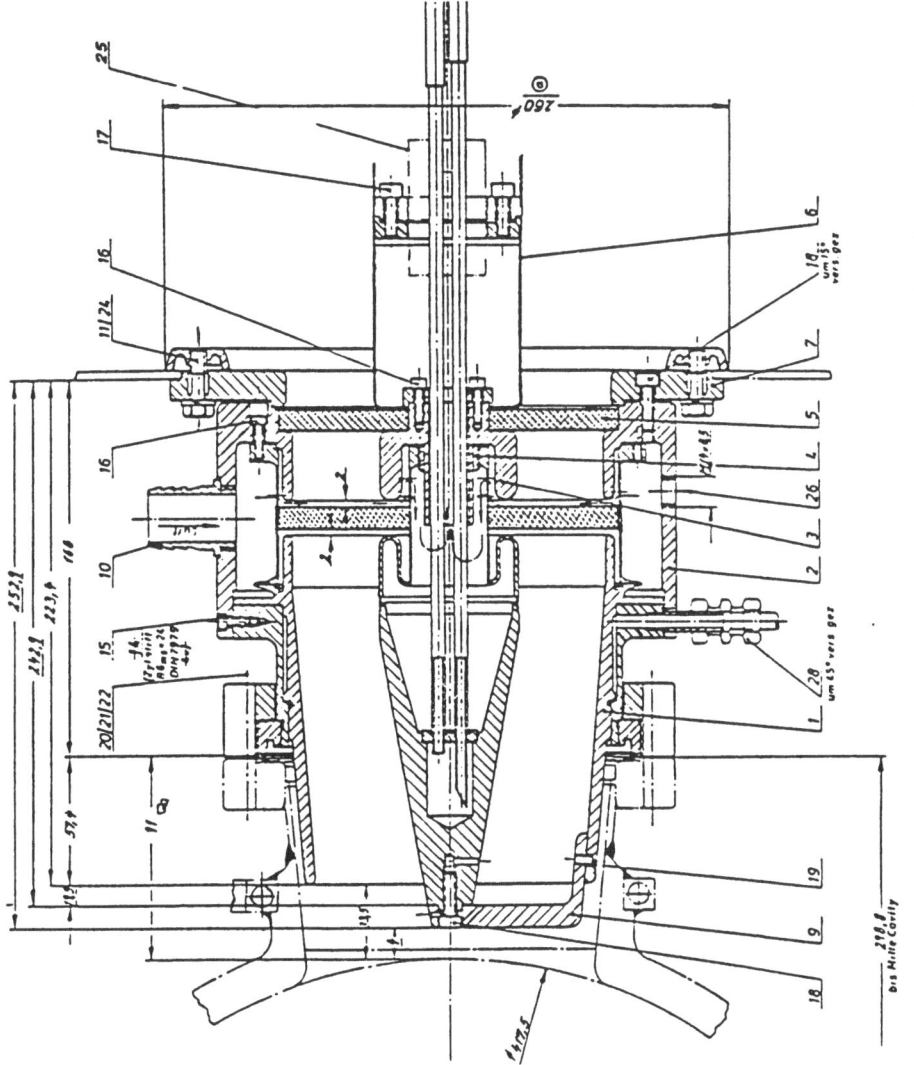

Fig. 23. Double disc alumina window of the PETRA cavity input coupler.

one or two orders of magnitude higher than with cavities alone, strong mutual RF coupling of the 16 accelerating units had to be abandoned immediately, in favour to weak coupling. Under weak coupling, a short by plasma discharge on a gap transforms into a high parallel (or low series-) impedance at the branch-off point, hence reducing automatically the RF power absorption. With this change, beam acceleration became never a RF (window) problem under poor vacuum, even up to 7 GeV (1964).

We conclude that window self-protection, or any other fast protection mechanism is useful or necessary.

9. Ferrites and their Applications

Ferrites consist of sintered oxide mixtures from iron, manganese, nickel, magnesium, zink, copper, aluminium, yttrium etc. Their relative permeability at low magnetising forces (μ_i) is between about 6 and 15000. The Curie temperature falls with higher permeability. Resistivity is between 10^2 and 10^7 Ohmcm, decreasing with higher temperature.

The following applications are of special interest for accelerator operation:

(1)   suppression of spurious microwave oscillations in high-power tetrode amplifiers by ferrite absorbers,

(2)   cavity frequency tuning by varying the permeability of ferrite toroids in coaxial HF and VHF cavities and tuners,

(3)   output (window) protection of klystrons against reflected waves by means of ferrite isolators and circulators.

The complex relative permeability of different ferrites is illustrated by fig.24. We show the frequency dependance of an absorber made of many U 60 elements as an example (see fig.25). The low-impedance absorber was intended to be placed between control grid and screen grid terminals of the 108 MHz tetrode amplifier discussed in chapter 11 in order to suppress spurious oscillations of the internal tetrode structure at frequencies above 600 MHz. In the final solution, ferrite absorbers were distributed inside the anode resonant circuit. The important point is here that the power absorption at the operational frequency stays within safe limits.

Cavity frequency tuning: Besides ferrites which are used for cavities in lower frequency ranges (eg CERN PS and booster up to about 10 MHz), newer products are on the market which look attractive for cavities in the VHF range, in particular for hadron facilities with rapid sequence of acceleration cycles. So far the Fermilab booster and main ring cavities are the only ones in operation with resonant frequencies around 50 MHz.

The classical method of biasing ferrite toroids in a coaxial cavity has been to apply the biasing magnetic field in the same direction as the magnetic RF field. This is usually done in double ended coaxial tuners by placing the conductors for the biasing current inside the field-free inner RF conductor.

A new method proposed by the LAMPF II design group is to apply perpendicular biasing i.e., either longitudinally or radially to the toroids. Perpendicular biasing reduces the RF losses in the ferrite material.

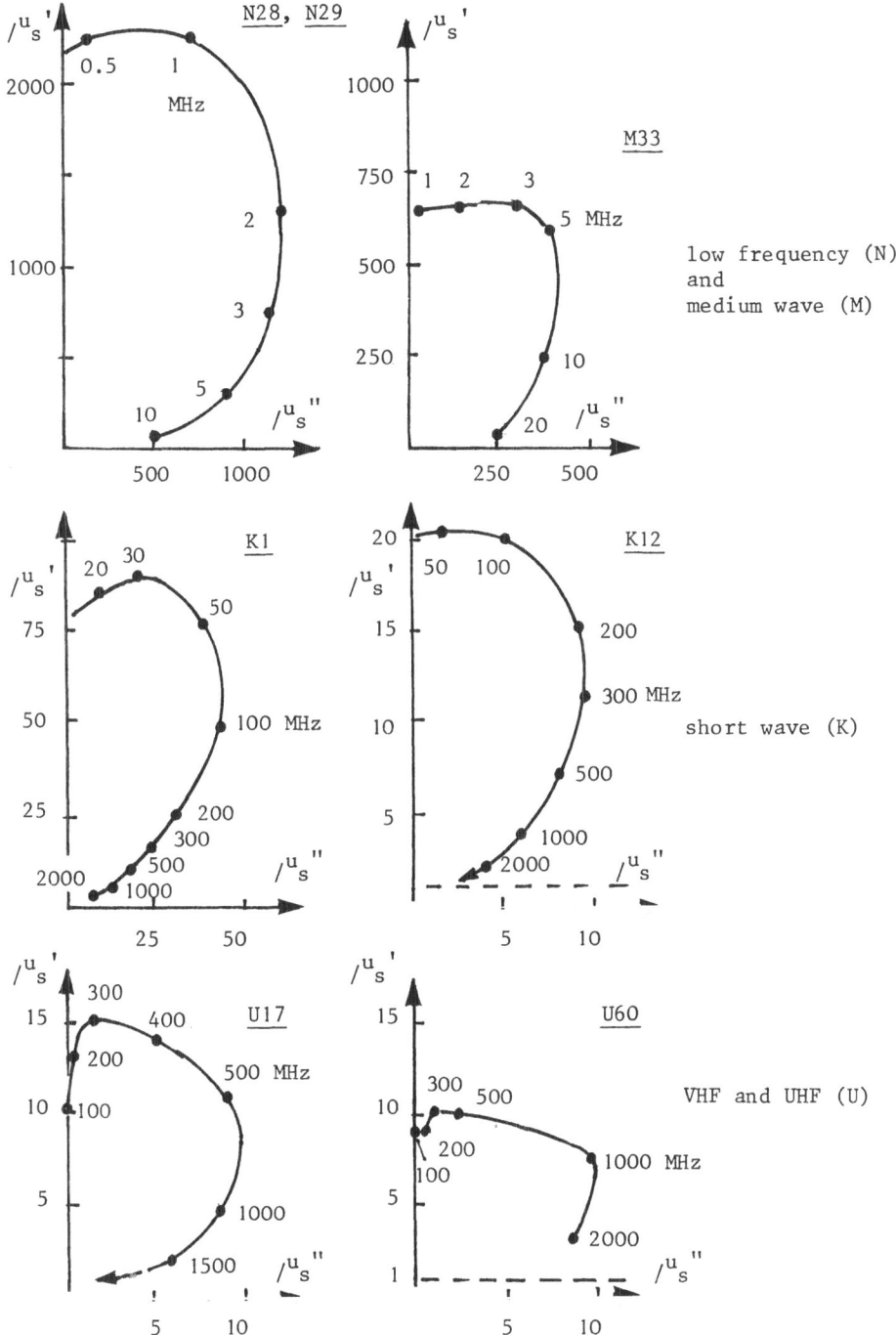

Fig. 24. Complex relative permeability of Siemens ferrites.
$/u_s'$ = inductive component, $/u_s''$ = loss component

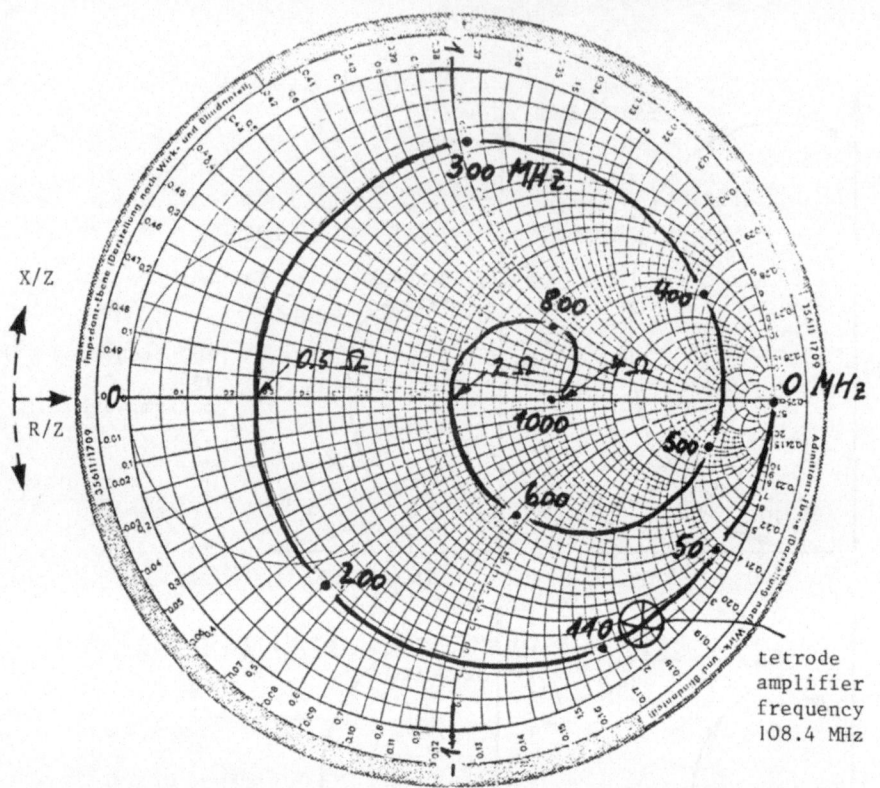

Fig. 25. Measured impedance of a coaxial ferrite absorber from 0 to 1000 MHz
(Smith diagram with Z = 2 Ohm).   Ferrite material U 60.

Fig.26 shows some measurement results on a VHF test cavity. The curve
labeled Q_r is the theoretical Q of the cavity resonator with a lossless ferrite, the other curves show Q values for perpendicular biasing, and in one
case parallel biasing for the purpose of comparison. The material which was
investigated is an aluminium-doped yttrium-iron-garnet ferrite Y 1 and a
magnesium-manganese-aluminium ferrite G 26 (both from TDK Co.).

Table 3. Calculated RF losses of the LAMPF II six-toroid prototype cavity
for different ferrites.

ferrite	electric Q	saturation magnetisation (Oe)	ferrite dissipation mag. mW/cm^3	ferrite dissipation elect. mW/cm^3	ferrite dissipation total mW/cm^3
TDK Y 300 Al garnet	>1000	300	370	97	467
TDK Y 5 Al garnet	>1000	600	82	97	179
TDK Y 1 Al garnet	>1000	1200	44	97	141
Trans Tech G 810 garnet	>1000	800	61	97	160

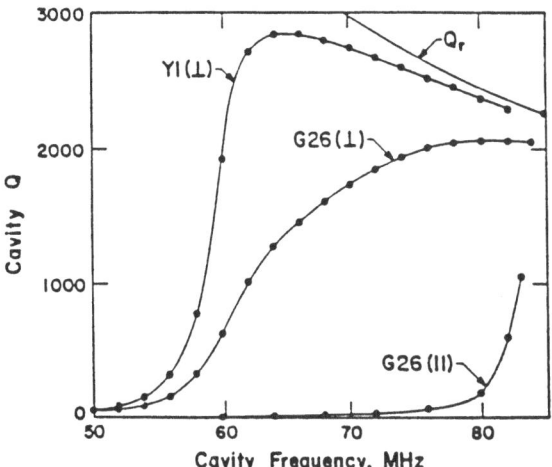

Fig. 26. Effect of perpendicular biasing of ferrite samples in a test cavity (LAMPF). The G 26 curves show the measured difference of cavity Q between parallel and perpendicular biasing.

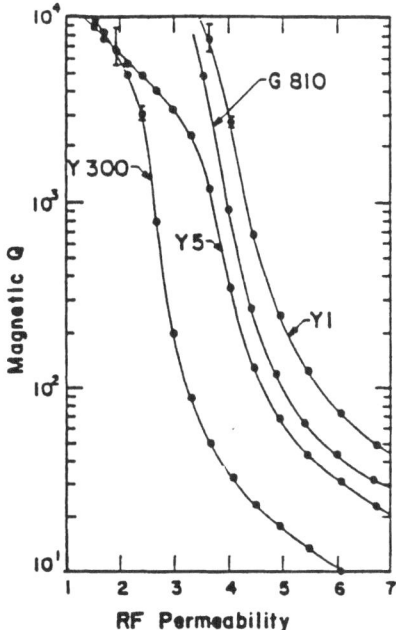

Fig. 27. Magnetic Q and RF permeability of the ferrites listed in table 3.

Dominant problems with ferrite-tuned cavities for hadron facilities are high gap voltage, typically 100 kV, wide tuning range in frequency, up to 25%, and efficient cooling of the ferrite. A permeability change occurs near the Curie temperature, for instance near 180°C for the Y 5 Al garnet. Mean power loss densities of about 0.2 W/cm^3 are considered to be the maximum values for stable operation. The magnetic Q decreases with higher power levels, for instance from 2300 to 1400 for a pulsed power density of 1.3 W/cm^3, as measured with a μ_{RF} of 3.1 at 71.5 MHz in the test setup of LAMPF.

The RF permeability $\mu_{RF} = \Delta B/\Delta H$, and the useful range for tuning are relatively small in comparison with other ferrites of higher μ but lower Q. Desirable is a maximum value of the product: magnetic Q x range of μ_{RF} x permissible increase in temperature.

10. Isolators and Circulators

The first high-power ferrite isolator for 500 MHz operation was specified and tested at DESY in 1964. It was designed for 400 kW pulse and 100 kW average RF power in the forward direction, and a maximum of 20 kW average dissipated power. Manufacturer was Raytheon Co./USA.

The isolator was installed in the waveguide run to the synchrotron RF cavities in order to protect the klystron output circuits against excessive reflections. Such reflections had to be considered as regular operating conditions, mainly due to high beam loading ratios in the high-Q cavities, poor vacuum pressure in the synchrotron vacuum chamber, and some other reasons for reflected RF power. 12 db isolation and 0.35 db insertion loss were achieved. The isolator was built in the form of a watercooled rectangular waveguide section (18" x 9") of about 3 m length, as shown schematically in fig.28. The nonreciprocal transmission characteristic was achieved by glued-on ferrite elements and permanent magnets at the proper distance from the middle of the waveguide i.e. near the area where circular polarisation of the RF magnetic field occurs.

The next step to handle more reflected RF power was envisaged at DESY in 1965, calling for a circulator which could possibly transmit 1 to 2 MW pulse and 300 kW average RF power, for synchrotron operation at 7.5 GeV. During the construction phase of the 3 GeV double storage ring DORIS, specifications were issued for 250 kW CW circulators. A phaseshifter circulator has been developed, at that time, by Telefunken.

The principle of using nonreciprocal phaseshifters for the construction of circulators is illustrated by fig.29. We do not have to add many comments on the functioning of the several possible arrangements of magic T's, 3 db couplers, etc. The schemes should be self-explaining. In practice it is also a question of how to orient the many waveguide input and output connections

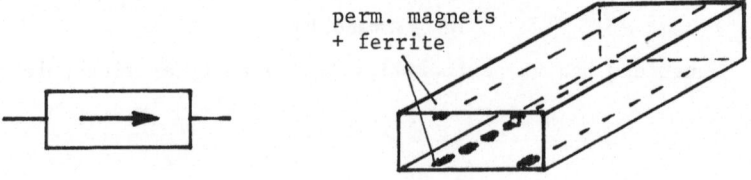

perm. magnets + ferrite

Fig. 28. Symbol of an isolator (left), and waveguide version (right).

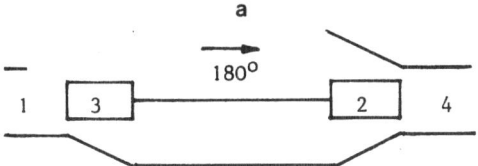

two magic T's and one nonreciprocal 180° phaseshifter

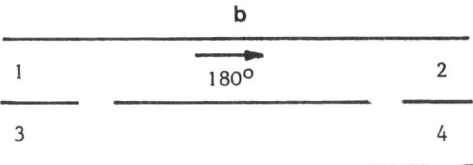

two 3 db couplers and one nonreciprocal 180° phaseshifter

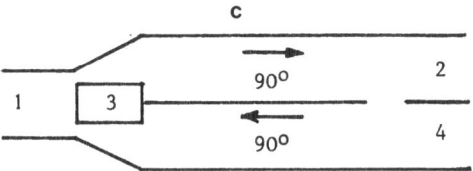

one magic T, one 3 db coupler and a nonreciprocal $\pm$ 90° phaseshifter

Fig. 29. Phaseshifter circulators, schematically (after Pivit).

since the dimensions of WR 1800 waveguides are still relatively large. A 500 MHz circulator including the necessary magnet for the ferrite is heavy.

The most interesting feature of circulators is the fact that the reflected RF power is no longer dissipated in the ferrite but transferred to a (preferably coaxial) water load. Such dummy loads can be built for several hundred kilowatt CW power.

We should come to the CERN-LEP specifications for an UHF circulator (1978). The performance data are

frequency	=	350 MHz
power	=	1.2 MW nominal CW
reverse power	=	25 % i.e. 300 kW maximum
insertion loss	=	0.15 db maximum
isolation	=	20 db nominal
waveguide	=	WR 2300 (23" x 11.5 ")

Symbolically, the flux of forward and reflected power is shown in fig.30.

Raytheon proposed a solution according to the scheme of fig.29c: 2 nonreciprocal 90° phase shifters (4 sections of 45° each) with reduced waveguide height ($\approx$1"), quarter wave transformer transitions on each phaseshifter, a folded H-plane magic T on the load side, and a 3 db coupler on the generator side (see fig. 31).

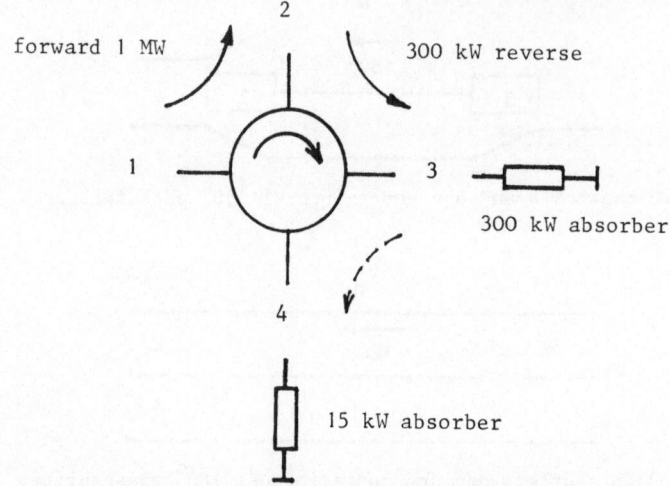

Fig. 30. Symbolic specification of the 350 MHz CERN-LEP circulator.

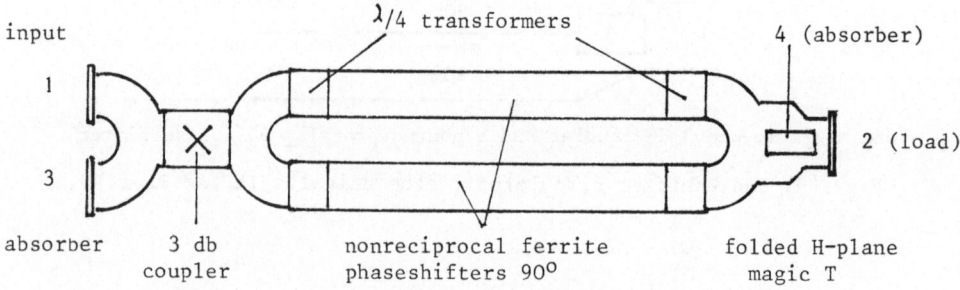

Fig. 31. Structure of the proposed CERN-LEP circulator (Raytheon).

The total length of this phaseshifter circulator is about 7 m, total weight about 3.5 tons. The ferrite is an yttrium-aluminium garnet (YAG) with about 1000 Gauss saturation magnetisation and a Curie temperature of about 200°C. The main contribution to the overall weight comes from the 1200 Gauss permanent magnet.

A much more compact solution are <u>junction type</u> (Y) high-power circulators which are fabricated for DESY by ANT Backnang (1985). They consist of a stack of ferrite discs at the center of a Y-type waveguide junction (see figs.32 and 33). CW operation up to 700 kW at 500 MHz has been possible.

It is immasing how much power can be handled by these relatively small Y-circulators. Some special protection measures for reliable operation are required, however. Useful are arc detectors to diagnose flashovers and to interrupt or reduce the supply of RF power if necessary. The insertion of thin Kapton foils to prevent dust particles to penetrate into the high-field ferrite region is proposed and studied by DESY.

The total number of high-power circulators required for CERN-LEP is approximately one hundred.

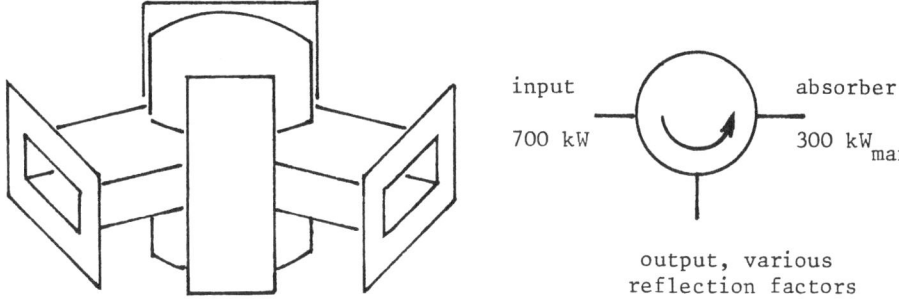

Fig. 32. Junction type circulator.

input
700 kW

absorber
300 kW$_{max}$

output, various
reflection factors

Fig. 33. Symbolic circulator
specifications.

11. Megawatt-Tetrode Amplifiers

Among the density modulated classical electron tubes, the high power tetrode has been developed to its ultimate physical limitation for RF power production in the VHF range.

At 202 MHz, a large number of tetrodes supply the necessary RF pulse and average power for the CERN-SPS. A typical 500 kW transmitter consists of 4 Siemens tetrodes which feed their output power into a set of coaxial 3 db couplers for combined transmission through a coaxial waveguide run to the accelerating units in the remote tunnel. Operation is long pulse (3s) with 50 % duty factor.

At 108 MHz, we find another interesting example, the Unilac heavy ion linac at GSI. The Alvarez RF structure of the Unilac requires pulses of up to 1.6 MW with a repetition rate of 50 Hz and a duty factor up to 25 %. A Thomson-CSF tetrode (TH 518) has been proposed and used for this purpose. Initially there were two 1.6 MW amplifiers. The number of tanks and amplifiers has been increased at a later stage.

Outstanding features of the chosen tetrode are the maximum ratings of anode dissipation (600 kW), screen grid dissipation (11 kW) and control grid dissipation (5 kW). Control grid and screen grid cages are made of pyrolythic graphite which has a very high melting point and at the same time an excellent conductivity.

The operational losses on control grid (g1) and screen grid (g2) by electron bombardment and RF currents are a critical point. RF displacement currents up to 1000 A must flow in the active anode/screen grid region (115 pF) at 108 MHz for an output power of 1.6 MW. This means an important additional stress to the screen grid. The situation for the control grid is also critical because of the very small distances between cathode, grid and screen grid. The interelectrode capacitances are 650 pF (cathode-grid) and 900 pF (grid-screen grid).

We may briefly consider the laws of current excitation in a tetrode. The electron emission from the cathode follows the space-charge law, i.e. $I_c = K \cdot V_{st}^{3/2}$, with the 'perveance' K and the 'steering voltage' V_{st}. The steering voltage is composed as $V_{st} = V_{g1} + D_2 V_{g2} + D_a V_a$. In this sum

V_a means anode-cathode voltage, V_{g2} screen grid-cathode voltage and V_{g1} grid-cathode voltage. The values of D are D_2 typically 20 %, D_a about 0.5 %. D is the 'inverse amplification factor', given by the capacitance ratios $D_2 \approx C_{g2-c}/C_{g1-c}$ and $D_a \approx C_{a-c}/C_{g1-c}$.

The cathode current $I_c(t)$ is the sum of anode current, screen grid and control grid current. As long as the anode-cathode voltage exceeds the voltage of the screen grid against cathode, and for small positive voltages between control grid and cathode, the main part of the cathode current flows into the anode. Hence, $I_a(t)$ is nearly equal to $I_c(t)$.

The RF input voltage is superimposed to a negative DC bias voltage on grid 1 and changes periodically the cathode current, as shown in fig.35. Here, we distinguish 3 different classes according to the value of the current conduction angle Θ. We discard class 'A' because of the high anode dissipation, and prefer to operate in class 'B' or 'C' which promises better efficiencies.

Typical current characteristics for the super power tetrode TH 518 are shown in fig.36, and a few operational parameters for 3 different pulse power levels are given in table 4. The required peak anode current reaches 300 to 400 A for pulses above 1 MW. The control grid voltage must be driven far into the positive range eg V_{g1} = +150 V, I_{g1} = 20 A, I_{g2} = 20 A.

A cross-section of the anode circuit of the Unilac Alvarez amplifier is sketched in fig.37. The anode of the tetrode is galvanically connected to the inner conductor of a folded full wave coaxial resonator. The screen grid terminal is on ground level, RF operation is based on grounded grid and grounded screen grid RF potential (theoretically) which enables the tetrode to operate as an amplifier without neutralisation. This means that the RF input voltage and tetrode anode current are in a mutual phase relation which results in a power transfer from input to output of the magnitude $V_{RFin} \cdot I_{a1}/2$, and hence, the output circuit loads the input circuit.

The RF voltage and current distribution along the anode resonator is also shown in fig.37. There is a first voltage node near the tube ceramic (screen grid terminal) and a second one in the outer part of the resonator. At this location, anode DC voltage and cooling water are introduced. In between the 2 nodes the RF amplitude reaches a maximum value of 35 kV, and the total voltage with a superimposed 20 kV DC voltage goes up to 55 kV.

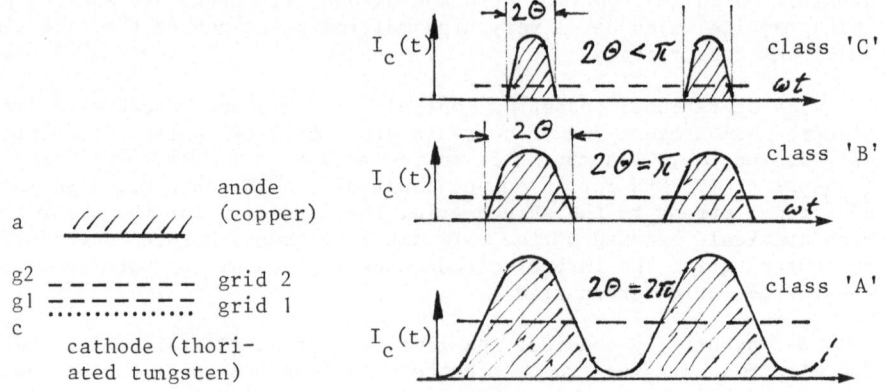

a   //////   anode (copper)

g2   --------   grid 2
g1   ========   grid 1
c

cathode (thoriated tungsten)

Fig. 34. Tetrode electrodes.   Fig. 35. Classes of operation.

Table 4. Megawatt-tetrode amplifier (TH 518),
        typical operational data for 100 Ohm anode load resistance.

desired pulse output power		0.5	1	2	(MW)
peak RF voltage, fundamental	$\hat{V}_{a1}$	10	14	20	(kV)
peak RF current, fundamental	$\hat{I}_{a1}$	100	140	200	(A)
required peak anode current	I_{apk}	200	280	400	(A)
anode pulse current	I_{ap}	64	89	128	(A)
DC anode current with 25 % duty cycle	I_{adc}	16	22	32	(A)
anode power dissipation with 25 % duty cycle		63	125	250	(kW)
efficiency with assumed RF/DC voltage ratio of 85 %		67	67	67	(%)

Note that for <u>class B</u> $I_{apk} = \pi I_{ap}$

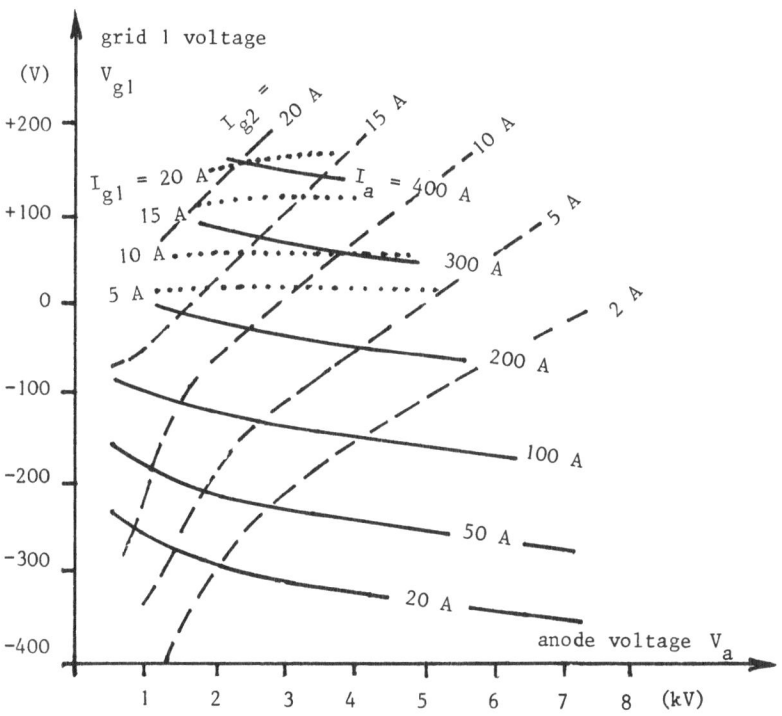

Fig. 36. Constant current characteristics for a 2 MW tetrode (TH 518)
        at V_{g2} = 1750 V.

303

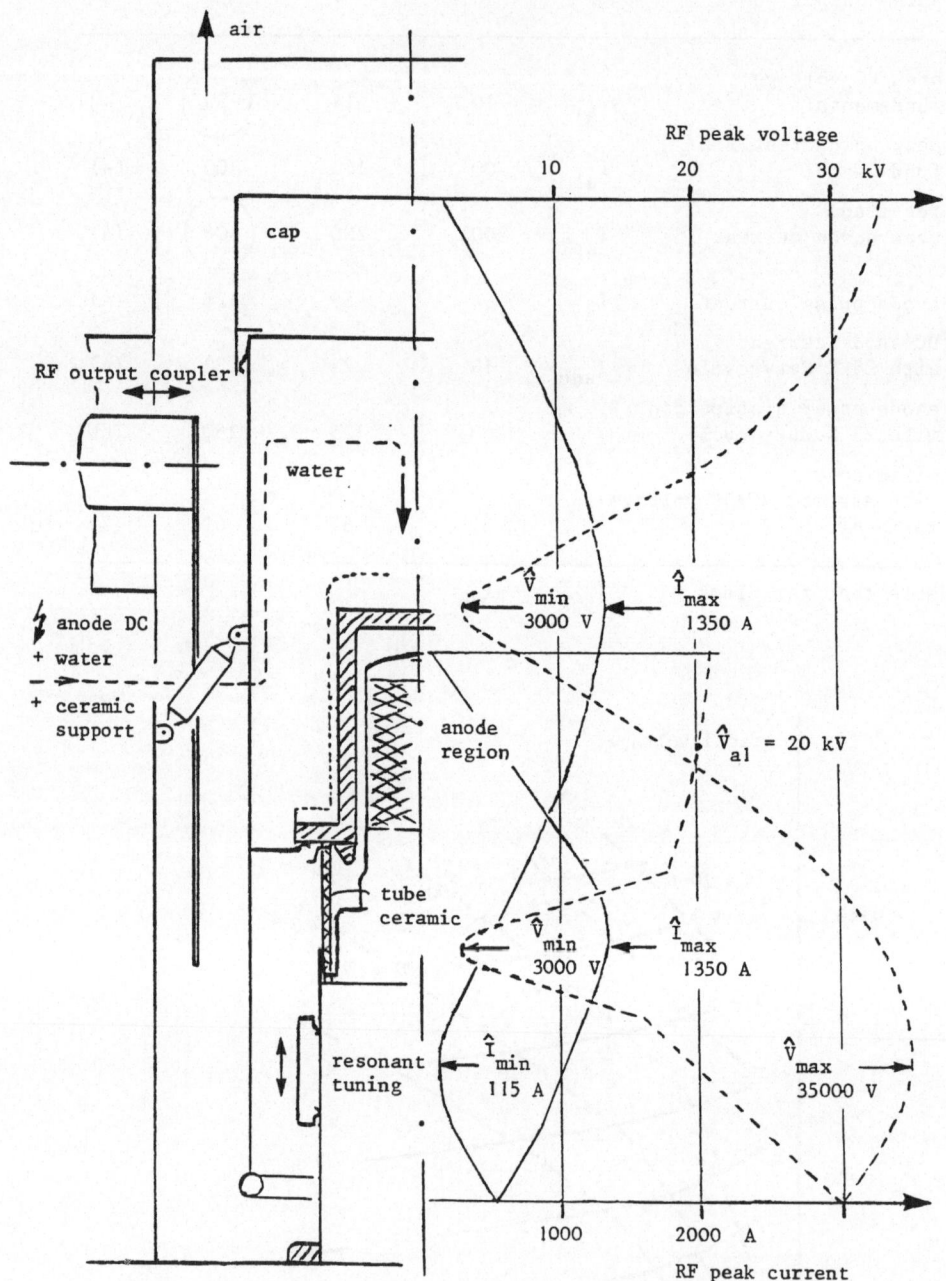

Fig. 37. Anode cavity of a 2MW tetrode amplifier for 108 MHz (Unilac), schematic cross-section, with calculated RF voltages and currents.

A cross-section of the RF input circuit is shown in fig.38, together with the active part of the tetrode i.e. cathode and control grid with their terminals. This coaxial structure is a half wave resonator until down to the 50 Ohm coaxial input line. A quarter wave falls entirely into the evacuated part of the tube. The RF voltage and current distribution along the resonator and inside the grid-cathode region is also shown in fig.38, based on 35 kW RF input power. The cathode impedance is transformed from its low value of 1 Ohm inside the tetrode into 250 Ohm and subsequently into 50 Ohm. Grid and screen grid are coupled by a half wave line section (not shown).

Operational troubles: The 'transconductance' dI_c/dV_{st} is proportional to $I_c^{1/3}$ and reaches relatively high values which enable the high Q internal structure of the tetrode to start microwave oscillations in circumferential modes. The circumference is about 45 cm, hence circumferential modes may become alive at this and at shorter wavelengths. Since the characteristic impedance of the cathode-, grid- and screen grid structure is extremely low, namely of the order of 1 Ohm but the inductances to the outer terminals are high, the 'grounded grid/screen grid' principle no longer applies and self-oscillations may result. Under such circumstances the anode current goes quickly up and the grid structure may be damaged. Because of the high inductances to the outer terminals, there is practically no chance left to couple external microwave absorbers efficiently to the microwave resonant circuits inside the tetrode. In practice, an improvement has only been possible by placing many ferrite absorbers into the anode resonator.

The RF power limitations of the tetrode amplifier can be stated as follows:

(1) the length of the active part of the tube must be short compared to a quarter wavelength,

(2) the diameter cannot be increased ad infinitum because of transverse mode excitation (circumferential resonances),

(3) length x diameter is therefore limited, this limits the maximum anode dissipation (not more than 0.5 to 1 kW/cm^2),

(4) the distances between cathode, grid and screen grid must be kept very small to keep the transit time of electrons within given limits,

(5) this in turn increases the risk of flashover and damages,

(6) also, the input capacitance becomes extremely high,

(7) spurious oscillations at UHF frequencies are possible (this is usually not sufficiently cured and may require an enormous amount of manpower - during operation).

But: good results with moderate power levels are obtained, at least up to 200 MHz. The possibility of coupling many amplifiers to a common load (eg at CERN SPS) to achieve 500 kW or more, is basically unlimited.

12. Cost Estimates

For large RF systems such as for LEP and for HERA, cost estimates were given in the related study reports from 1978, 1979 (LEP) and 1981 (HERA), based on experience at that time.

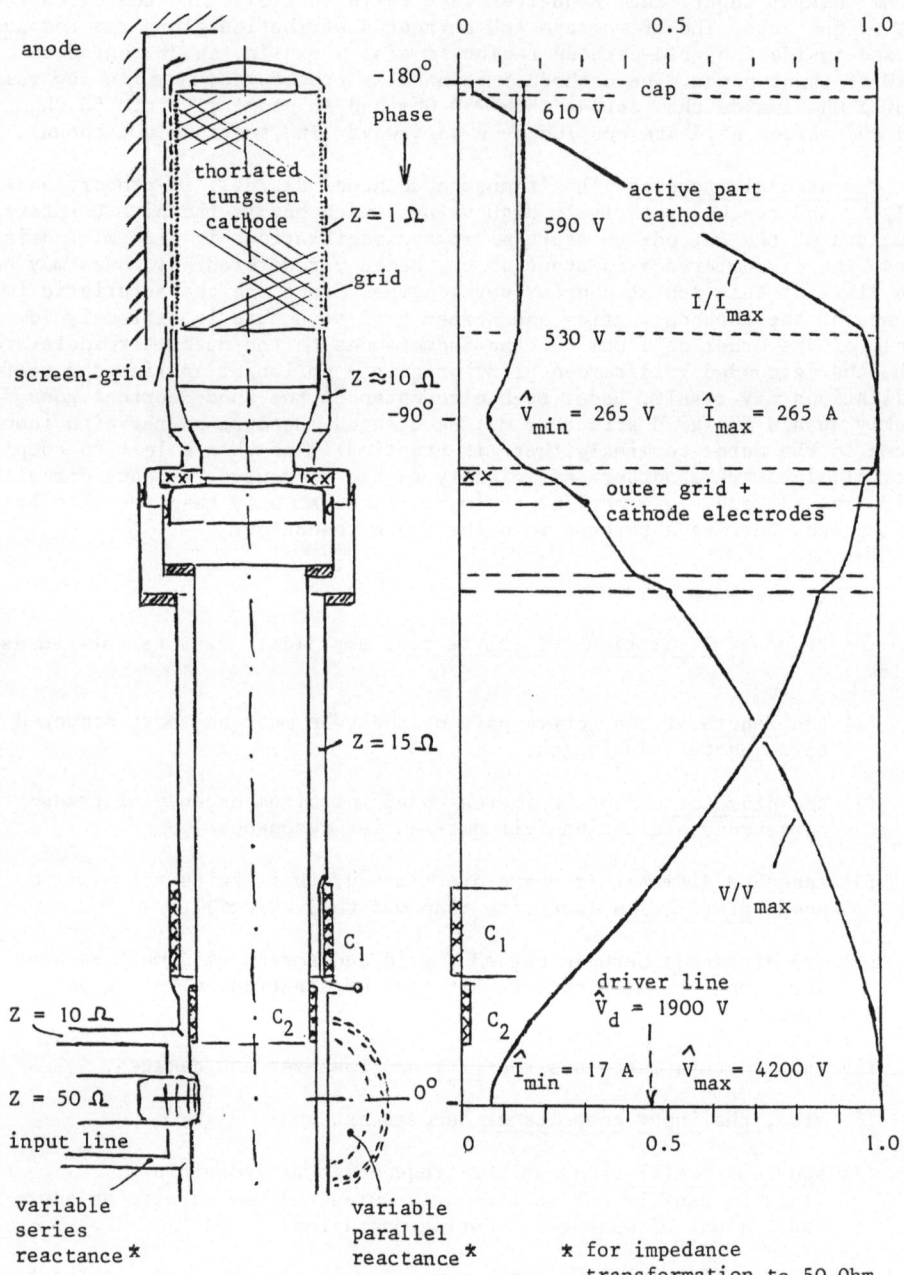

Fig. 38. Input circuit of the 2 MW tetrode amplifier, with RF voltage and current distribution for 35 kW input power. Ceramic capacitors C_1 and C_2 separate the DC potentials of grid and cathode from screen grid (= ground) potential.

For the electron ring of HERA, half of the RF equipment will be taken out from PETRA. This is 6.6 MW RF power (12 klystrons and 112 cavities). The cost of 12 additional klystrons (500 MHz CW), complete with high-voltage part, was estimated at 9.5 MDM, the additional HV power supplies (3 x 3 MW) at 8.6 MDM, 80 additional five-cell cavities with waveguides, circulators + controls at 17 MDM. This gives a total investment of 35.1 MDM for a 6.6 MW CW system at 1981 prices. In the meantime, the material index has risen by about 19% and the global index by about 23%. An approximate idea of present day cost would be around <u>6.5 MDM per MW</u>.

For LEP, the RF system was originally designed for operation at 357 MHz fundamental frequency and 1071 MHz third harmonic operation. The cost estimate in 1978 prices for the RF system was the following: (a) 73.5 MSF for 80 klystron units, 1 MW each, including modulating tubes, circulators, all waveguides and coaxial loads, (b) 75.9 MSF for 640 five-cell cavities at 357 MHz, (c) 12.6 MSF for the low-power 357 MHz system, (d) 35.9 MSF for the third harmonic system consisting of 16 klystrons, 365 kW each, 512 five-cell cavities, and other components, (e) 31 MSF for 80 klystron power supplies, each 1.6 MW DC, and (f) 3.4 MSF for 16 klystron power supplies (3rd harm.), each 600 kW DC. This gives a total of 232.3 MSF for an 85.7 MW RF system or around 2.7 MSF/MW plus about 25 % index increase, resulting in approximately <u>4.1 MDM/MW</u>.

The subsequent cost estimate from 1979 for a 353 MHz LEP RF system included 96 klystrons of 1 MW each, 768 five-cell cavities + 768 storage cavities (353 MHz), drive and control system, a third harmonic system (16 klystrons, 400 kW each, 256 seven-cell cavities), 96 klystron power supplies, each 1.44 MW DC, and 16 klystron power supplies, each 600 kW DC. The total estimate on capital investment resulted in 287.9 MSF for 102.4 MW, or 2.8 MSF/MW, which is practically unchanged compared to the previous figure.

SPECIAL LITERATURE

E. Demmel, UHF-Hochleistungsklystrons für Teilchenbeschleuniger und Plasmaheizung, NTG Fachberichte, IEEE Fachtagung Mai 1980

E. Pivit, Ein Hochleistungszirkulator für das Deutsche Elektronen-Synchrotron DESY, Intern. Elektron. Rundschau <u>25</u> (1971)

W. Hauth, S. Lenz, E. Pivit, Entwurf und Realisierung von Hohlleiter-Verzweigungszirkulatoren für höchste Leistungen, Frequenz <u>40</u> (1986) 1

BOOKS

C.G. Montgomery et al., ed., Principles of Microwave Circuits, Radiation Lab. Series Vol.8, Mc Graw-Hill 1948

E.L. Ginzton, Microwave Measurements, Mc Graw-Hill 1957

Th. Moreno, Microwave Transmission Design Data, Dover Publ. 1958

Ph.H. Smith, Electronic Applications of the Smith Chart, Mc Graw-Hill 1964

L. Young, ed., Advances in Microwaves, Vol.I and IV,
    Acad. Press 1966 and 1969

P. Grivet, The Physics of Transmission Lines at High and Very High
    Frequencies, Vol.1 and 2, Acad. Press 1970 and 1976

H. Meinke u. F.W. Gundlach, Hrg., Taschenbuch der Hochfrequenztechnik,
    Springer Verlag 1962 (and later ed.)

H.H. Meinke, Einführung in die Elektrotechnik höherer Frequenzen,
    2.Bd., Springer Verlag 1966

K. Küpfmüller, Einführung in die theoretische Elektrotechnik
    Springer Verlag 1965

E. Stadler, Hochfrequenztechnik kurz u. bündig,
    Vogel Verlag Würzburg 1973

NOTE

Various data on commercial products have been mentioned in this contribution. By nature, such information (on historical development and actual
trends) cannot be complete nor fully accurate. It is therefore recommended
that readers contact the manufacturers concerned whenever more detailed de-
scriptions or specifications are of interest.

The author wishes to express his thanks for friendly permission by individual persons, firms and institutions to present part of their work in
this technical review.

RADIO FREQUENCY SYSTEMS FOR PRESENT

AND FUTURE ACCELERATORS

Eugene C. Raka

Brookhaven National Laboratory
Upton
Long Island, NY 11973, USA

INTRODUCTION

In order to bunch, accelerate and store charged particle beams it is
necessary to install one or more "accelerating" gaps on the circumference
of the machine. These gaps are usually part of a resonant structure
driven by a source of radio frequency power. The frequency, phase and
amplitude of the voltage appearing at the gap(s) is controlled by servo
loops that use programmed inputs as well as error signals derived from the
particle beam itself in many instances.

The design specifications for any rf system would include the initial
(injection) and final (at maximum beam energy) frequencies and thus the
harmonic number $h = f_{rf}/f_o$ where f_o is the particle rotation frequency; the
peak gap voltage and the number of gaps; the acceleration rate in keV or
MeV per turn and hence the maximum power to be delivered to the beam; the
amount of power needed to make up for the losses in the resonant structure
and the impedance presented to the beam image current by the structure at
the principal resonant frequency and at the resonant frequencies of any
higher order modes that produce significant longitudinal or transverse
field at the gap. There are also other parameters of interest to the
designer and many of these will be referred to in our discussion of present and future rf systems.

FNAL MAIN RING

Some of the parameters for this rf system are listed in Table 1, and a
sketch of the resonator is shown in Figure 1. It consists of a pair of
$\lambda/4$ standing wave resonators folded back on themselves to conserve linear
space. The outside diameter is 0.68 m and the length 1.78 m.[1] Tuning is
accomplished by varying the bias field (in the same direction as the
coupled rf magnetic field) in stacks of ferrite rings loop coupled to the
cavity. A pair of loops are installed on each cavity as can be seen in
Figure 2. Each loop can change the stored energy in the cavity by 0.7%
thus giving a tuning range of about 400 kc (at 53 MHz).

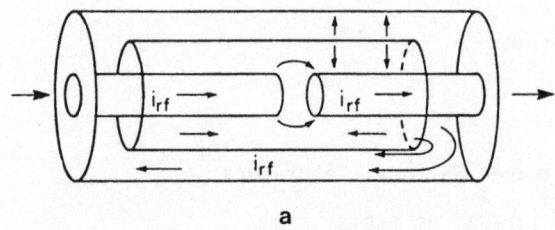

a

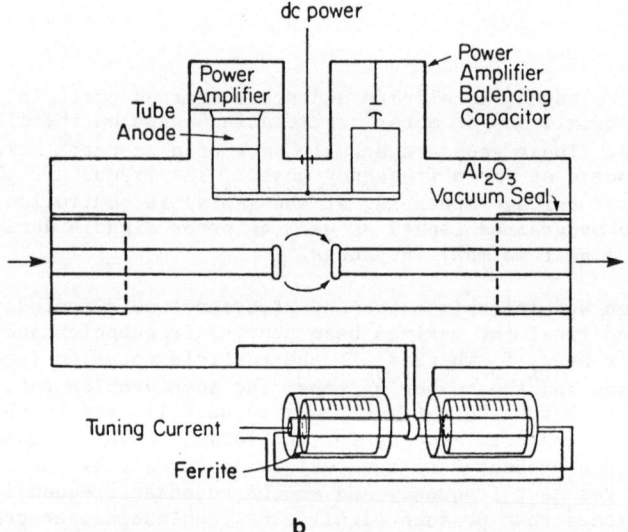

b

Fig. 1. Sketch of FNAL main ring rf cavity.
(a) Folded (14) TEM resonators.
(b) Method of coupling rf power and tuning.

Table 1. FNAL Main Ring rf System Parameters

Injection Energy	8 GeV
Final Energy (pre tevatron)	400 GeV
f_∞ (f_0 at $\beta = 1$)	47.7 kc
f_{rf}	52.81-53.104 MHz
h (harmonic number)	1113
Acceleration Rate (Φ_s 45° & 135°)	2.6 MeV/turn
Peak rf voltage (18 cavities)	3.67 MV
Peak rf voltage per cavity	>200 kV
Beam current at 3.25×10^{13} protons	≈250 mA
Cavity R_s/Q	104
Q_ℓ	5800
R_s (including PA)	600 KΩ
Excitation power @ 204 kV/gap (total)	624 kW
Maximum power to beam @ 3.25×10^{13}	650 kW

Now in addition to tracking the change in the rf frequency during acceleration, the tuning loop must also correct for changes in beam loading. Usually the phase of the total voltage appearing across the gap is compared to the phase of the drive current applied to the final stage of the power amplifier. Zero error signal corresponds to a real load impedance (shunt impedance of the resonator plus beam impedance). In general the resonator will be detuned by an amount $\Delta\omega = (\omega_{rf} - \omega_r) = -\omega_r \tilde{I}_b R_0 \cos\Phi_s / 2V_c$ where $\tilde{I}_b$ is the component of the beam image current at ω_{rf}, $R_0 = R_s/Q$ and ω_r is the nominal resonant frequency of the cavity. In the FNAL main ring the transition energy occurs at ≈17.6 GeV so the fractional detuning due to beam loading changes sign. Hence a change of $2|\Delta\omega|$ must take place. Since normally the tuning loop response is quite slow compared to the time it takes the cavities to respond to a phase jump of $2(\pi/2 - \Phi_s)$ the power amplifier will not see a real load for a brief period and hence must supply some reactive current. We note that for 3.5×10^{13} in the main ring $\Delta\omega/\omega_r \cong 0.9 \times 10^{-4}$ or $\Delta f \approx 4.7$ kc.

Fig. 2. FNAL main ring cavities (July 1976).

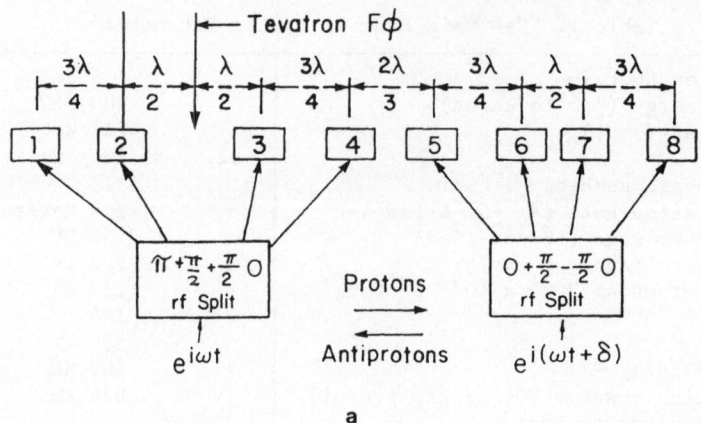

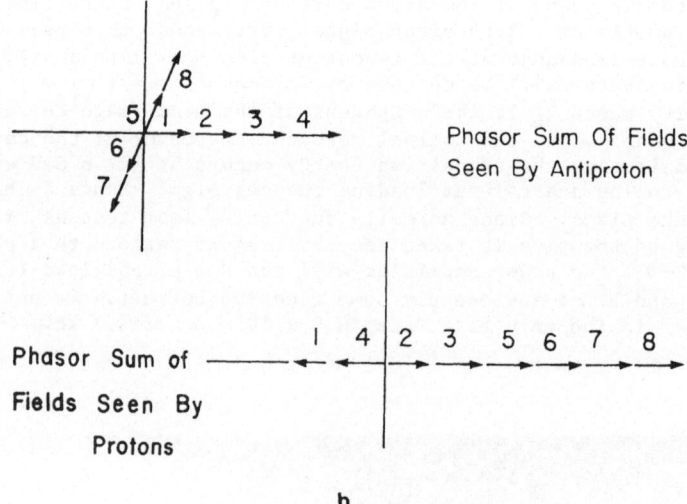

Fig. 3. Tevatron rf power distribution system.

In order to damp higher order resonant modes in the structure, nickle zinc ferrite was installed in enclosures located at the end walls of the cavity which are iris-coupled to the resonator.[1] It was later found necessary to install a ferrite loaded waveguide at the central low impedance point of the cavity in order to couple to some modes which have a large magnetic field there but not at the end walls.[2] This addition eliminated the coupled bunch longitudinal instabilities observed at 8×10^{12} protons/pulse and no further longitudinal instabilities were observed up to a peak intensity of 3.25×10^{13}.

The FNAL Tevatron ring is a superconducting accelerator and storage ring. In the fixed target mode it accelerates protons to 900 GeV and in the colliding beam mode it accelerates three protron and three antiproton bunches to 900 GeV and then stores them at this energy for several hours. Eight resonators are installed on the ring in two groups of four, each spaced in such a manner that by proper phasing, protons and antiprotons can be accelerated simultaneously (see Figure 3). In Table 2 we summarize the parameters of this system. Typical acceleration time for $p,\bar{p}$ is ≈25 seconds with 1.4 MV available for each group of bunches. The acceleration rate for the fixed target mode is not much greater since the limiting factor is the ramp rate of superconducting magnets. At 2×10^{13} protons/pulse the maximum power delivered to the beam is ≈20 kW per station while the resonator excitation power alone is 54 kW. Thus static beam loading will never be severe in this machine (this can occur when the power delivered to the beam is equal to the cavity excitation power plus the rf power dissipated in the internal impedance of the power amplifier).

TABLE 2. FNAL Tevatron rf System Parameters

Injection Energy	150 GeV
Final Energy	0.8-0.9 TeV
f_ω	47.7 kc
f_{rf}	53.104 MHz
h	1113
Accelerator Rate (6 stations @ 1.8 MV peak)	785 keV/turn
Peak voltage (8 cavities)	>2.4 MV
Peak voltage/cavity (two gaps)	360 kV
Cavity power at peak voltage	54 kW
R_s (two gaps)	1.2 M
Φ_o	7100
τ_c (unloaded cavity time constant)	42 μ/sec
τ_c (loaded)	12 μ/sec

In Figure 4 we show a drawing of the resonator[3] and in Figure 5 a cross section of the accelerator tunnel where the power amplifiers and resonators are located. The power tube is an Eimac ¥576B (special 4CW 200,00E) whose output capacitance along with its enclosure is part of a quarter wave resonator tuned independently to 53.104 MHz. A 9-3/16 inch 50Ω transmission line is connected to the anode end of the resonator and it feeds the tube output to the cavity resonator on the ring. The length of this line is adjusted to be 2 wave lengths at 53.104 MHz so that the real and reactive component of the load (beam and cavity) are transmitted to one to the tube circuit.[4]

The two halves of the resonator are driven in parallel while the electrical length of the resonator at 53.104 MHz is 170° so that V_{eff} sin (170°/2) = 0.995 V_{gap}. There are two sets of modes that can be excited in the resonator; one set is characterized by both gaps oscillating in phase and these modes couple to the transmission line and hence are controlled by the source impedance; the other set is characterized by the gaps oscillating 180° out of phase and these result in a current maxima at the center of the resonator. The Hipernom resistor located at the center of the drift tube serves to damp these resonances.

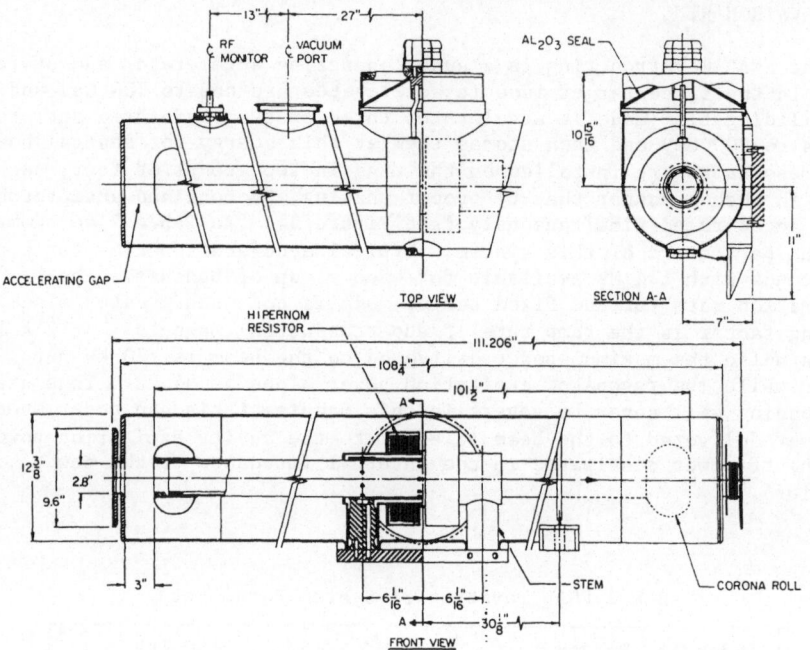

Fig. 4. Prototype tevatron resonator.

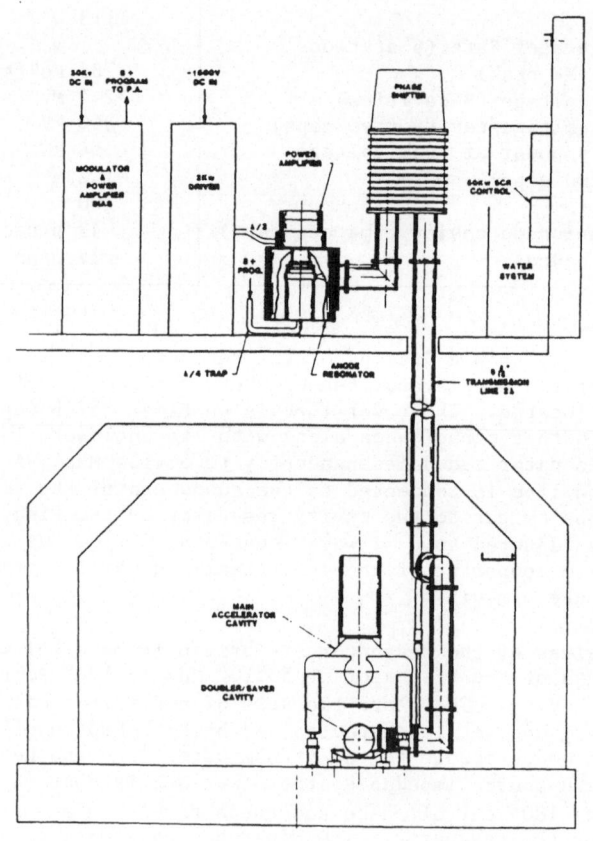

Fig. 5. Tevatron rf station installation.

Tuning is accomplished by controlling the cooling water applied to the drift tube and coils on the outer walls. There is no beam controlled turning loop since at 180 kV/gap and 2×10^{13} protons, the required detuning is ≈3.75 kc or only one half of the natural bandwidth of the resonator.

Now Figure 3(a) shows the relative location of the eight resonators and Figure 3(b) demonstrates how they act on the proton and antiproton bunches in the collision mode for an arbitrary phase angle δ. The angle δ can be used to move the collision point of the bunches but the normal value for the spacing shown is $\pi/6$. In Figure 6 we see a cavity installed on the ring along with the 9-3/16" coaxial feed from the power amplifier.

CERN SPS

Originally the CERN Super Proton Synchroton was designed only to accelerate protons. In the early 1980's it was adapted to accelerate and store simultaneously protons and antiprotons. Before 1990 it will function as an injector for the CERN LEP accelerator by providing both electrons and positrons at 20 GeV. In order to perform these functions, it has been necessary to augment the initial 200 MHz travelling wave system for pp̄ physics and it will be necessary to add 32.200 MHz standing wave cavities for lepton acceleration.

Travelling Wave System

In designing the SPS rf system, it was decided to employ travelling wave structures.[5] This permitted the location of the power source at ground level while the accelerator tunnel was located 60 m underground. The structures always present a matched load to the source independent of the frequency and beam loading. Since the required frequency swing (see Table 3) was less than 5×10^{-3} it was possible to design a structure with the desired bandwidth, thus removing the need to tune a resonator.

The rf system specifications are shown in Table 3. One parameter not listed is $\tau = (\omega - \omega_o) \, 1/V_g$ the total phase slip between the travelling

Fig. 6. Tevatron Cavity on Ring with 9-3/16" coaxial feed.

Table 3. SPS p$\bar{\text{p}}$ Acceleration System Parameters

Injection energy p's	10 GeV (f_{rf} = 199.526 MHz)
Injection energy p$\bar{\text{p}}$	26 GeV
Final energy p's	400 GeV (f_{rf} = 200.396 MHz)
Final energy p$\bar{\text{p}}$	270 GeV
Transition energy	24 GeV (f_{rf} = 200.222 MHz)
Peak acceleration voltage	4 x 2 MV
f_∞	43.2 kc
h	4631
Beam current at 3 x 10^{13} p's	210 mA
Acceleration rate (nominal)	≈2.5 MeV/turn
Beam power at 3 x 10^{13}	525 kV
Incident power required (4 cavities at 0.9 mV peaks ϕ_s = 45°, 135°)	678 kW
Power dissipation in 4 cavities @ 0.9 mV	13 kW
R_s/cavity (6 MΩ/m)	121 MΩ
Beam cavity coupling impedance r_c	1.4 MΩ
τ_c (cavity filling time)	0.7 μ/sec
ϕ	19,650
V_g (group velocity)/c	0.0946 (backward wave)
Interaction length (54 cells)	20.196 m

wave of group velocity V_g and the proton bunches along the structure length 1. Here ω_o is the rf frequency at which there is perfect synchronism between the particles and the wave. In order to be able to jump the stable phase angle at transition without changing the amplitude of the incident, wave ω_o is chosen to be equal to ω_{rf} at the transition energy.[5]

The accelerating structure which consists of identically spaced drift tubes supported by horizontal bars is shown in Figure 7. Since the passband of the bar structure is of the backward wave type, the power is fed into the downstream end of the cavity. It operates in the π/2 mode at ω_o i.e. the phase shift from bar to bar is 90°. A cavity consists of five sections of 11 cells, each of which is 3.74 mm long, including two half cells for input and output coupling. Fine tuning of the overall structire to ω_o is achieved by controlling the cooling water temperature.

There are many higher order modes of the structure that are synchronous with the beam. Since these couple poorly to the input and output loops two addditional loops are installed on the ends of the cavity to damp these modes without extracting significant power from the main resonant mode. Transverse deflecting modes have also been identified in the structure and additional loops were installed to damp them.[6] Because of the wide bandwidth of the cavities, there is significant impedance at frequencies ±nf_o from the driving frequency hf_o where n = 1,2.....20. Longitudinal coupled bunch instablities driven by this impedance with mode numbers of 1 to 17 have been observed and damped with feedback.[6]

Now the beam induced voltage is zero at the drive end of the cavity, so the power source does not see the beam. The beam power is subtracted from the power incident on the terminating load at the upstream end and it is only necessary to supply additional power as a function of beam current to maintain the same total cavity voltage. This required power can be written as ($\tau = 0$):

$$P_{max} = 1/8 \left[\frac{|V|^2}{r_c} + r_c I_b^2 + 2|V| I_b \cos \Phi_s \right]$$

where I_b is the component of beam current at ω_{rf}, V is the total cavity voltage and Φ_s is measured from the peak of the rf wave.

Fig. 7. Drift tube and bar assembly inside of the cavity.

Fig. 8. Cavity Installation in SPS tunnel.

There are now four travelling wave cavities installed on the accelerator, each with its own 500 kW power amplifier. A second set of power amplifiers of 2 MW capability has also been installed in order to insure that power is always available. In the $p\bar{p}$ mode one changes the direction of power input of two of the cavities so that the $\bar{p}$'s are accelerated by one pair of cavities and the p's by the other set. Since one is dealing with only six bunches in this mode the effect of the beam induced signal in either set of structures can be ignored. Since the cavity filling time including power amplifier is ≈1 µsec, while the spacing for three $\bar{p}$ or 3p bunches is 7 µsec, individual phase and amplitude control is possible for each one.

It should be noted that three cavities now have four sections each so that the four cavities total 17 sections[6] versus 15 for the original three.[5] Figure 8 shows one of the cavities in the SPS tunnel.

200 MHz Standing Wave System

In order to accelerate electrons and positrons to the 20 GeV energy necessary for injection into LEP about 30 MV of rf at 200 MHz is required. This will be provided by 32 single cell copper cavities each fed by a power amplifier capable of 60 kW CW and 110 kW peak power for a 1 second period and 20% duty factor.[7] At the design value of $|MV|$ gap (288 mm long) 60 kW is required to excite the cavity which has an R_s of ≈8.5 MΩ and a Q_o ≈49,000. The peak surface electric field is ≈12 MV/m at this voltage. A piston tuner which is servo controlled, compensates for thermal effects, mechanical variations and beam loading conditions. The total range is 400 kc which is also sufficient to cover the variations during p̄p acceleration, since in the latter case the injection energy is 26 GeV. The power tube is a Siemen's RS 2058 CJ tetrode operated class AB with a grounded screen and 10 kV anode voltage at an efficiency of 64%.

In addition to the power tube and tuning piston, both of which are mounted on the cavity, Figure 9, there are also higher order mode suppressors[7] and a fundamental mode damper. The former are two resonant loops that are water cooled, mounted so as to couple to two field polarizations. When the SPS is operating in the fixed target mode with 3 x 10^13 protons the standing wave cavities must not perturb the 4000 bunches due to beam induced voltages. Hence the 200 MHz mode must be strongly damped. This is accomplished by a magnetic coupling loop connected to a shorting stub and a λ/4 transformer terminated in a 50Ω load as shown in Figure 10. The stub compensates for the circuit reactance so that the resulting real impedance is transformed to the load which is water cooled.[7] The loop which is rectangular is plunged into the cavity during proton acceleration producing a damping factor of 500 and is retracted for lepton acceleration so that both types of machine cycles can be interleaved.

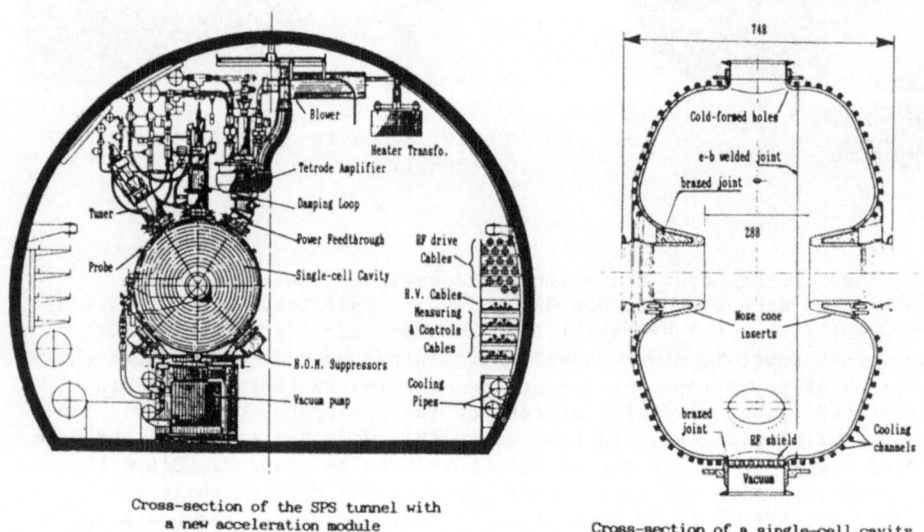

Cross-section of the SPS tunnel with
a new acceleration module

Cross-section of a single-cell cavity

Fig. 9. SPS 200 MHz standing wave cavity and ring installation.

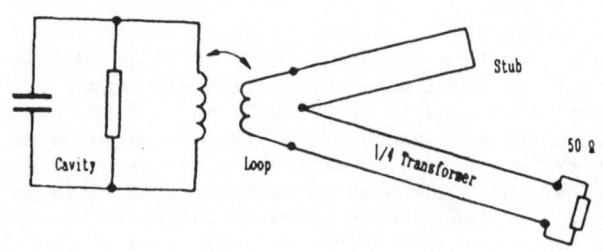

Fig. 10. SPS 200 MHz cavity damping loop schematic.

100 MHz Standing Wave System

In addition to the two 200 MHz systems mentioned above, a third set of six cavities operating at 100 MHz are also being installed on the SPS ring. These will be used to capture the larger $\bar{p}$ bunches expected from the new antiproton collector ring ACOL. During acceleration, both the 100 MHz and the 200 MHz systems (travelling wave) will be used. In the storage mode the 100 MHz system can also be used depending upon bunch length requirements.

Six cavities each giving 400 kV with 16 kW excitation are required. A mechanical tuner provides the necessary 70 kc frequency swing for $\bar{p}$ acceleration. The cavities are also equipped with higher order mode suppressors and a tuned loop that is plunged in to reduce the cavity Q by 200 during fixed target proton running and lepton acceleration.

CERN LEP rf SYSTEM

In phase I of LEP the large electron positron collider the maximum energy will be 55 GeV. The necessary rf voltage is to be supplied by 128 accelerating structures driven by 16 Klystrons, each capable of furnishing 1 MW of rf power at ≈350 MHz.[8] Figure 11 shows the power distribution system for one of the eight units, while in Figure 12 is shown their location on the LEP ring. Thirty-two units are placed on each side of two of the eight intersecting regions determined by the four e^+ and four e^- bunches circulating in the machine. Now the spacing between adjacent cavities is $\lambda/2$ while the spacing between cavities 1,16 or 2,15 etc. is an odd number of half wavelengths. Thus the rf must be 180° out of phase between each pair so that with an e^- bunch entering from the left will always see the same accelerating voltage while an e^+ bunch coming from the opposite direction will also always see the same accelerating voltage.

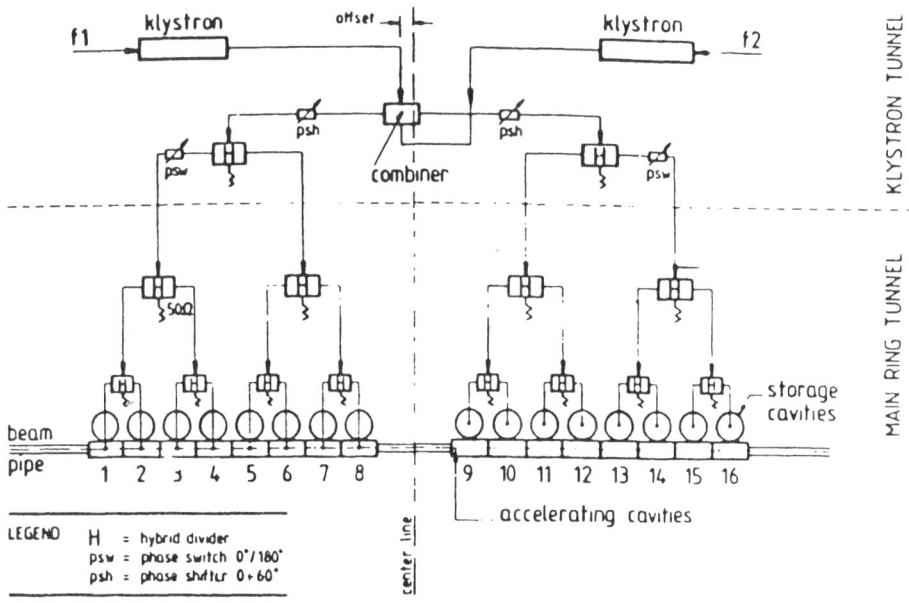

Fig. 11.   rf power distribution for one LEP unit.

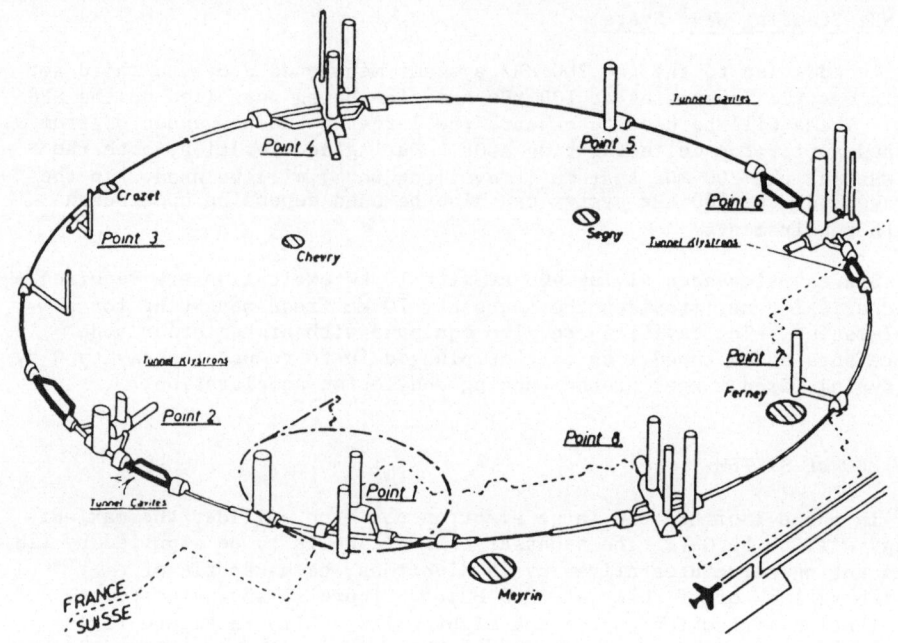

Fig. 12. LEP tunnel showing regions equipped with rf power in Phase I.

The accelerating structures are five cell π mode slot coupled cavities side coupled to a spherical storage cavity operated in the H_{011} mode. The combined structure has two resonant frequencies separated by $\Delta\omega = 2\pi \times 8f_o$ or

$$\omega_1 = \omega_{rf} = \Delta\omega/2 \qquad \omega_2 = \omega_{rf} - \Delta\omega/2$$

and if one excites them with the sum of two voltages:

$$V_i = V_o\cos\omega_1 t + V_o\cos\omega_2 t$$

the two cavity voltages will be given by[9]

$$V_1 = 2V_o\cos\omega_{rf} \cos \Delta\omega/2 \qquad V_2 = -2V_o \sin \omega_{rf} \sin \Delta\omega/2$$

Thus the amplitude of the accelerating cavity voltage will be modulated at $4f_o$ the e^- or e^+ bunch repetition frequency. Energy will be transferred between the cavities at the frequency Δf and the power dissipation is reduced by the factor q where

$$q = 1/2 \frac{(Q_1 + Q_2)}{Q_2} \quad \text{and} \quad P = 1/2 \left[\frac{V^2}{2R_1} + \frac{V^2}{2R_2}\right] = \frac{V^2 q}{2R_1}$$

Now $Q_1 = 40,000$ for the accelerating cavity and $Q_2 = 160,000$ for the storage cavity so that one has q = 0.625. Figure 13 is a drawing of the combined structure showing the storage cavity which is ≈ 1.23 m in diameter. In Table 4 we list the rf system parameters.

In phase I the reflected power due to beam loading can produce a missmatch that will result in a VSWR 1.3 at design intensity. It is expected that this can be tolerated by the klystrons and hence initially

TABLE 4. LEP rf SYSTEM PARAMETERS[8]

Frequency ($\omega_{rf}/2\pi$)	352.209 MHz
h	31320
f_∞	11.2425 kc
Effective shunt impedance of accelerating cavities (5 cells)	27.66 MΩ
Effective shunt impedance with storage cavity	42.56 MΩ
Maximum voltage gap	625 kV
Average excitation power/cavity	114.7 kW
Accelerating cavity Q	$\approx$40,000
Storage cavity Q	$\approx$160,000
τ_c for coupled system	56.5 μsec
Cavity input coupling factor	1.21
Cavity active length (5 x $\lambda/2$)	2.1279 m

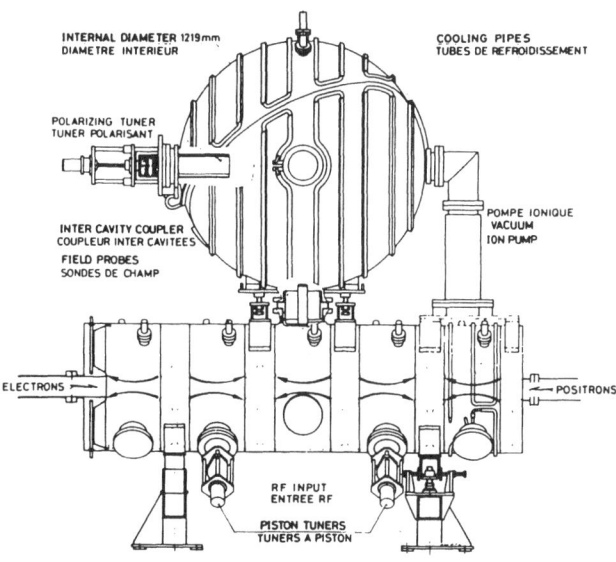

Fig. 13. LEP storage cavity and accelerating cavity assembly.

no isolators were to be installed though space was reserved for them. However during preliminary system testing, one of the klystrons suffered damage and it was decided to purchase and install isolators for all units prior to start up.

HERA PROTON ACCELERATION SYSTEM

The HERA[10] project located at the DESY laboratory in Hamburg will collide 30 GeV electrons on 820 GeV protons. The latter particles will be accelerated from 40 GeV injection energy in a ring containing super-conducting magnets. The injector will be the PETRA ring which has been

adapted to accelerate protons as well as electrons (the latter are
accelerated to 14 GeV before being injected into the HERA electron ring).
We will only describe the proton rf systems in PETRA and HERA here but
note that for electron acceleration the 500 MHz cavities used in PETRA
will be employed in both rings. Superconducting 500 MHz cavities will be
added to the storage ring in order to obtain the design energy of 30
GeV.

In Table 5 we list some of the parameters for proton acceleration.
The 2.4 MV at 208 MHz in the HERA ring will be provided by four single
cell copper cavities of the same design as the CERN SPS 200 MHz standing
wave cavities. The nose cones will be slightly longer to increase the
resonant frequency and they will be operated at 600 kV/gap. Since the
shunt impedance will be essentially the same only 21.5 kW excitation
power per cavity will be needed. The acceleration rate will be even
slower than in the FNAL Tevatron ring so that the beam power at the
design current will be only a fraction of the excitation power.

The tuning range required for acceleration is ≈50 kc which is well
within the range of the system developed for the CERN design. We note
that in the storage mode and during acceleration as well, since the
stable phase angle is very small, the cavities will be detuned by ≈4 kc,
or about twice the half bandwidth, due to reactive beam loading. It may
become necessary in the future to introduce feedback around the cavity
and driver to lower the voltage induced by the component of the beam
current at 208 MHz.

At injection this component of beam current is not expected to be
appreciable since the bunches will be too large to fit into 208 MHz
buckets. First the 52.033 MHz rf system must capture three groups of 70
bunches from PETRA (leaving a 10 bunch gap) into every fifth bucket.

Table 5. HERA Proton rf System Parameters

PETRA II injection energy (p)	7 GeV
PETRA II final energy (p)	40 GeV
HERA final energy (p)	820 GeV
rf frequency PETRA II (h = 400)	52 MHz
rf voltage PETRA II	≈200 kV
HERA rf frequency at injection (h = 1100)	52.033 MHz
Peak 52 MHz voltage	310 kV
Number of bunches	210
rf frequency for acceleration (h = 4400)	208.13 MHz
Peak voltage	2.4 MV
Average beam current	130 mA
γ_{tr} HERA	27.6
η at 40 GeV	7.63×10^{-4}
η in PETRA at 40 GeV	0.0249
f_∞ HERA	47.3 kc

We can write for matched transfer of the bunches from buckets at the same
rf frequency in both machines

$$V_H = \frac{R_H}{R_p} \frac{\eta_H}{\eta_p} V_p$$

where H and p refer to the separate rings. Now the factor $\eta = (1/\gamma_{tr}^2) - (1/\gamma^2)$ at 40 GeV is positive in both machines as this is above the transition energy in each one (since $\gamma_{tr} = 6.265$ in PETRA and the proton injection is at 7.5 GeV transition is not crossed in either machine). However as shown in Table 5 there is a large difference between them and this results in $V_H = 0.084\ V_p$ so that at 200 kV in PETRA one needs only 16.8 kV in HERA to match the phase space trajectories in both machines.

Now the second function of the 52 MHz system is to compress the bunches so that they will fit into the high frequency rf buckets. Since the bunch length for fixed bunch longitudinal phase space area is proportional to $V^{1/4}$ this process requires much less voltage in the HERA ring. The peak voltage will be raised to 310 kV giving a bunch width reduction of greater than 50%. During this process it will be necessary to insure that the induced voltage at 208 MHz is minimized so that each of the 210 bunches is captured in only one of the high frequency buckets.

The 52 MHz rf systems are being developed at the Chalk River Laboratory in Canada.[11,12] A sketch of the PETRA II cavity is shown in Figure 14. It is based on the FNAL design described in Figure 1(a). Here the intermediate cylinder is supported by a single post at the center and the cavity is in air with a ceramic seal at the gap. The power amplifier is coupled to the cavity as shown in Figure 1(b) and higher order modes are damped with a system described in reference (12). In Figure 14 we have shown only the external resonant tuning cavity. This is a $\lambda/2$ resonator which uses the coupling loop as part of the half wavelength line. It is loaded with perpendicular biased microwave ferrite at one end.[11] At high bias current the resonant frequency is far above that of the main cavity and there is little interaction between them. At low bias current the external resonant frequency is lowered which then also lowers the accelerating mode frequency. A tuning range of 500 kc is required and by using the perpendicular biased ferrite it is possible to obtain this with a single tuner and in a more efficient manner than in the original FNAL tuning system (400 kc tuning range).

The 52 MHz HERA cavities will be of a similar design but the entire cavity will be evacuated. There will be no ceramic seals and no tuning loop. The intermediate cylinder will again be supported on a post.

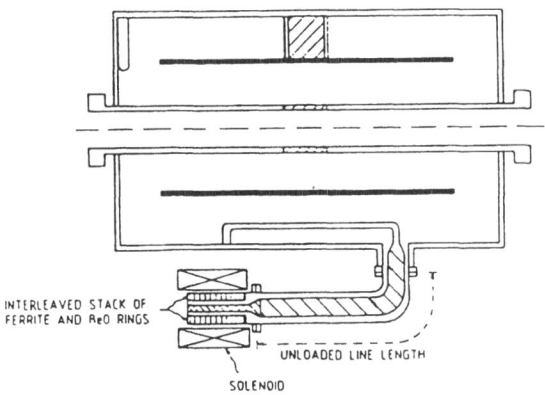

INTERLEAVED STACK OF
FERRITE AND ReO RINGS

UNLOADED LINE LENGTH

SOLENOID

Schematic layout of the PETRA II 50 MHz cavity
with the loop coupled, external perpendicular
biased ferrite loaded frequency tuner.

Fig. 14. Proposed PETRA II cavity.

Since a tuning range of at least 25 kc will be needed to compensate for beam loading some form of electrical tuning is necessary. Higher mode dampers will also be required. Because the peak voltage per gap will be about 160 kV the rf power required is about three times greater than for the PETRA cavities.

CERN PS e^+e^- ACCELERATION SYSTEM

As mentioned earlier the CERN SPS will accelerate e^+ and e^- beams for injection into LEP. However prior to this, these beams will also be accelerated in the Proton Synchrotron. The rf system will be required to accelerate these particles from 600 MeV to 3.5 GeV. Two high Q resonators operating at 114.511 MHz with a gap voltage of 500 kV have been constructed and tested.[13] Figure 15 shows the cavities and auxiliary equipment while the system parameters are listed in Table 6. They are driven by power amplifiers rated at 50 kW cw and 200 kW pulsed, through 40 m of coaxial cable.

Slow tuning is accomplished with the piston tuner. Fast tuning which is required to position the bunches prior to extraction is accomplished with two coaxial lines partially loaded with perpendicularly biased microwave ferrite and loop coupled to the cavity Figure 16(a). The rf power loss in each tuner is below 1 kW at 500 kV gap voltage which is estimated to be 1/10 the loss of ordinary ferrite with parallel bias. Four higher order mode dampers (0 in Figure 15) are employed to suppress resonances above the accelerating frequency. This triaxial structure is shown in Figure 16(b). The 50Ω load is coupled to the resonator through a notch filter tuned to 114 MHz, that consists of the

Table 6. CERN PS e^+e^- Acceleration System Parameters

Injection Energy	600 MeV
Final Energy	3.5 GeV
rf frequency (h = 240)	114.511 MHz
Peak rf voltage	1 MV
Q_ℓ	56,000
R_s/Q_ℓ	180 Ω
Cavity excitation power	12.5 kW
Tuning range (slow)	260 kc
Tuning range (fast)	20 kc

outermost enclosure and the outer conductor of the coaxial line that connects the loop to the load. In Figure 17(a) is shown the resulting parasitic mode attenuation.

It was not possible to reduce the very high shunt impedance of these cavities at 114 MHz to the few kΩ required for proton acceleration, using coupling loops and external loads as was done with the 100 MHz and 200 MHz cavities installed on the CERN SPS. Thus two shorting bars that can be inserted across the gap were designed and installed (S in Figure 14). This moves the lowest resonant mode to 180 MHz where the mode dampers produce additional attenuation [Figure 17(b)]. Under normal operation the shorting arms absorb 3% of the total cavity dissipation. At full cw power it will be necessary to provide water cooling for the shorting bars.

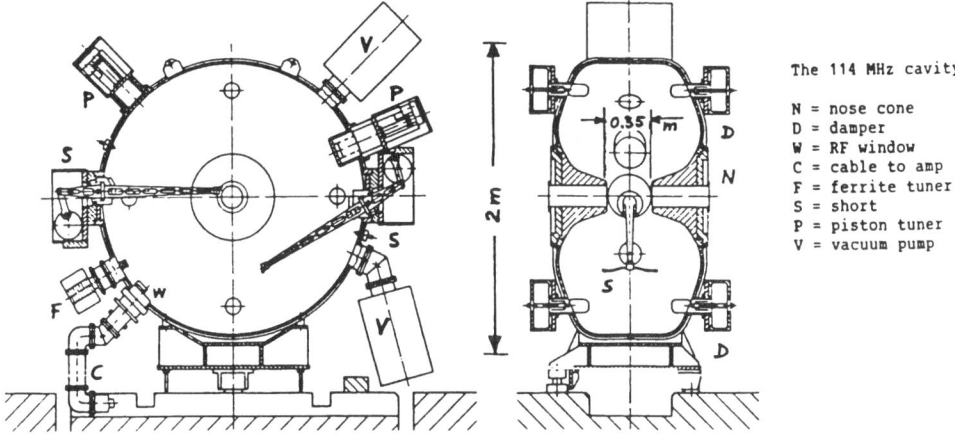

Fig. 15. CERN e⁺e⁻ cavity for PS accelerator.

The 114 MHz cavity

N = nose cone
D = damper
W = RF window
C = cable to amp
F = ferrite tuner
S = short
P = piston tuner
V = vacuum pump

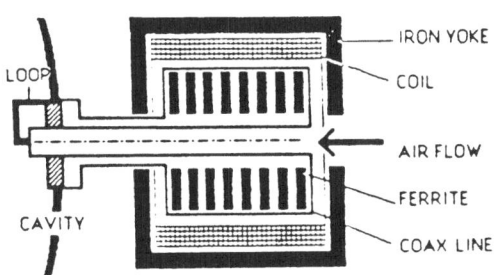

The ferrite tuner

Fig. 16(a). 114 MHz cavity
fast tuner.

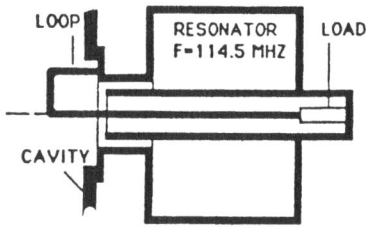

The higher mode damper

Fig. 16(b). 114 MHz cavity
mode damper.

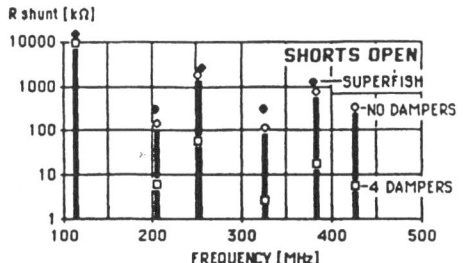

R shunt with and without damper
(for electrons and positrons)

Fig. 17(a). Parasitic mode attenuation without gap shorts.

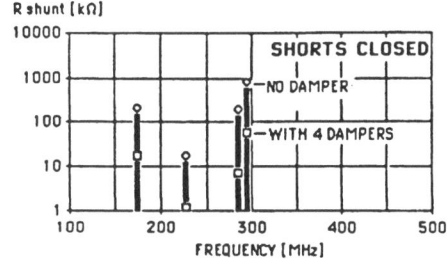

R shunt with and without damper
(for protons)

Fig. 17(b) Parasitic mode
attentuation with gap shorts.

CERN EPA MONOCHROMATIC CAVITY

Another link in the series of injectors for the CERN LEP project is the Electron Positron Accumulator ring. It takes particles from a fast cycling (100 Hz) linac at 600 MeV and accumulates them for injection into the PS. The rf system[14] operates at 19.08 MHz (h = 8) and must provide a peak voltage of 50 kV. A sketch of the model cavity used to test novel methods of damping higher order modes is shown in Figure 18. The cavity is a single capacity loaded $\lambda/4$ coaxial resonator. The outer cylinder is 2 m long by 1 m in diameter. A ceramic gap bridges the vacuum chamber which passes through the stem of the resonator.

In order to damp higher order modes provision is made to place ferrite rings at either end of the gap; to place ferrite in "disc" at the end of the stem and also inside the stem; and to put water in a box located at the far end of the cavity. Radial slits in the "disc" and longitudinal slits in the stem expose the ferrite to higher order "dipole" modes. Tests showed that with just one ring near the gap plus the stem and disc ferrite reduced the Q value of the measured modes by 10-80%. The final cavity incorporated two ferrite rings and the water box which together produced further attenuation and essentially eliminated some of the modes completely. It was decided to use tap water rather than demineralized water since the former resulted in greater damping. The additional 3 kW loss was easily made up by the 50 kW power amplifier.

Mechanical tuning is employed and no provision is made for beam loading compensation. The fully loaded Q is $\approx$ 3400 and the R_s/Q is 35-40Ω. Operation at the design intensity of 2.2×10^{10} particles/bunch and 500 MeV has already been achieved.

LOW IMPEDANCE RF SYSTEMS

Very low impedance rf systems are required in machines where one has both dc beams and bunched beams. This was the case in the CERN Intersecting Storage Rings, and would have been the case in the Brookhaven CBA

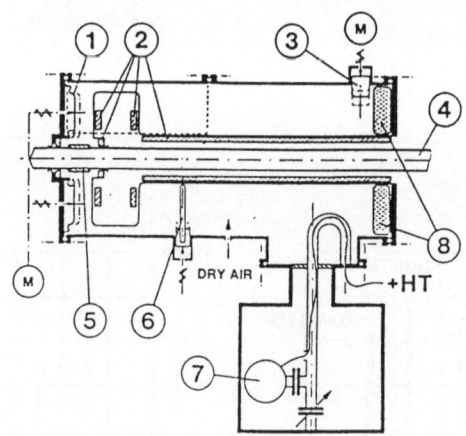

Cavity Layout; (1) Coarse tuner
(2) Ferrite absorbers (3) Fine tuner
(4) Vacuum chamber (5) Ceramic cylinder
(6) Ceramic Support (7) Power Tube (8) Water Box

Fig. 18. CERN Monochromatic test cavity.

had it been completed. It is also the situation in the CERN Antiproton Accumulator ring. The impedance requirements depend upon the intensity and the square of the momentum spread in the dc beams. If this limit is exceeded then the beam would tend to self-bunch at some harmonic of the revolution frequency in an uncontrolled manner.

CERN ISR System

In the CERN ISR low impedance ($< 50\Omega$ at $\approx$ 10 MHz) per cavity (six cavities) was achieved by a feedback loop between the cavity and the power amplifier including the driver.[15] The resonant impedance of each cavity was 3 kΩ, so an overall reduction of greater than 60 was obtained with the feedback. Higher order modes were damped with ferrite rings that were an integral part of the distributed 100 pF fixed tuning capacitor placed across the accelerating gap.[15]

Operationally twenty bunches from the PS were injected into the ISR which was 3/2 larger so that at h = 30 it had the same rf frequency as the PS. The bunches were accelerated slowly and deposited into a stack of dc beam. During this process the voltage per cavity could vary from 3 kV to less than 15 volts. Hence a low impedance presented to the beam was also necessary to reduce the signal induced at the rf frequency and its harmonics. In addition a complementary or feed forward compensation system was employed to further reduce the beam voltage appearing at the gap. A beam derived signal from a wideband pick-up electrode was fed into the driver stage with the correct phase (or time delay) and amplitude to produce an output current in opposition to the bunch beam current appearing at the gap. This open loop correction further reduced the dynamic impedance at f_{rf} to $< 5\Omega$/gap.

It should be noted however that the feedback loop and/or a wideband feed forward system can put severe peak current limitations on the power tube driving the cavity. A tightly bunched beam can have peak currents of several amperes and this must be supplied by the tube if the overall system is to function in a linear manner. In the case of the ISR the original system had to be augmented with a separate power amplifier per cavity as the injected currents from the PS were increased.[16]

Proposed CBA System

For the Brookhaven Colliding Beam Accelerator it was proposed to inject bunched beams and stack them as was done in the CERN ISR. However after accumulating up to 8 amperes this beam was to be rebunched and accelerated from 30 GeV to 200 GeV. Thus in addition to a stacking rf system that would have been similar to the ISR, an acceleration system operating at h = 3 ($\approx$ 240 kc) and capable of providing 36 kV (12 kV/gap) was required. In addition the impedance per gap had to be 15Ω or less so that in the presence of the 8 A beam the beam would not rebunch of its own accord.

A solution to the design problem was obtained by developing a cathode follower power amplifier.[17,18] The inherent feedback present in this configuration guarantees a low output impedance if stable operation of a large high power tube can be achieved. In addition the peak current requirement mentioned above is easily satisfied if the tube is large enough. This can be understood by referring to Figure 19 which is a typical plot of beam current pulses. Once the beam is bunched the tube quiescent current must always be greater than the average or dc component of the circulating current in order to avoid cutoff (i.e. for any very low impedance system the power tube must always operate class A). With

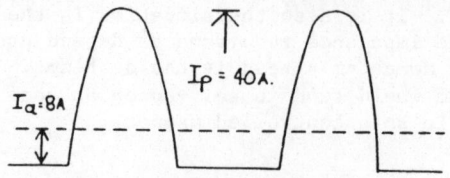

Typical Beam Current

Fig. 19. Beam current pulses in proposed CBA.

the cathode connected to an accelerating gap which is upstream of the ground plane, the peak beam image current will always tend to increase the tube current and hence preserve class A operation. If one attempted to excite the cavity in push pull then one tube at least (whether cathode follower or plate coupled) would always have to operate with a quiescent current $\left(I_{peak} - I_{average}\right)$.[18]

SSC RF SYSTEM

The Superconducting Super Collider is a two ring proton-proton accelerator, colliding beam machine that has been proposed for

Table 7. SSC rf System

Injection	1 TeV
Final energy	20 TeV
f_∞	3.614 kc
rf frequency	374.74 MHz
h	103,680
Bunch frequency	62.4 MHz
Acceleration rate	5.26 MV/turn
Peak rf voltage	20 MV
Acceleration time	≈1000 sec
Beam current	73 mA
Peak beam current	2 A
Maximum power to beam	384 kW
Cavity Excitation power (total)	1 MV

construction in the United States in the early 1990's. In Table 7 we list the rf system parameters for the present design.[19] The choice of frequency was primarily determined by the well developed technology for efficiently producing large power and obtaining high gap voltages around 350 MHz. Examples being the PEP electron storage ring at SLAC[20] and the CERN LEP project mentioned above. The peak voltage requirement (20 MV) was based on the need to provide bunches with sufficient momentum spread in the storage or colliding mode to ensure a minimum growth rate due to intra-beam scattering. Again the acceleration time is long as in the Tevatron or the HERA proton ring but the energy gain per turn is necessarily larger because of the machine size and required energy increase.

It is assumed that eight full cell π mode cavities similar to the PEP cavities (and to the LEP accelerating cavity shown in Figure 12) but made from copper will be employed. The unloaded Q will be ≈40,000 with an R_s = 5 MΩ/cell and 40 cells at 500 kV/gap will provide the necessary voltage. At present it is proposed not to detune the cavities to compensate for beam loading. This is because the maximum detuning which would be $\Delta f \cong 6.8$ kc (occurring during storage) or almost twice the rotation frequency of 3.614 kc could result in the cavities driving coupled bunch instabilities.[19]

Since the cavities will be resonant with the drive frequency there will always be reflected power whether one is accelerating or storing the beam. The coupling coefficient β, which is defined as the ratio of the power radiated out of the coupling loop to the power dissipated in the cavity walls if excited by the beam, can be chosen to minimize the reflected power using the following expression.[21]

$$P_i = \frac{V_c^2}{8\beta R_s} \left[\left\{ (\beta+1) + \frac{I_b R_s}{V_c} \sin \Phi_s \right\}^2 + \left\{ \frac{I_b R_s \cos \Phi_s}{V_c} + \tan \psi \right\}^2 \right]$$

Now $\tan \psi = (\beta+1) \tan \Theta$ where Θ is the detuning angle of the cavity given by

$$\tan \Theta = \frac{- I_b R_s}{(\beta+1)} \cos \Phi_s$$

I_b is the rf component of the beam current and Φ_s the stable phase angle. Here we have assumed that $\Phi_b = \Phi_s$ where Φ_b is the phase of the rf component beam current relative to the rf frequency (this is true if the bunch is short compared to the rf wavelength). Thus for the SSC one assumes $\Theta = 0$ and taking $\Phi_s = 164.75$ the minimum value for β is found to be 1.92. The required incident power P_i would be 1. 65 MW for the eight cavities. If the coupling is increased to β = 3 then the required power is 1.79 MV during acceleration and 1.51 MW during storage while the impedance seen by the beam at f_{rf} is reduced to 5 MΩ ÷ (β+1) or 1.25 MΩ/gap.[19]

Since it takes only 1 MW to excite the cavities there is considerable reflected power (≈250 kW/klystron) which must be absorbed before reaching the klystrons. Now transient beam loading is present due to a 2.7 μsec gap required in the chain of bunches for extraction (as well as during injection when only fractions of a turn are stacked sequentially). In order to compensate for this it is necessary to feed the cavities in such a manner that precludes the power distribution system originally proposed for SSC.[19] Thus in order to absorb the reflected power circulators similar to those employed at LEP will be required at every cavity.[22]

REFERENCES

1. J.E. Griffin, Q.A. Kerns, NAL main-ring cavity test results, IEEE Trans. Nucl. Sci. NS-18, No. 3, 241-243 (1971).
2. R.F. Stiening, J.E. Griffen, Longitudinal instabilities in the Fermilab 400 GeV main accelerator, IEEE Trans. Nucl. Sci. NS-22, No. 3, 1859-1861 (1975).
3. Q. Kerns, M. May, H.W. Miller, J. Reid, F. Turkot, R. Webber, D. Wildman, Energy saver prototype accelerating resonator, IEEE Trans. Nucl. Sci. NS-28, No. 3, 2782-2784 (1981).
4. Q. Kerns, C. Kerns, H. Miller, S. Tawzer, J. Reid, R. Webber, D. Wildman, IEEE Trans. Nucl. Sci. NS-32, No. 5, 2809-2811 (1985).

5.   G. Dome, The SPS acceleration system, Proc. 1976 Proton Linac
     Conf. Chalk River, Canada, 138-147 (1976).

6.   D. Boussard, G. Dome, T.P.R. Linnecar, Acceleration in the CERN SPS.
     Present status and future developments, IEEE Trans. Nucl. Sci.
     NS-26, No. 3, 3231-3233 (1979).

7.   P.E. Faugeras, H. Beger, J.P. Kindermann, V. Rodel, G. Rogner, A.
     Warman, The new rf system for lepton acceleration in the CERN SPS,
     Proc. 1987 IEEE Particle Accel. Conf. Vol. 3, 1719-1721 (1988).

8.   H. Frischolz, Generation and distribution of radio-frequency power
     in LEP, IEEE Trans. Nucl. Sci. NS-32, No. 5, 2791-2793 (1985).

9.   P. Brown, H. Frischholz, G. Geschonke, H. Henke, I. Wilson,
     Development and first results with a storage resonator, IEEE Trans.
     Nucl. Sci. NS-28, No. 3, 2707-2710 (1981).

10. B.H. Wiik, Progress with HERA, IEEE Trans. Nucl. Sci. NS-32, No. 5,
     1587-1591 (1985).

11. R.M. Hutcheon, A perpendicular-biased ferrite tuner for the 52 MHz
     PETRA II cavities, Proc. 1987 IEEE Particle Accel. Conf. Vol. 3,
     1543-1545 (1988).

12. R.M. Hutcheon, J C. Brown, R.J. Burton, R.A. Vokec, A simple higher
     order mode damping system for the PETRA II cavities, Proc. 1987 IEEE
     Particle Accel. Conf. Vol. 3, 1546-1548 (1988).

13. B.J. Evans, R. Garoby, R. Hohbach, G. Nassibian, P. Marchand, S.
     Talas, The 1 MV 114 MHz electron accelerating system for the CERN
     PS, Proc. 1987 IEEE Particle Accel. Conf. Vol. 3, 1901-1903 (1988).

14. S. Bartalucci, M. Bell, F. Caspers, K. Hubner, P. Marchand, A.
     Susini, R. Power, A monochromatic rf cavity, Proc. 1987 IEEE
     Particle Accel. Conf. Vol. 3, 1791-1793 (1988).

15. F.A. Ferger, W. Schnell, The high power part of the rf system for
     the CERN ISR, CERN-ISR-RF/70-34 (1970).

16. H. Frischholz, W. Schnell, Compensation of beam loading in the ISR
     rf cavities, IEEE Trans. Nucl. Sci. NS-24, No. 3, 1683-1685 (1977).

17   T.W. Hardek, W.E. Chyna, Common-anode amplifier development, IEEE
     Trans. Nucl. Sci. NS-26, No. 3, 3959-3961 (1979).

18. S. Giordano, M. Puglisi, A cathode follower power amplifier, IEEE
     Trans. Nucl. Sci. NS-30, 3408-3410 (1983).

19. SSC Central Design Group. Conceptual design of the SSC,
     SSC-SR-2020, (March 1986).

20. M. Allen, I. Korvonen, J.L. Pellegrin, P.B. Wilson, rf system for
     the PEP storage ring, IEEE Trans. Nucl. Sci. NS-24, No. 3, 1780-1782
     (1977).

21. P.B. Wilson, High energy electron linacs, AIP Conf. Proc. No. 87,
     Physics of high energy particle accelerators, 474 (1982).

22. E. Raka, Transient beam loading and rf power distribution in the
     SSC, Proc. 1986 Summer Study on the Physics of the SSC, 542-544
     (1987).

RF CAVITY PRIMER FOR CYCLIC PROTON ACCELERATORS

James E. Griffin

Fermi National Accelerator Laboratory
Batavia, Illinois 60510 U.S.A.

INTRODUCTION

The purpose of this note is to describe the electrical and mechanical properties of particle accelerator rf cavities in a manner which will be useful to physics and engineering graduates entering the accelerator field. The discussion will be limited to proton (or antiproton) synchrotron accelerators or storage rings operating roughly in the range of 20 to 200 MHz. The very high gradient, fixed frequency UHF or microwave devices appropriate for electron machines and the somewhat lower frequency and broader bandwidth devices required for heavy ion accelerators are discussed extensively in other papers in this series.

While it is common practice to employ field calculation programs such as SUPERFISH, URMEL, or MAFIA as design aids in the development of rf cavities, [1,2,3] we attempt here to elucidate various of the design parameters commonly dealt with in proton machines through the use of simple standing wave coaxial resonator expressions. In so doing, we treat only standing wave structures. Although low-impedance, moderately broad pass-band travelling wave accelerating systems are used in the CERN SPS, [4] such systems are more commonly found in linacs, and they have not been used widely in large cyclic accelerators.

Two appendices providing useful supporting material regarding relativistic particle dynamics and synchrotron motion in cyclic accelerators are added to supplement the text.

ACCELERATOR REQUIREMENTS

Before proceeding with a detailed description of rf cavities, it will be useful to develop some ideas of the requirements and limitations imposed on the designs by the general properties of the accelerators in which they are to be used.

Physical Size, Apertures

Proton synchrotron accelerator complexes usually consist of a series of accelerators operating over a sequence of energy ranges. Protons or H⁻ ions are injected into the first synchrotron from a linear accelerator at an energy between 100 and 1000 MeV. Because of practical limitations on the range over which useful bending magnetic fields can be developed, the momentum range of each successive synchrotron is limited to about 30:1. A further limitation on the momentum range of the first ring may arise because of the cost balance between providing the bending magnet length required for the highest momentum and the magnet size required to provide the physical aperture for the injected beam.

We assume, for example, that the first ring in the sequence spans a momentum range of 12.5:1, carrying protons from 1.696 to 21.2 GeV/c. (Useful relationships between kinetic energy, total energy, velocity, momentum, β, γ, and electrical forces on protons are developed in Appendix A.) This ring uses iron core magnets with a maximum useful bending field of 1.5 T developing a bending radius at maximum momentum of 47.1 m. A total bending magnet length of 296 meters is required to deflect the beam by 2π radians. The actual orbit length must be increased substantially to allow space for focussing quadrupole magnets, higher order correction magnets, injection and extraction devices, diagnostic equipment, and finally, rf accelerating cavities. The total circumference of such a machine would be near 500 m, with, perhaps, 120 m available for rf cavities.

The next ring in the sequence could carry protons from 21.2 to about 500 GeV/c with a circumference of about 7500 m. At least 100 m of circumference would be available for rf equipment. Following this ring one might find a large ring of 8 - 10 T superconducting magnets sustaining a maximum energy of 30 TeV and a circumference approaching 80 km.

The required apertures, or beam pipe sizes, in these machines might reasonably start at 6 by 10 cm at the lowest energy, decreasing to a 4 cm circular aperture at the highest energy.

Frequency Range, Frequency, Harmonic Number

At the lowest injection energy, 1 GeV, $\beta = 0.875$. With the proposed circumference of 500 m, protons require 1.9 microseconds to complete one turn; the rotation frequency is 524.65 kHz. At the extraction momentum, 21.2 GeV/c, β has increased to 0.99902, and the rotation frequency is 599 kHz. The rf system could, in principle, operate at exactly these frequencies, swinging about 15 percent during acceleration. In fact, there are compelling reasons to operate at much higher frequencies, related to the rotation frequencies by an integer harmonic number h. Modern proton facilities have been operating at rf frequencies between 30 MHz (Fermilab, Petra-II) and 200 MHz (SPS, HERA). The choice is related to the frequency modulation requirements, the method of transfer of beam from one machine to the next, and the beam bunch composition desired in the final very high energy ring. In the case at hand, it is reasonable to choose for the first ring a harmonic number h=100 so that the rf frequency must swing from 52.46 to 59.9 MHz during acceleration.

The second ring in the sequence would operate over the very narrow frequency range 59.9 to 59.96 MHz at a harmonic number h=1500. The third ring operates at constant frequency 59.96 MHz, or possibly twice or four times that frequency at h=16000, 32000, or 64000.

In the lowest energy ring the magnetic guide field ramp during acceleration will probably be the independent variable and the rf frequency will have to be tuned to keep the beam at some mean radius within the aperture. This means that the rf system will have to be electrically tunable, almost certainly using the variability of ferrite magnetic permeability to tune the rf cavities. Until recently Ni-Zn Iron spinel ferrite would have been used, limiting the maximum frequency to about 60 MHz. Recently there have been developments[5,6] using very low loss yttrium garnet ferrite for this purpose. The frequency range of cavities tuned in this manner may be increased to a few hundred MHz.

We are considering here rf systems whose only function is to develop a time varying electric field along the direction of motion of protons. Because these fields are varying (usually sinusoidally) synchronously with the rotation period, the energy gain of a particular proton may be positive, negative, or zero, depending upon its time of passage through the rf field. (We assume here that the rf fields are developed across gaps in the beam pipe which are short with respect to an rf wavelength and that the proton gap crossing time is short with respect to an rf period. The "transit time factor" is unity.[7] This is not true in all accelerators.) The change in energy resulting from a single passage through the rf accelerating system will cause a change in rotation period of the particle so that subsequent crossings will occur at progressively different times. A detailed discussion of this "phase motion" is given in Appendix B.

The alternating fields confine the circulating proton beam into localized groups or "bunches." The bunches contain protons spanning a range of energy around a "synchronous energy" and span a phase or time period which is a fraction of one rf period. The bunches are centered around some phase angle, ϕ_s which depends on the static or dynamic conditions of the synchrotron bending magnetic field.* A machine with harmonic number h can sustain h such bunches although there need not be a bunch located on each rf wave (i.e., all bunches need not contain the same number of particles. See Figure 1.). If the magnetic bending field

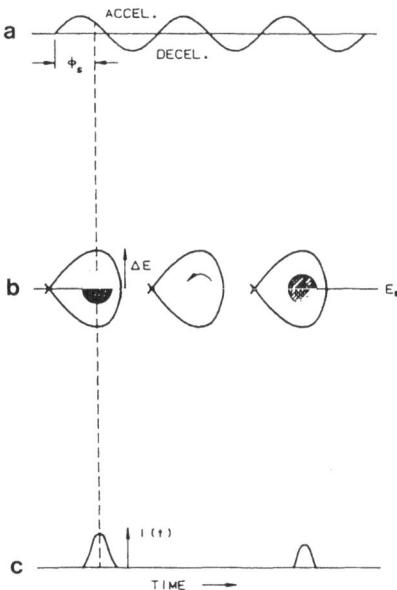

Figure 1 a) RF voltage wave. b) Phase space boundaries generated by the rf voltage for synchronous phase angle ϕ_s=135 degrees, i.e., above transition. Coordinates are time and $\pm\Delta E$, particle variations from a synchronous energy E_s. Shaded areas represent populations of particles within the "buckets." The second bucket contains no particles. The arrow shows the direction of motion of particles above transition. c) Projection of bucket population on the time axis. Beam "line charge distribution" per unit time, or current i(t).

---

*This confinement principle of phase stability was first stated by V. I. Veksler in 1944 and independently by E. M. McMillan very shortly thereafter.

is increased or decreased, the average phase angle at which particles cross the accelerating gaps must be changed so that the average particle momentum will be that required to maintain an orbit centered in the beam pipe. Here net energy must be delivered to or removed from the beam by the rf system.

In order to calculate the required rf voltage and the power which the rf system must deliver to the proton beam, we must propose more details for the hypothetical machines under consideration. Let us assume that the 1 GeV linac delivers 24 ma of H⁻ ions with an energy spead of ±1 MeV to a stripping foil in the injection channel of the Booster accelerator. If injection is allowed to proceed for seven turns, or 13.3 µs, a total of. 2 x 10¹² particles will be injected on each acceleration cycle.

The product of the total injected energy spread and the rotation period yields longitudinal emittance of 3.8 eV-s or 0.038 eV-s per rf wave, since h=100. Using equations 8B and 8C, Appendix B, we find that an rf voltage slightly larger than 61 kV must be applied just following injection in order to create the required "bucket area" to bunch the beam. This voltage should be developed "adiabatically"[8] in order to minimize an unwanted increase in the longitudinal emittance during bunching. Alternatively, the rf voltage might be applied in advance of injection and previously bunched bursts of beam, created in the linac, may be "painted" into the existing buckets to provide a desired beam charge distribution. In such a case the required voltage would be substantially larger, a few hundred kV, depending on details of painting. During this injection period the rf system is delivering no net energy to the beam.

For acceleration the rf system must, in addition to creating bucket area, provide accelerating voltage and deliver energy to the beam. The accelerating voltage and power depend on details of the rate of acceleration. For the example being developed we assume that the Booster magnetic field time dependence (equivalent to the beam momentum time dependence) will consist of a 15 Hz Sine wave. The beam momentum dependence is

$$cp(t) = 11.44 - 9.75 \cos(30\pi t) \quad \text{GeV/c} \tag{1}$$

where t is in seconds. The beam momentum is changed from 1.69 to 21.1 GeV/c in 33.33 ms. The rate of change of momentum reaches a maximum of 919 GeV/cs at 16.67 x 10⁻³s. The rate of change of momentum is equal to the net force exerted by the rf electric field which, averaged over one turn, is the rf voltage per turn seen by a particle, divided by the orbit length

$$\frac{dp}{dt} = \frac{V \sin \phi_s}{2\pi R}$$

$$V \sin \phi_s = \frac{2\pi R}{c} \frac{d(pc)}{dt} = \frac{1}{F_\infty} \frac{d(pc)}{dt} \tag{2}$$

where F_∞ is the rotation frequency of a particle having velocity c, (in the Booster case, ~ 600 kHz). The required rf accelerating voltage at the maximum slope points is 1.5MV per turn.

The required peak voltage depends on the synchronous phase angle ϕ_s which, in turn, depends on the required bucket area through the moving bucket factor.[9,10] For constant rf voltage, the bucket area is minimum at 1.732 times the transition point. For the example at hand we may take $\gamma_t = 7$ so the bucket area is minimum at 16.66 ms, the maximum slope point. We may assume that a bucket area of 0.06 eV-s is required at this time in the cycle yielding a synchronous phase angle of 67 degrees and a peak rf voltage of 1.63 MV.

Note that by adding about 18 percent of second harmonic to the magnetic field at the proper phase, the momentum slope could be reduced to a minimum, requiring only about 1 MV of rf voltage for the above Booster example.

The maximum energy gain of 1.5 MeV per turn per particle is 1.2×10^{-7} Joules per second per particle or 240 kW for 2×10^{12} particles. This is the peak power which must be delivered to the beam by the booster rf system. The rf system must operate for about 35 ms out of each 67 ms acceleration period delivering an average power of 100 kW to the beam.

We assume that the next ring in the series is filled with 14 booster batches (2.8×10^{13} protons) and accelerates at a rate of 100 GeV/s. This requires the rf system to deliver 448 kW peak power to the beam at an accelerating voltage of 2.5 MV per turn. If the ring is assumed to have γ_t near 18, a few kV of additional rf voltage is required to create an adequate bucket area, about 0.08 eV-s. During injection and extraction, when the synchronous phase angle is zero, a minor difficulty may rise in creating sufficiently low rf voltages to avoid the generation of excessively large buckets which might result in undesired beam phase space dilution.

The 80 kM ring will require about 5 MV of rf voltage at constant frequency in order to accelerate to 30 TeV in about 1500 seconds. The required bucket area can be generated with a few tens of kV so the synchronous phase angle will be quite large. At a total beam intensity of 2×10^{14} protons, the rf system must provide about 600 kW to the beam during acceleration.

Beam Loading

The beam passing through the rf cavities, bunched as shown in Figure 1, will have a component of rf current at frequency $f_{rf} = hF_0$ between one and two times the dc beam current. This factor depends on the bunch shape, or line charge density distribution, and it may be determined by evaluating the normalized Fourier transform of an average bunch distribution at frequency hF_0 and multiplying by twice the dc current. In the example Booster, at 590 kHz, the rf beam current at h=100 would be about 0.3A.

The beam current will generate an rf voltage at the accelerating cavity gap just as the rf power generator does. This situation is shown in Figure 2 where the rf cavity is represented near resonance by a lumped component RLC circuit. The current generator delivering the rf power source current i_g to the circuit is assumed to represent the rf power source transformed in amplitude to the gap impedance (i.e., the actual rf voltage delivered to the cavity at some point may be 10 kV which is transformed by the cavity geometry to 100 kV at the accelerating gap or gaps. The excitation current transformed to the gap would then be decreased by a factor of 10, for instance 10A from the power source is represented by 1A at the gap. The transformed rf power source internal impedance is combined with the accelerating cavity impedance). The phasor diagrams accompanying the circuit of Figure 2 show the beam current i_b on the positive real axis. The current, which actually excites the cavity, is the image of the beam current on the vacuum chamber walls $i_i = -i_b$. Using this convention the rf power generator delivers current at the required synchronous phase angle (ϕ_s-90) degrees. In the diagram shown, $\phi_s \rangle 90°$, indicating operation above transition, also indicated in Fig. 1.

In the absence of beam current, with the rf cavity tuned to resonance, a generator current $i_g = i_0$ is required to develop the desired gap voltage V_g, and this voltage will have the phase angle of the generator current, ϕ_s -90°. The introduction of beam image current

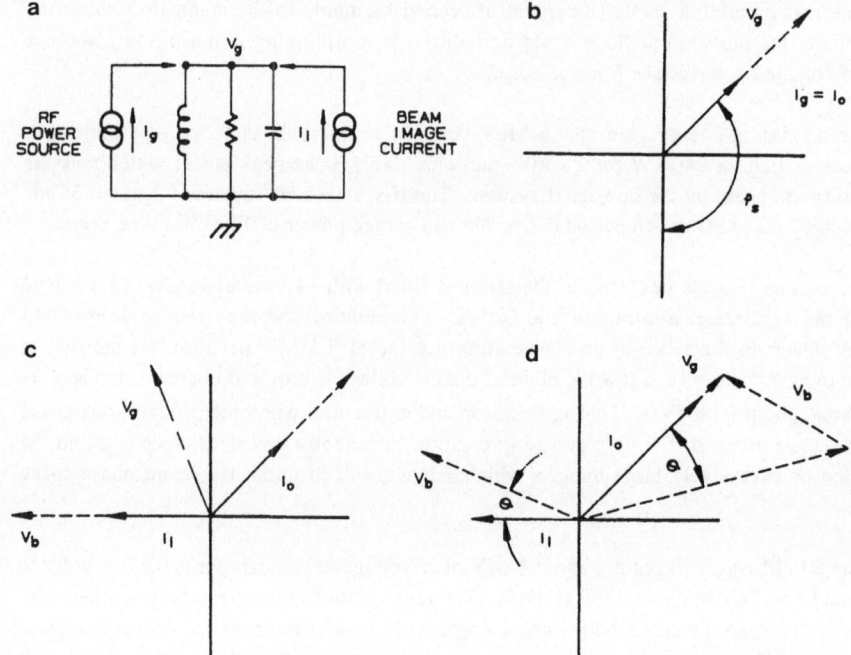

Figure 2 a) Rf cavity represented by a lumped RLC circuit near resonance. The circuit is excited by two current generators, i_g the rf power source, and i_i the beam image current. b) Phasor diagram showing rf voltage V_g (dashed) resulting from generator current i_o at $\phi_s =$ 135 degrees. The cavity is tuned to resonance, and there is no beam current. Voltage phasors are arbitrarily doubled in length for clarity. c) The effect of introduction of beam current with Fourier amplitude equal to i_0 The resultant gap voltage phasor V_g is moved to a phase angle 202 degrees resulting in a net decelerating voltage. d) Phasor diagram with cavity detuned below the excitation frequency by 35 degrees and the generator current increased by a factor 1.7. The accelerating voltage phasor is restored to the original amplitude and phase.

will generate an additional component of gap voltage which will move the resultant gap voltage phase away from that of the generator current. The desired gap voltage phase could be re-established by adjusting the phase of the generator current, but such a procedure results in the rf power generator delivering energy to a load with an effective reactive component, an inefficient situation.

By detuning the rf cavity appropriately, the load presented to the power generator can be made to appear real.[11,12] The rf cavity complex impedance is expressed

$$Z(\omega) = R_s \cos \theta \, e^{-j\theta} \tag{3}$$

$$\text{where } \theta = \tan^{-1} 2Q\frac{(f-f_0)}{f} \, .$$

The detuning required to compensate for the beam current is expressed

$$\tan \theta = 2Q\frac{\Delta f_1}{f} = \frac{i_b \cos\phi_s}{i_0} = \frac{i_b R_s \cos\phi_s}{V_g} \tag{4}$$

$$\Delta f_1 = \left(\frac{R_s}{Q}\right) \frac{hF_0 \; i_b \; \cos\phi_s}{2 \, V_g}.$$ (5)

In order to preserve the magnitude of the required gap voltage, the generator current must be increased in the presence of beam current,

$$i_g = i_0 + i_b \sin \theta_s.$$ (6)

The beam loading compensation scheme described here requires two feedback systems operating on the rf cavity and power source system. One feedback loop adjusts the delivered power such that the generator current develops the desired gap voltage while the other loop controls the cavity tuning (or detuning), such that the load presented to the power source appears real under all conditions of beam current.

Beam Stability

The rf cavities will inevitably present impedances at the accelerating gap, in which rf frequency components of the beam current can induce voltages. Such induced voltages will react back on the beam in a manner which may cause unwanted synchrotron motion, usually referred to an longitudinal instability. [13] Such instabilities may cause an unwanted increase in the longitudinal emittance, (essentially bunch length and momentum spread), resulting in undesirable beam configuration for injection into a succeeding machine or experimental use. In severe cases such instability may result in beam loss. Instabilities caused by rf system impedances can be broken into two classes; those that are associated with the fundamental operating frequency, and those that result from spurious impedances at other frequencies. Instabilities associated with the fundamental rf operating frequency are usually termed Robinson instability.[14,15] Coherent longitudinal oscillation of the beam bunched about the synchronous phase angle cause the appearance of frequency modulation sidebands adjacent to the rf frequency component of beam current. The largest of these sidebands are separated from the rf frequency by the synchrotron frequency (cf. Appendix B), usually a few hundred Hz to a few kHz. These current components induce voltages in the rf cavity impedance. If the cavity is tuned slightly off resonance, the real component of impedance will be different for the upper and lower sidebands, and the difference in amplitude of these two excitations creates a driving term or a damping term for the coherent synchrotron motion. Stability against this form of coherent bunch motion can be established by tuning the rf cavity system off resonance in such a way that the damping term dominates. If a machine is being operated below transition, stability is obtained by tuning the rf system slightly above the excitation frequency so that the resistive component at the upper sideband is larger than that at the lower sideband frequency. Above transition, detuning below the excitation frequency is required. Fortunately, the direction of detuning to establish Robinson stability is the same as that required to compensate for beam loading.

There is another aspect to Robinson instability which amounts physically to the effect that coherent bunch synchrotron oscillation reacts on the rf system in such a way as to reduce the bucket area. The amount of reduction is governed by the ratio of power delivered to the beam to that dissipated in the rf system. When this ratio is equal to one, the bucket area is reduced to zero and all beam will be lost if this happens during acceleration. Of course, the power delivered to the beam is proportional to the amount of beam in the machine so this effect is a beam self-limiting one, in that beam may be lost due to bucket area reduction until the power ratio is reduced to unity. This aspect of the Robinson instability can be eliminated by adjusting the shunt impedance of the rf system, and hence the internal power dissipation, until the internal dissipation is equal to or greater than the power delivered to the beam. In the Booster example the rf system would be required to dissipate 240 kW at a total gap voltage (all cavities) of 1.5 MV. The total shunt impedance should then be less than 4.7M-Ohm.

The Robinson instability relationships were developed considering rf cavities with no feedback control on the phase or amplitude of the gap voltage. The internal power dissipation requirement can be reduced substantially by application of various forms of feedback around the rf cavity-power-amplifier system, the effect of which would be to reduce the effective gap, or output, impedance of the rf system as seen by the beam. With appropriate feedback in place the beam to cavity power ratio can, in principle, be increased by an order of magnitude. Ratios of 2:1 have been achieved in practice. While this technique can reduce the total power dissipation and operating cost, it may not substantially reduce the size and cost of the rf power amplifier because, in order to develop the required low output impedance, the amplifier must be capable of delivering substantial power in transient situations to counteract the effects of beam excitation as it happens.

Coupled Bunch Instability

The bunches in a machine are capable of individual non-coherent longitudinal oscillations within their individual buckets. Frequently the motion is such that all bunches oscillate with roughly the same amplitude with a progressive phase difference advancing through the successive bunches, completing $2m\pi$ radians around the machine. In other cases, a few bunches oscillate with large amplitude while others remain stable. In any case such motion, when Fourier analyzed, can display a spectrum of beam rf currents at harmonics of the rotation frequency plus or minus the synchrotron frequency. The amplitude of these spectral lines is again governed by the shape of the Fourier transform of an average individual bunch and substantial amplitudes may be generated up to about 1 GHz by bunches a few ns in length. Real components of impedance at any frequency within the spectral range may be excited by these beam currents, and if the voltage generated is sufficiently high, the bunch motion responsible for a particular frequency may grow in amplitude. Therefore, it is expedient to minimize the amplitude of gap shurt impedance created by spurious resonances in all rf cavities. Specific requirements depend on machine details and anticipated beam intensity, but generally, reduction of spurious shunt impedances to a few tens of kOhms is sufficient. These reductions are frequently done by special traps or networks within the cavities which are designed to avoid coupling large amounts of energy from the fundamental operating mode. Such traps are usually terminated in resistances or composed of resistive material which transforms a low impedance to the accelerating gap at selected frequencies.

The magnitude of the real component of impedance above which an instability may arise is related to machine parameters, beam current and beam momentum spread roughly as [16,17]

$$\kappa = \frac{R}{n} < \frac{(const.)E_s \, \eta}{i_b(n)} \left(\frac{dp}{p}\right)^2 \qquad (7)$$

where the constant, (around 3), is related to the beam charge distribution within the bucket and $i_b(n)$ is the beam current component at frequency nF_0. The limiting value of R/n may range from a few to a few hundred ohms.

RF Cavity R/Q

Beam stability requirements may place a limit on the ratio of shunt resistance to Q (quality factor) for the entire rf system at the fundamental operating frequency. In a large machine with a very low rotation frequency and large harmonic number, the two rotation harmonics adjacent to the rf frequency $(h\pm1)F_0$ are very close to the operating frequency hF_0. At large beam intensity the beam loading cavity detuning, Eq. 5, will cause the impedances at the adjacent sidebands to become unequal, and it may raise the larger of the two to an amplitude capable of causing an $m=1$ coupled bunch instability. Therefore, the real component of cavity impedance must be kept below $\kappa(h\pm1) \cong \kappa h$ (h is large) at frequencies $F_0(h\pm1)$. From Eq. 3, the real component of cavity impedance is

$$R(f) = R_s \cos^2 \theta = \frac{R_s}{1+[2Q\frac{\Delta f}{f}]^2}$$

$$\cong \frac{R_s}{\frac{4Q^2}{h^2} \frac{(\Delta f)^2}{F_0^2}} . \tag{8}$$

From Eq. 5, the cavity is already detuned by Δf_1, so Δf in Eq. 8 becomes

$$\Delta f = F_0 - \Delta f_1 = F_0 \left[1 - \frac{hi_b \cos \phi_s}{2V_g} \right]. \tag{9}$$

Because the detuning condition, Δf_1, is largest at injection or extraction, when V_g is small and $\phi_s=0$, we set $\cos\phi_s = 1$. Combining Eqs. 8 and 9 we arrive at a sort of limiting condition on R_s/Q.

$$R_s/Q < \frac{2}{\left(\frac{h}{\kappa R_s}\right)^{1/2} + \frac{i_b h}{V_g}} \tag{10}$$

In the 7500 m ring example, if the beam cavity power ratio is 2:1, the injection voltage 200 kV, and the beam stability limit $\kappa=3$ Ohms, then R_s/Q is limited to about 186 Ohms. For Kaon factory rings, where the beam current may be an order of magnitude higher, the R/Q limitation becomes more severe.

Transient Beam Loading

During the bunch-to-bucket beam transfer of several batches of beam, there are periods when the larger ring is partially full. In order for the beam loading compensation system to operate correctly in this situation, the feedback loops would require bandwidth sufficient to generate corrections in a few rf periods, usually not feasible. Frequently the rf cavity time constant, $\alpha = \omega/2Q$, is several rotation periods so the voltage developed by bursts of current in the partially filled ring can be written

$$v(t) = (\Delta i) Z_c (\omega) \left[1-e^{-(\alpha+j\omega)t} \right]$$

$$\cong (\Delta i)Z_c (\omega) (\alpha+j\omega)t$$

$$\cong (\Delta i)R_s \alpha t = \frac{(\Delta i)\omega}{2} \left(\frac{R_s}{Q}\right)t. \tag{11}$$

Since these voltage excursions are, during the early transient period, in phase with the beam image current, they are in quadrature with the required gap voltage, when ϕ_s is either zero or π. The phase excursions, easily obtained from the gap voltage and Eq. 11, will cause selective phase space dilution if not compensated. Frequently some form of feed-forward compensation is employed, but high precision is difficult. It is clear that reduction of R/Q is beneficial.

Vacuum and Radiation Environment

The design of each rf cavity must be such that appropriate vacuum can be maintained in the accelerating gap region. The required pressures would be 10^{-8} - 10^{-9} T in the first two rings and perhaps 10^{-10} - 10^{-11} T in the very large ring.

The rf equipment which is to be operated within the accelerator enclosure must be designed to operate satisfactorily for long periods in a region of substantial ionizing radiation flux. Accumulated radiation doses of 1 MRad per year of x-ray and light charged particles and 10^{13} neutrons/cm^2 per year should be anticipated. Operational dose rates may be 300-500 Rad/h. In addition to satisfactory operation, rf equipment should be designed so that maintenance can be accomplished easily and expeditiously since it must be done in a significant residual radiation environment.

The insulating material used to isolate the accelerator vacuum from other parts of the rf system, thus serving as an rf window, must be made of material whose electrical loss properties are very good initially and will not deteriorate excessively under the radiation flux anticipated.

In addition to the above requirements, components of the rf cavities, such as insulating material, should not be made of materials which may emit noxious smoke into the accelerator enclosure in the event of electrical breakdown.

RF CAVITY GEOMETRIES

Simple Coaxial Structure Properties

The longitudinal electric fields required of the rf system can be developed by enclosing the beam pipe and its accelerating gap by a shorted coaxial structure as shown in Figure 3a. The inner conductor of the coaxial structure may be larger than the accelerator beam pipe and need not be made of the same material. The electrical properties of coaxial structures are widely known,[18,19] and we intend only to develop a few useful relationships here.

The inductance per unit length (ignoring magnetic fields within the conductors) is

$$L_l = \frac{\mu_r \mu_0}{2\pi} \ln \frac{r_2}{r_1} \qquad \text{Henrys per meter.} \qquad (12)$$

The capacitance per unit length between inner and outer conductor is

$$C_l = \frac{2\pi \varepsilon_0 \varepsilon_r}{\ln (r_2/r_1)} \qquad \text{Farads per meter.} \qquad (13)$$

The resistance per unit length is:

$$R_l = \frac{\rho_s}{2\pi} \left(\frac{1}{r_1} + \frac{1}{r_2} \right) \qquad \text{Ohms per meter} \qquad (14)$$

where $\rho_s = (\pi f \mu/\sigma)$ is the surface resistivity resulting from skin effect, and σ is the bulk conductivity of the material. At 60 MHz ρ_s is about 2 x 10^{-3} Ohms per square for high

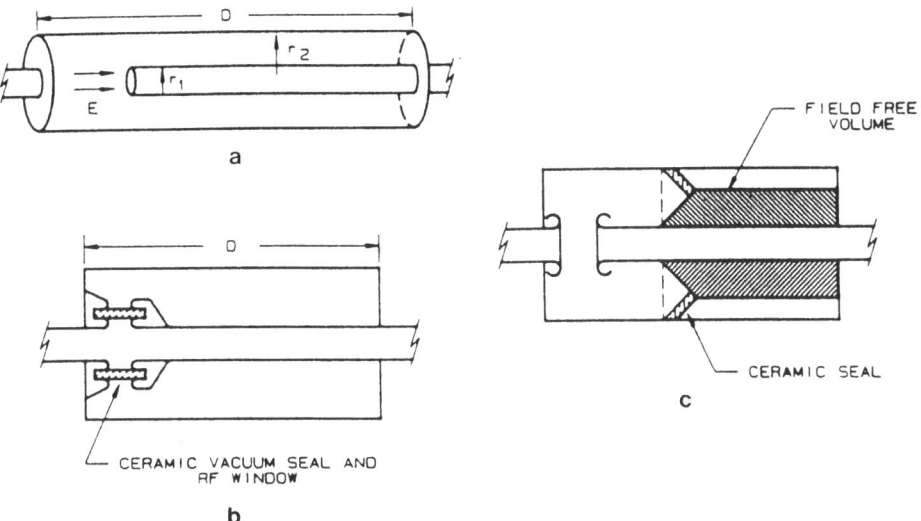

a

b

CERAMIC VACUUM SEAL AND
RF WINDOW

FIELD FREE
VOLUME

CERAMIC SEAL

c

Figure 3 a) a gap in the metallic beam pipe is formed into an accelerating resonator by enclosing the region with a shorted outer coaxial conductor. b) the physical length of the structure may be reduced without raising the frequency by adding lumped capacitance at the accelerating gap. In this case the capacitance is further increased by the presence of a ceramic cylinder which serves as an rf window and vacuum seal. c) The impedance transformation ratio between the low impedance excitation point and the accelerating gap can be affected by varying the characteristic impedance of the structure along its length. This also may facilitate removal of the ceramic window to a point of lower electric field, as shown.

conductivity copper and about 25 percent higher for aluminum. We will ignore real conductance, G, between inner and outer conductor since the structure is usually air filled or evacuated.

Coaxial transmission lines support travelling TEM waves of the form

$$A(z,t) = Ae^{j\omega t - \gamma z} \tag{15}$$

where A is the amplitude of a voltage, current, electric, or magnetic field, and $\gamma = \delta + j\beta$ is a propagation constant consisting of a spatial attenuation term and a spatial phase constant. For the low loss structures under consideration ($R \ll \omega L$, $G \equiv 0$)

$$\delta = \frac{1}{2} R \left(\frac{C_1}{L_1} \right)^{1/2} \tag{16}$$

$$\beta = \omega [L_1 C_1]^{1/2} = \omega [\mu_0 \varepsilon_0]^{1/2} = \frac{2\pi}{\lambda} \tag{17}$$

where λ is the free space wavelength of waves with angular frequency ω. In regions of a structure containing ferrite, or large amounts of dielectric material, the definition of β should include the relative permeability or permitivity of the materials.

The travelling waves in the structures can be combined to form standing waves whose amplitudes and phases combine to meet the boundary conditions imposed by open circuits (at the accelerating gap) or short circuits at the shorted end of the structure. These boundary conditions are, of course, idealizations, since the open circuit will consist of some shunt capacitance in parallel with a real conductance and the "short" may have series resistance and inductance. For the idealized boundary conditions the spatial and time variations in voltages and currents, or electric and magnetic fields, will be shifted in phase with respect to each other by 90 degrees when the cavity is operated at, or near, resonance. Given this condition, it is no longer necessary to include $e^{j\omega t}$ in each expression. Maximum amplitudes of voltage and current are related through the characteristic impedance of the structure

$$Z_c = \left[\frac{R_l + j\omega L_l}{j\omega C_l}\right]^{1/2} \cong \left[\frac{L_l}{C_l}\right]^{1/2}\left(1 - j\frac{R}{2\omega L}\right) \tag{18}$$

If voltage and current are known at any point in the system, they may be calculated at any other point using the transfer matrix expression

$$\begin{pmatrix} v(z) \\ i(z) \end{pmatrix} = \begin{pmatrix} \cosh(\gamma z) & Z_c \sinh(\gamma z) \\ \dfrac{\sinh(\gamma z)}{Z_c} & \cosh(\gamma z) \end{pmatrix} \begin{pmatrix} v(0) \\ i(0) \end{pmatrix}. \tag{19}$$

It is almost always valid to neglect losses when relating voltages and currents within an accelerating cavity. So, defining $R_c = [L_l/C_l]^{1/2}$ as a real characteristic impedance, the expression can be written

$$\begin{pmatrix} v(z) \\ i(z) \end{pmatrix} = \begin{pmatrix} \cos\beta z & jR_c \sin(\beta z) \\ j\dfrac{\sin\beta z}{R_c} & \cos(\beta z) \end{pmatrix} \begin{pmatrix} v(0) \\ i(0) \end{pmatrix}. \tag{20}$$

If the radical E field and thereby, the voltage, at the shorted end of the line is assumed to be zero, and a current I_0 is assumed, then the voltage and current distributions along the line are

$$I(z) = I_0 \cos\beta z,$$

and

$$V(z) = jI_0 R_c \sin\beta z. \tag{21}$$

For line length D the gap voltage and the short end current are related by

$$\frac{V_g}{I_0} = jR_c \sin\beta D. \tag{22}$$

Ignoring losses, the impedance at the open end is

$$Z_{in} = jR_c \tan\beta D. \tag{23}$$

Moving the open end gap around onto the beam pipe and additional structure in that region, such as a ceramic vacuum seal with field limiting corona rolls, will create a concentration of electric field in the gap region which is accounted for by a lumped gap capacitor C_g. The admittance of the parallel combination of the lumped gap capacitance and the shorted coaxial line is

$$Y = \begin{bmatrix} \text{Real} \\ \text{Part} \end{bmatrix} + j(\omega C_g - Y_c \cot\beta D). \tag{24}$$

where $Y_c = R_c^{-1}$ and the real part of Y is created by losses in the line and possibly in the gap capacitance. Parallel resonance (or "anti-resonance") occurs when the imaginary part of the input admittance is zero, at which point the magnitude of the input impedance is maximized and the time averaged electric and magnetic stored energies in the system are equal. From Eq. 24 this condition is

$$\omega R_c C_g = \cot\beta D$$

or

$$D = \frac{\lambda}{2\pi} \cot^{-1}\left(\omega R_c C_g\right). \tag{25}$$

clearly the shortest resonator length is less than $\lambda/4$, reaching $\lambda/4$ when $C_g = 0$.

Stored Energy

The time averaged stored magnetic energy in any incremental length, dz, of the resonator is

$$dW_e = \frac{1}{4} I(z) I^*(z) L_1 dz = \frac{1}{4} I_0^2 L_1 \cos^2(\beta z) dz. \tag{26}$$

Similarly, the incremental stored electrical energy is (using Eg. 22),

$$dW_e = \frac{1}{4} V(z) V^*(z) C_1 dz = \frac{1}{4} I_0^2 L \sin^2(\beta z) dz. \tag{27}$$

The total stored energy per unit length $dW_m + dW_e$, is a constant,

$$dW(z) = \frac{1}{4} I_0^2 L_1 dz \tag{28}$$

independent of position along the line. So the total stored energy in the coaxial line is $I_0^2 L_1 D/4$. This is not the total stored energy in the resonant system unless $C_g = 0$, because the stored energy in the gap capacitor has been ignored.

Problem: for a resonator with gap capacitance C_g, calculate separately the time averaged magnetic and electric stored energies and show that the difference is exactly the electric energy stored in the gap capacitance.

Assuming that the magnetic energy stored in the gap capacitance is negligible, the total stored energy is twice the time averaged magnetic energy in the line.

$$W_s = \frac{I_0^2 L_1}{2} \int_0^D \cos^2\beta z dz$$

$$= \frac{I_0^2 L_1}{2} \left[\frac{D}{2} + \frac{1}{2\beta} \sin\beta D \cos\beta D\right]. \tag{29}$$

Using Eg. 22, this energy can be written in terms of the gap voltage

$$W_s = \frac{\pi V_g^2}{8R_c \omega} \left[\frac{2\beta D}{\pi \sin^2\beta D} + \frac{2}{\pi} \cot\beta D\right]. \tag{30}$$

The leading factor is the stored energy in a line of R_c with $C_g = 0$ and voltage V_g. It is evident that the stored energy is inversely proportional to the characteristic impedance of the structure. The term in brackets is unity for $C_g = 0$ and increases (for constant gap voltage) as the line is shortened by adding gap capacitance. When the reactance of the gap capacitance equals R_c ($\cot\beta D = 1$) the line physical length is $\lambda/8$ and the stored energy is increased by 1.64. At 60 MHz, a $\lambda/8$ line of $R_c = 80$ Ohms operating at $V_g = 100$ kV has stored energy 0.21 Joules.

Problem: It is frequently useful to taper or step the structure characteristic impedance from a small value near the high current end to a larger value near the accelerating gap, holding the outer dimension constant. An example is shown in Figure 3c. Consider a structure of total physical length $\lambda/4$, stepped at the midpoint from impedance R_c to kR_c ($k \geq 1$).

Show that:

a) For constant frequency a gap capacitance Cg is required where

$$C_g = \frac{1}{\omega R_c k} \left(\frac{k-1}{k+1} \right) \tag{31}$$

b) The gap voltage is related to the maximum current by

$$V_g = j \frac{I_o R_c}{2} (1+k). \tag{32}$$

c) The energy stored in the high impedance sector is larger than that stored in the low impedance sector (not including energy stored in C_g) by the factor

$$W_h = \frac{1}{2} (k + \frac{1}{k}) W_l. \tag{33}$$

d) The <u>total</u> stored energy at constant gap voltage decreases as k increases

$$W = \frac{\pi V_g^2}{8 R_c \omega k} \left[1 + \frac{2}{\pi} \left(\frac{k-1}{k=1} \right) \right]. \tag{34}$$

Q, Shunt Impedance, R_s/Q

The incremental energy dissipated in a coaxial structure is

$$dW_D = \frac{1}{2} ii^* R_l dz. \tag{35}$$

where R_l is the total rf resistance per unit length from Eq. 14. The total energy dissipated in a shorted line of length D becomes

$$W_D = \int_0^D dW_D = \frac{R_l I_o}{2} \left[\frac{D}{2} + \frac{1}{2\beta} \sin\beta D \cos\beta D \right]. \tag{36}$$

The quality factor Q is defined

$$Q = \frac{2\pi W_s}{W_D} = \frac{\omega W_s}{P}. \tag{37}$$

where P is the power absorbed by the structure. Using Eq. 29 for W_s we find that the Q of a coaxial structure of constant R_c is

$$Q = \frac{\omega L_1}{R_1}. \tag{38}$$

This implies that the Q of a coaxial resonator is independent of length, and therefore, independent of the amount of shortening resulting from the presence of a gap capacitance.

The power P in Eq. 37 can be expressed in terms of the peak voltage at some point in the structure and an effective resistance across the electric field at that point. Usually the gap voltage is chosen and the effective resistance is referred to as the shunt resistance of the cavity, R_s.

$$P = \frac{V^2}{2R_s}. \tag{39}$$

For a structure with $C_g = 0$, using Eqs. 34 and 37, we find

$$R_s = \frac{V^2}{2P} = \frac{V^2 Q}{2\omega W_s} = \frac{4}{\pi} QR_c. \tag{40}$$

This is the real component of the imput impedance of a shorted quarter-wave resonator at resonance.

Problem: Use Eqs. 18 and 19 to calculate the complex input impedance of a shorted coaxial line. Verify Eq. 40 by examining the real part of the expression when the line length is $\lambda/4$

For a coaxial structure foreshortened by a gap capacitance C_g (let $\omega C_g R_c \equiv q$) the real part of the input impedance at resonance becomes

$$R_{in} = \frac{2R_c Q}{q + (1+q^2) \cot^{-1} q}. \tag{41}$$

If a particular structure is shortened to length $\lambda/8$ by making q=1, the shunt impedance is reduced by a factor 0.61, therefore, the power dissipated in the structure at constant gap voltage is increased by a factor of 2.7

It is evident from Eqs. 40 and 41 that R/Q is just $4Q/\pi$ for $C_g=0$ and it is reduced by the same factor as is R_s when the line is shortened by C_g, since the Q does not change.

In general, R/Q can be thought of as the reactance of some representative structure capacitance C_r and R/Q may be lowered by increasing some shunt capacitance. This can be seen by finding that capacitance which represents R_s/Q if $C_g = 0$.

$$C_r = \frac{Q}{R_s \omega} = \frac{\pi}{4R_c \omega} = \frac{\pi c C_1}{4\omega} = \frac{\lambda}{8} C_1 \tag{42}$$

Since the resonator is $\lambda/4$ in length this represents just half of the total capacitance of the structure. For a shortened line, R/Q is roughly the capacitive reactance of the gap capacitor in parallel with half that of the line capacitance.

The above expressions are guidelines from which boundaries on various parameters can be extracted. Many sources of energy dissipation have been ignored, such as energy dissipated in the end-wall and energy resulting from dielectric loss in a vacuum seal ceramic insulator located near the gap. Energy is also lost in input coupling loops and tuner components, etc.

Using the above expressions, a resonator made of copper, with inner radius 0.1 m and $R_c = 80$, foreshortened to length $\lambda/8$ operating at 60 MHz, would have Q about 24000. The shunt impedance would be about 1.5 MOhm, and R/Q is near 63 Ohms. For an actual structure the Q and R_s might be reduced by one-half, while R/Q would remain roughly unchanged. The rf power required to develop 100 kV at the gap would be less than 10 kW. Since this will probably be far less than the power required by the beam, some problem with Robinson stability might be anticipated. This problem will probably be relieved by several mechanisms. If the cavity requires substantial tuning, the Q and shunt impedance will be lowered substantially by energy dissipated in the tuning system. If little tuning is required, a higher gap voltage would probably be developed and cavity dissipation will increase by the square of the voltage increase.

Tuning

In order to tune the rf cavity over the required range it is necessary to change the ratio of electric to magnetic stored energy by an appropriate amount. This can be done by placing a variable capacitor in a region of high voltage or a variable inductor in a region of high current. Because precise control over the cavity tune is required with substantial bandwidth, tuning is usually done by creating an inductor containing ferrite, where magnetic properties can be changed by application of an external magnetic field. In Figure 4a we show a line of constant impedance R_c shortened to length $\lambda/8$ by a gap capacitor and augmented by a lumped variable inductor at the high current end. The variable inductance required to tune a particular structure over the required range may be approximated through the use of a simple tuning expression (probably due to R. M. Foster),

$$\frac{\Delta W}{W} = 2\frac{\Delta f}{f}. \tag{43}$$

where ΔW is the change in peak magnetic energy necessary to develop the required frequency change. Since one cannot reduce the stored energy in a ferrite inductor to zero, the energy stored in the inductor at the highest frequency may be about $2\Delta W$. Using the above expression and Eqs. 22 and 30, a relation between the resonator properties and the reactance of the lumped inductor can be developed.

$$\omega L \cong R_c \left(\frac{\Delta f}{f}\right) (\beta D + \sin\beta D\cos\beta D). \tag{44}$$

Tunable Cavity Example

Now we can expand on the 80 Ohm $\lambda/8$ structure previously discussed by requiring that it be tunable from 52 to 60 MHz. At 60 MHz, Eq. 44 suggests a tuning inductance 4×10^{-8} H for tuning range $\Delta f/f = 0.15$. The gap capacitance must be 23 pF to re-establish resonance. With V_g 100 kv the maximum current is about 1500 A and the tuning inductor stores 0.045 J. To reach the lower end of the tuning range the tuning inductance must be increased to 9×10^{-8}

a

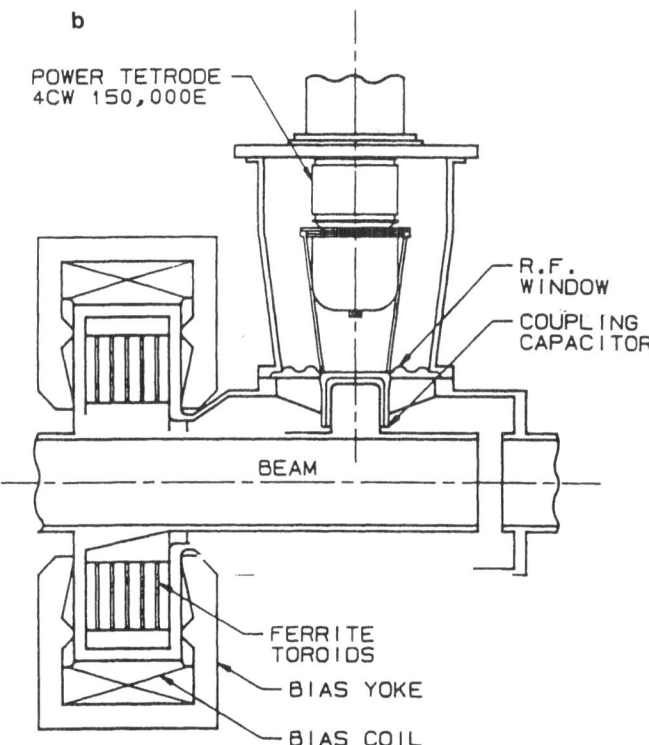

b

POWER TETRODE
4CW 150,000E

R.F. WINDOW

COUPLING CAPACITOR

BEAM

FERRITE TOROIDS

BIAS YOKE

BIAS COIL

Figure 4 a) Schematic representation of a coaxial structure of length $\lambda/8$ with lumped gap capacitance and variable tuning inductance. The system can be excited by an rf source current generator, output capacitance, and coupling capacitance as shown. Also shown is a series inductance in the rf power source which represents the short transmission line created by the physical size of a large power amplifier tube. b) Possible physical realization of the schematic representation of a tunable accelerating structure.

H. For 100 kV gap voltage the maximum current is 1370 A, and the peak stored energy in the tuning inductor is 0.085 J so the magnetic stored energy is increased by 0.04 J.

The required change in tuning inductance can be obtained by changing the relative permeability of ferrite by the required factor of about 2.25.[5] If yttrium garnet ferrite is used, it is reasonable to consider a permeability range 1.4 - 3.2 The tuner configuration will be determined by the volume of ferrite required which, in turn, is set by the rate at which energy dissipated in the ferrite can be removed. A maximum dissipation of 0.1 W per cm^3 is reasonable. (With this type of ferrite dielectric losses due to stored electric fields must also be considered.) We assume here that the ferrite has Q=2000, and that the time averaged stored energy, at 52 MHz, is 0.06 J. The power dissipation approaches 10 kW so that 10^5 cm^3 of ferrite is required.

A coaxial volume with inner radius 38 cm and outer radius 48 cm is added to the end of the cavity and filled with ferrite (along with some heat conducting material, such as BeO). By Eq. 12, the inductance per unit length of this section is 1.5 x 10^{-7} H/m at μ=3.2. In order to develop the required inductance of 8 x 10^{-8} H at 52 MHz, the length of the ferrite filled section is 53 cm. The volume of ferrite is 145 x 10^3 cm^3, so the dissipation will be about 70 mW/cm^3, which meets the dissipation requirement. Transverse field bias for the ferrite can be provided by encircling the ferrite region with bias windings and providing a flux return path, as shown in Figure 4b.

In order to provide a large tuning system bandwidth the coaxial region containing the ferrite should be provided with radial and longitudinal slots to minimize eddy currents resulting from changes in bias current. The bandwidth can be further improved by using thin iron bearing glass tape (METGLAS[*]) for the flux return path material.

The cavity described is similar in concept to a prototype cavity built at Los Alamos Laboratory.[6]

Geometry Variations

The simple geometry developed in the preceeding section may not always offer the input power coupling, tuner location, rf vacuum properties, or ease of maintenance desired. Other cavity geometries, usually with more than one accelerating gap, are possible. The boundary conditions imposed on structures described in Figure 3 can be met by eliminating the shorting plane end wall and extending the coaxial line to another symmetrically located accelerating gap as shown in Figure 5a. Here the rf current is maximum at the center, and the rf voltages developed at each end will be π radians out of phase with each other. This ensures that the accelerating fields developed at the gaps will be in time phase with each other, as shown. Since the distance 2D is slightly less than λ/2 (for a structure slightly shortened by gap capacitances), protons travelling through the cavity will require nearly one-half of an rf period to go from gap to gap, during which time the phase of the gap fields will reverse so the structure can deliver almost no net accelerating voltage to the particles. If the gap spacing 2D is mλ/2 where m is less than 1, the net accelerating voltage will be $V_g(1+\cos m\pi)$, i.e., approaching zero for m near 1 and approaching 2Vg as m approaches zero. While the gap spacing can be reduced somewhat by capacitive loading, it is impractical to improve such an accelerating geometry simply by gap loading. The gap spacing can, however, be reduced to zero by reconfiguring the same geometry as shown in Figure 5b. Here the electrical length from mid-plane to the end and back into the centrally located gap is λ/4. Each half may assume the properties of a graded structure with R_c on the outside perhaps 25 Ohms and R_c of the center part perhaps 80 Ohms. This makes coupling to a relatively low impedance rf power source convenient while allowing development of very large gap voltages. The "intermediate cylinder" may be supported at each end by cylindrical alumina vacuum seals allowing the entire center section, where high voltages are developed, to be evacuated, while allowing access to the outer, high current region for coupling loops, etc. Since there is no fundamental frequency electric field at the outer region mid-plane,

---

[*] ™Allied Chemical Company, Morristown, New Jersey.

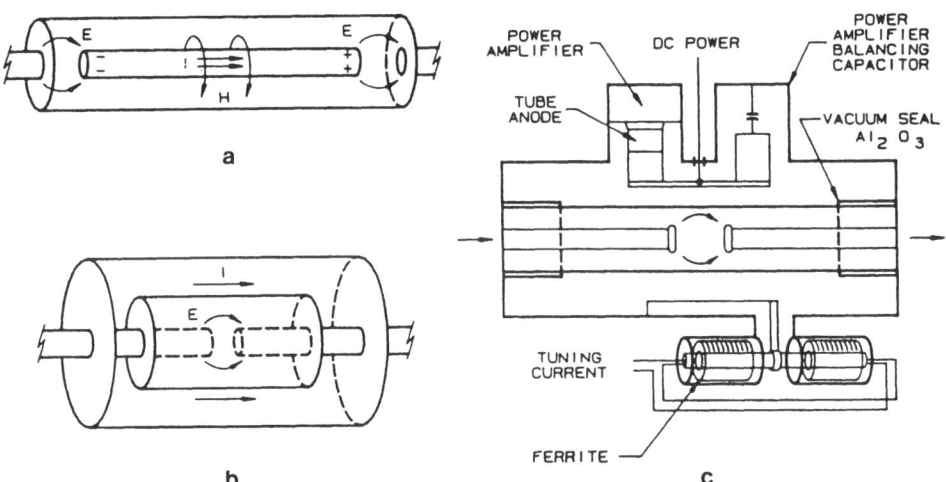

a

b   c

Figure 5 a) Double gap coaxial resonator. RF current is maximum at center and voltages at the ends oscillate out of phase causing accelerating fields which are always in time phase with each other. b) Re-entrant double gap structure. The two gaps are combined into a single gap at the center of the structure. c) Re-entrant cavity geometry showing loop coupling of tuner and power amplifier to the high magnetic field region. Also shown are ceramic vacuum seals supporitng the intermediate cylinder.

water cooling to the intermediate cylinder can be conveniently delivered and a dc potential may be applied to the cylinder to inhibit multipactoring. Because the stored energy and the losses scale in the same way, the Q of a re-entrant cavity will be just that of a simple graded line, (nominally 12000 at 60 MHz). The shunt impedance of the cavity, and R/Q, will both be twice that of a single gap cavity since the gap effects appear in series.

This re-entrant geometry is the basis of the Fermilab Main Ring rf cavity design.[18] In that design the cavities are tuned about 300 kHz near 53 MHz by ferrite tuners which are coupled to the outer, high magnetic field, region of the cavity by loops, as shown in Figure 4c. RF power is easily coupled into the re-entrant geometry by loop coupling to the high current region, as shown in Figure 5c.

Accelerating cavities with separated gaps, as in Figure 5a, can be used effectively for acceleration if the gaps can be made to oscillate in phase so that the developed accelerating fields are out of phase. This can be done by connecting some impedance, for instance a ferrite tuner, from the center of the line to ground, as shown in Figure 6a. In this case current flows from the tuner into each side of the structure equally causing the opposite ends of the line to oscillate in phase. This results in accelerating fields 180 degrees out of phase. This amounts to tuning two cavities with one tuner, so the tuner must be capable of storing roughly twice as much energy as in the tuned cavity example. Since the impedances of the two line sections are presented to the tuner in parallel, it must be capable of developing an inductive reactance lower by half than the previous example. The maximum accelerating voltage developed by this geometry is

$$V_{acc} = 2V_g \sin\frac{2\pi D}{\lambda}. \tag{45}$$

where D is the line half length. Maximum voltage is developed when D=λ/4 and this may be

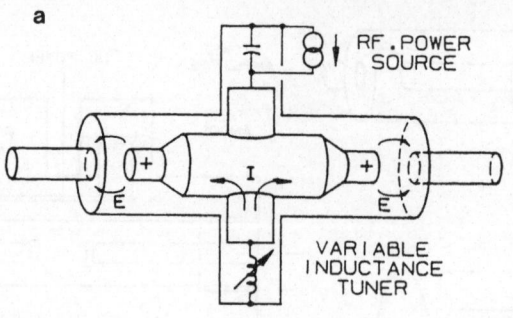

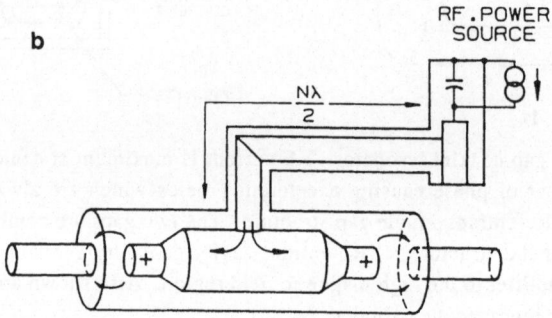

Figure 6 a) Double gap structure with tuner and power amplifier connected at center. The operating mode is now one in which rf current supplied from the external tuner is split symmetrically between the two halves of the cavity so that the accelerating fields developed at the gaps oscillate 180° out of phase. b) Separated gap accelerating structure in which the current is supplied to the center from a relatively long transmission line from a remotely located power source. The transmission line supports standing waves, and it is really part of the accelerating cavity.

achievable with an appropriately graded R_c, but the result may not achieve as low an R/Q value as desired.

The Q of such a cavity is calculable just as a ferrite tunable cavity. The shunt impedance presented to the beam, as compared to that measured in the laboratory, depends on the length. A current generator delivering resonant frequency current i to one of the gaps will develop a voltage v_g at that gap. One can infer from this a resistance $R = v_g/i$ at that gap, and the power dissipated in the cavity (using peak values of voltage and current) is

$$P = \frac{iv_g}{2} = \frac{ii^*R}{2} = \frac{v_g v_g^*}{2R}.$$

(46)

An additional current generator, at some other phase angle, applied to either gap, will develop an additional voltage at the new phase angle at <u>each</u> gap. The two voltages must be added vectorially and the power calculated using Eq. 46.

We again consider structures with half-length D = mλ/4 (m < 1). Beam current entering the cavity from the left can be considered to excite the cavity at phase angle ϕ = mπ/2 (with

350

respect to $\phi = 0$ at cavity center plane). The same current, <u>leaving</u> the cavity at the right gap, excites the cavity with current of <u>opposite sign</u> delayed by $-\phi$. The beam generated cavity voltage is

$$V_b = I_b \left(e^{jm\frac{\pi}{2}} - e^{-j\frac{m\pi}{2}} \right) R$$

$$= 2jIR \sin m\frac{\pi}{2} \tag{47}$$

The power delivered to the rf cavity by beam excitation is

$$P_b = \frac{V_b V_b^*}{2R} = \frac{I_b^2}{2} \left(4R \sin^2 \frac{m\pi}{2} \right). \tag{48}$$

For beam loading, stability, and R/Q considerations, the shunt impedance of such a cavity must be considered to be $R_s = 4R \sin^2(m\pi/2)$ where R is the shunt impedance inferred from external measurement at one gap.

This geometry allows several ferrite tuners to be connected to the cavity in parallel at the mid-point so that they do not occupy space along the beam pipe, and they are easily accessible for cooling, application of tuning bias current, maintenance, etc. External place-ment of the tuning inductance establishes the center point of the cavity at voltage and im-pedance levels convenient for input power coupling. The wide range of beam loading and tuning impedance usually encountered do not establish this point as one of constant real load impedance suitable for termination of a long transmission line, however, it is usually a suitable point to couple to the anode of a large power amplifier tube located very close to the cavity.

The in-phase spearated gap cavity can be coupled through a long transmission line to a remotely located power amplifier when operated at fixed frequency. In such a case the transmission line will usually have standing waves and, in fact, becomes part of the resonant accelerating structure. This is the basis for the Fermillab TEVATRON rf system.[19]

Mechanical Considerations

Accelerator rf cavities must operate reliably and stably for long periods in a remote, hostile, and, only occasionally accessible environment. To meet these requirements they should be built as simply and ruggedly as possible. The very high rf currents encountered at the low impedance regions of the cavity are capable of damaging loose or casually assembled joints and the very high electric fields at other points can result in damaging arcs or dielectric failure of vacuum seals. These hazards can be minimized by care and ingenuity in the selection and placement of rf clamp joints, dielectric and metallic materials, ferrites, etc. Many unexpected malfunctions will be exposed through the development of a complete and operable prototype for each accelerating system. Frequently such prototypes evolve into spare systems which can be kept in operation outside of the accelerator, providing immediate access to a known reliable component in case of accelerator failure. Accelerator access time is extremely expensive time, and the cost of additional operating systems is easily offset by the reduction of true maintenance time they generate.

CONCLUSION

These notes are a collection of a few ideas developed over a number of years of engagement with a variety of proton accelerator rf systems. It is hoped that they may be useful to persons entering the field.

The author wishes to thank collectively all of the many people with whom he has collaborated or from whom he has assimilated or stolen ideas, with the sole purpose of making the best possible instruments for doing the best possible physics.

APPENDIX A
Relativistic Mechanics

The charged particles (protons) under consideration here are acted upon by electric and magnetic fields in their motion around closed orbits in the accelerator. The force acting on the particles is the Lorenz force (MKs units)

$$\mathbf{F} = e[\mathbf{v} \times \mathbf{B}(s,t) + \mathbf{E}(s,t)]$$

$$\text{where } = \mathbf{v} = \frac{ds}{dt} \hat{s}$$

and s is the distance along the particle orbit. In a proton synchrotron the ideal orbit is a closed path of definite length, or circumference C. The path is usually not quite circular but one can define an average radius such that $2\pi R = C$.

The forces resulting from magnetic fields B are always perpendicular to the direction of motion, hence, they cannot affect the particle energy, but serve primarily to establish the path curvature required for a closed orbit. It is the electric fields E(s,t) provided by the rf system and directed along the path s which can affect the particle energy.

The particles under consideration are relativistic, i.e., their speeds are reasonably near the free space speed of light c. Since these speeds cannot exceed c, momentum changes resulting from the longitudinal fields E(s,b) result in changes in both particle speed and mass. Some useful relationships can be derived from the energy-momentum triangle shown in Figure A1.

If a particle energy is increased slightly by passage through an rf accelerating cavity electric field, the increase in particle momentum dp (or dp/p) is composed partially of an increase in speed $d\beta/\beta$ and partially an increase in mass. For a fixed average orbit bending magnetic field the two changes will have opposite effects on the period of time required for a particle to complete one orbital path. An increase in speed will, of course, shorten the period while an increase in mass will cause the particle to pass through the bending field with slightly larger average radius, increasing its path length and orbital period. In strong focussing synchrotrons magnetic bending and focussing fields are composed so that changes in orbit length resulting from increases in momentum are reduced somewhat by a momentum compaction factor α.

$$\frac{dc}{c} = \alpha \frac{dp}{p} \qquad \alpha \equiv \frac{1}{\gamma_t^2} \tag{1a}$$

$$\text{Also} \qquad \frac{dv}{v} = \frac{d\beta}{\beta} = \frac{1}{\gamma^2} \frac{dp}{p} \tag{2a}$$

$$\text{Total Energy} = \begin{bmatrix} \text{Kinetic} \\ \text{Energy} \end{bmatrix} + \begin{bmatrix} \text{Rest Mass} \\ \text{Energy} \end{bmatrix}$$

$$E = T + m_0c^2 \qquad\qquad \text{For Proton}$$
$$m_0c^2 \cong 938.28 \text{ MeV}$$

Define $\beta \equiv v/c$

$$\gamma \equiv \frac{E}{m_0c^2}$$

$c = \text{vel. of light}$
free space

Energy Momentum Triangle

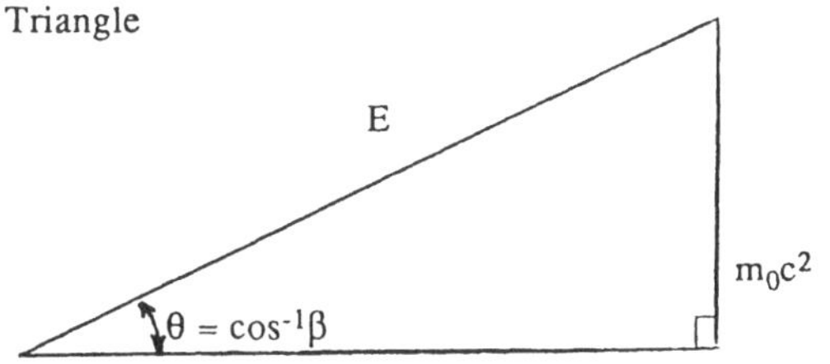

$$E^2 = p^2c^2 + m_0^2c^4$$

$$pc = \beta E$$

$$\frac{dE}{E} = \frac{d\gamma}{\gamma} = \beta^2 \frac{dp}{p} \qquad \gamma = \left[1 - \beta^2\right]^{-1/2}$$

$$\frac{dE}{E} = (\beta\gamma)^2 \frac{d\beta}{\beta} = (\gamma^2 - 1) \frac{d\beta}{\beta}$$

$$\frac{d\mathbf{P}}{dt} = \mathbf{F} = e[(\mathbf{V} \times \mathbf{B}) + \mathbf{E}]$$

Figure A1) Some useful relationships between various kinematic properties of particles.

The fractional change in rotation period can now be related to a change in momentum,

$$\frac{dT}{T} = \frac{dC}{C} - \frac{dV}{V} = \left(\frac{1}{\gamma_t^2} - \frac{1}{\gamma^2}\right)\frac{dp}{p} = \eta\frac{dp}{p} = \frac{\eta}{\beta^2}\frac{dE}{E} \qquad (3a)$$

The constant γ_t (transition gamma) is determined by the detailed design of the bending and focussing magnets in the ring. η, referred to as the <u>revolution</u> <u>frequency</u> <u>dispersion</u> or <u>mixing</u> <u>factor</u>, is a function of particle energy (or gamma). We see that at $\gamma = \gamma_t$ a change in momentum results in no change in rotational period (or frequency). For energies such that $\gamma < \gamma_t$ η is negative and the change in particle <u>speed</u> dominates, so an increase in energy results in a decrease in rotation period (increase in rotational frequency), while for $\gamma > \gamma_t$ the converse is true.

These relationships between rotation period and momentum need not be related uniquely to changes in energy associated with the presence of rf longitudinal fields. They very generally relate the periods of particles of different momentum moving within the <u>momentum</u> <u>acceptance</u> or <u>momentum</u> <u>aperture</u> of the ring. Frequently the incremental energy ΔE or momentum Δp are defined with respect to a particle on a <u>central</u> <u>orbit</u> or <u>synchronous</u> <u>particle</u>. Deviations from the arrival time or energy of a synchronous particle are most conveniently represented by pairs of variables which are <u>cannonically</u> <u>conjugate</u> in the sense of Hamiltonian mechanics. For this illustration we use two such variables, the energy deviation ΔE and the arrival time increment τ. Other useful pairs of variables are momentum and path length (Δp) and (Δs), or $W = \Delta E/h\Omega$ and $\Delta\phi$, the angle on the rf wave corresponding to the time increment τ.

An excellent review of Hamiltonian single particle dynamics may be found in Ref. 9.

APPENDIX B
Synchrotron Motion

There are many excellent and detailed treatments of synchrotron motion in the literature.[1,2,3b] We propose to develop here a simple and intuitive model for such motion with only that mathematical detail necessary to motivate the design of rf accelerating cavities. We examine the motion of two particles and their interaction with an rf field as shown in Figure B1. Initially each of the particles has the same energy and each requires a time T_0 the <u>synchronous</u> <u>rotation</u> <u>period</u> to make one complete orbit turn. The rf period is shorter than T_0 by an integral factor h, the <u>rf</u> <u>harmonic</u> <u>number</u>. Particle 1 passes the accelerating gap at a time when the accelerating voltage is zero so its momentum remains unchanged, consequently, after one turn (exactly h rf periods) it arrives at the accelerating gap when the voltage is again zero and the process continues with no change in particle 1 momentum. (We assume that this particle loses negligible energy to its environment during its orbital travel; true for protons at presently attainable energy, but not true for electrons.) Particle 2 arrives at the accelerating gap a few nanoseconds later than particle 1, and it experiences a non-zero <u>accelerating</u> field, increasing its momentum. Here we considered the <u>velocity</u> increase to be dominant, $\gamma < \gamma_t$, η negative. Particle 2 traverses its orbit in a time slightly less than T_0, arrives slightly earlier, where it again receives an acceleration, and the process continues, particle 2 moving toward particle 1 in time. When particle 2, arriving at successively earlier times, arrives coincident with particle 1, its momentum has been substantially increased so it continues to arrive at earlier times, now experiencing a repeated <u>decelerating</u> field, until it again reaches the synchronous momentum, now arriving <u>earlier</u> by the same time increment that it initially had for late arrivals. The particle now continues to be decelerated, arriving successively <u>later</u> and the process continues, with particle 2 executing <u>synchroton</u> <u>time</u> or <u>phase</u> oscillation about the positive slope zero field point on the rf wave.

The rotational angular velocity of a hypothetical particle moving with velocity c

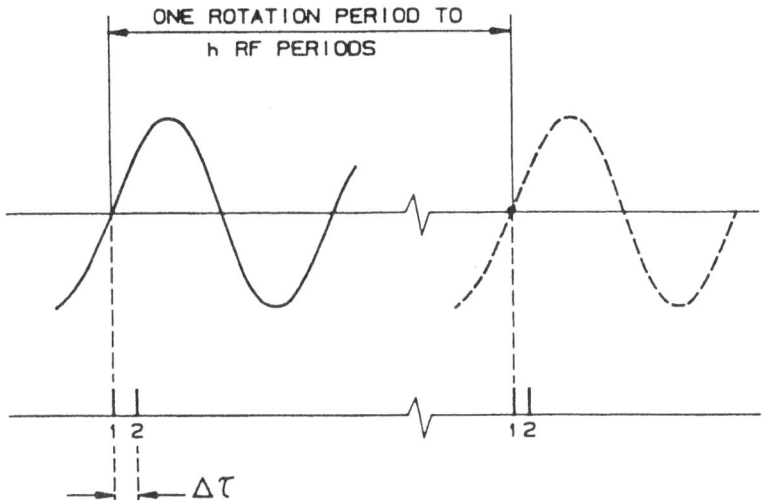

Figure B1) Two segments of rf accelerating fields are shown separated by one rotation period or h rf periods. Two particles are shown displaced slightly in time. Particle 1 receives no acceleration from the field and arrives at the same phase angle (0 degrees) after one turn. Particle 2 is accelerated on the first passage and consequently arrives at a smaller phase angle after one turn.

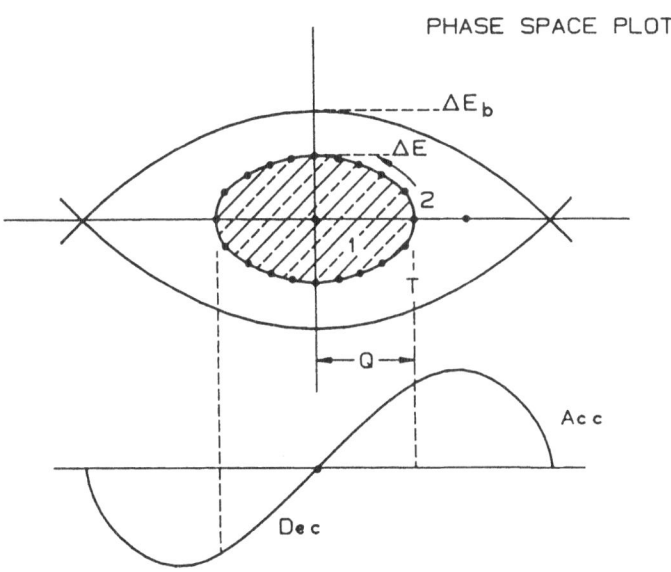

Figure B2) A single rf wave is shown with its relationship to the phase space boundary it creates in an environment of constant accelerator bending field ("stationary bucket").

(infinite energy) would be $\Omega_c = c/R$ and that of a finite energy particle characterized by β is $\Omega = \beta\Omega_c$.

The rf voltage developed at an accelerating gap by an rf system operating at harmonic number h is

$$v(t) = V \sin(h\Omega t). \tag{1b}$$

The change in energy, ΔE of a particle crossing the gap at time τ during one turn will be

$$\frac{(\Delta E)}{\text{turn}} = eV \sin (h\Omega\tau)$$
$$\cong eV(h\Omega\tau) \tag{2b}$$

for τ small with respect to one rf period. The average time rate of change of particle 2 becomes

$$\frac{d(\Delta E)}{dt} = \left(\frac{\Delta E}{\text{turn}}\right) \left(\frac{\text{turns}}{\text{second}}\right) = \frac{eV\Omega^2\tau}{2\pi}. \tag{3b}$$

Using Eq. 3A the time derivative of τ can be written

$$\frac{d\tau}{d\tau} = \frac{\eta}{\beta^2} \frac{(\Delta E)}{E}. \tag{4b}$$

For the case under consideration $\gamma<\gamma_t$, η is negative and Eqs. 3B and 4B can be combined to yield a small amplitude harmonic oscillator equation

$$\frac{d^2 (\Delta E)}{dt} + \omega_s E = 0. \tag{5b}$$

The energy deviation ΔE and the time increment τ oscillate sinusoidally (with a $\pi/2$ phase shift) with the small amplitude synchrotron angular frequency ω_s

$$\omega_s = \frac{\Omega}{\beta} \left[\frac{eVh\eta}{2\pi E}\right]^{1/2}$$
$$\text{and } \Omega/\beta = \Omega_c^- = c/R \tag{6b}$$

In proton machines the factor in brackets is typically extremely small so that the synchrotron frequency is much lower than the rotation frequency. At the low energy end of a Booster synchrotron (~1 GeV) the synchrotron frequency might be a few kHz, while in a 1 TeV storage ring the frequency may be a few tens of Hz.

For larger amplitude oscillations the motion is described by simple pendulum equations, and the oscillation frequency becomes lower, approaching zero for a particle with τ approaching one-half of a synchrotron period.

In Figure 2b the motion of particle 2, starting at maximum time displacement τ is shown on a plot with coordinates ΔE and τ. One cycle of the rf accelerating field is shown also. Particle 1, arriving always at the positive slope zero voltage point remains at the center of the phase space plot. Particle 2 moves on an elliptical path around the plot. Any other particles in the machine, starting at any point on the ellipse, will follow the same path at the same rate.

Particles with smaller initial energy or time increments will follow smaller nested ellipses so that the entire contour defined by the motion of particle 2 may be filled with particles, all moving with nearly the same angular velocity. Such a distribution of particles, confined to a region ±τ on the rf curve, is a beam bunch. The area of the ellipse defined by the maximum amplitude particles is referred to as the longitudinal emittance of a single bunch.

$$S = \pi \Delta E \tau \text{ eV-s.} \tag{7b}$$

Consider now a hypothetical particle which starts incrementally prior to the negative slope zero rf voltage point, i.e., $\tau = \pi/2 - \delta$. Such a particle would, at first, receive an extremely small acceleration per turn, and it would move very slowly toward earlier time where the acceleration per turn becomes larger until it reaches a maximum at $\tau = \pi/4$. Such a particle would oscillate very slowly over the entire rf period, generating the dashed path shown. It would receive the maximum possible acceleration and deceleration each cycle. The dashed curve represents a separatrix, and it defines the boundary of the rf system capability for confining particles, commonly referred to as an rf bucket. Particles with variables within the separatrix will move on closed contours within the bucket while particles outside of the separatrix will move along paths outside of the series of buckets and are not confined by the rf system.

Some useful relationships are presented here without derivation.

$$\text{Bucket Height } \Delta E_b = \beta \left[\frac{2E_s V}{\pi h |\eta|} \right]^{1/2} \text{ eV} \tag{8b}$$

$$\text{Bucket Area } A_b = \frac{8\Delta E_b}{h\Omega} \text{ eV-s} \tag{9b}$$

It is convenient to define an angle ϕ on the rf wave,

$$\phi = h\Omega\tau = \omega_{rf}\tau. \tag{10b}$$

The energy deviation ΔE of a particle starting from angle ϕ is

$$\Delta E = \Delta E_b \sin\frac{\phi}{2}. \tag{11b}$$

References

1. K. Halbach and R. F. Helsinger, Part. Accelerators, 7 (1976) 213
2. T. Weiland, NIM 216 (1983) 329 and DESY-M-82-024, (1982)
3. T. Weiland, Part. Accelerators, 17 (1985)
4. G. Dôme, Proc. 1976 Proton Linear Accel. Conf., Chalk River, Ontario, Canada (1976) 138-147
5. W. R. Smythe, IEEE Trans. Nucl. Sci. NS-30, (1983) 2173
6. W. R. Smythe and T. G. Brophy, IEEE Trans. Nucl. Sci. NS-32, (1985) 2951
7. P. B. Wilson, AIP Conf. Proc. No. 87 (Physics of High Energy Accel., Fermilab Summer School), (1981) 452
8. R. Johnson, S. van der Meer, F. Pedersen, and G. Shering, IEEE Proc. Nucl. Sci, NS-30, (1983) 2290
9. F. T. Cole, AIP Conf. Proc. 153 (Physics of Particle Accelerators, SLAC, 1985) 45

10. C. Bovet, R. Gouiran, I. Gumowski, R. Reich, CERN Report DL/70/4, (1970) Appendix D

11. J. Griffin, IEEE Proc. Nucl Sci. NS-22, (1975) 1910

12. P. B. Wilson, Proc. IX Internat. Conf. on High Energy Accelerators, SLAC, (1974) 57

13. V. Vaccaro, Comments on Beam Instability, Eloisatron Workshop, Ettore Majorana Center, (1987)

14. K. Robinson, Cambridge Electron Accel. Lab Rep. 1010, (1964)

15. A. Hofmann, Proc. Internat. School of Particle Accelerators, Ettore Majorana Center, (1977) 139

16. A. Hofmann, ibid.

17. E. Keil and W. Schnell, CERN Rep. ISR-TH-RF/69-48, (1969)

18. J. E. Griffin and Q. A. Kerns, IEEE Trans. Nucl. Sci. NS-18, (1971) 241

19. Q. Kerns et al., IEEE Trans. Nucl. Sci. NS-28, (1981) 2782

CONCLUDING REMARKS

It seems to me that this Workshop has achieved the goal indicated in the "Opening Address": to speak and to think of radiofrequency and microwave problems (in the beautiful frame of Erice).

This is not a trivial fact, for at least two important reasons. First, it is the first time that a workshop "for accelerator people" has centered entirely upon RF. Second, it confirms that RF and microwave techniques are - finally - considered a relevant part of any electrodynamic accelerator.

The vivid presentations of various topics have clearly shown that the pure and applied sciences, together with the most recent technological advances, have created a lot of very sophisticated and powerful devices. In less than 30 years RF and microwave techniques have changed drastically, their rate of progress unmatched by other areas of the field of particle accelerators.

At times, during this pleasant week, I thought back to the early days of synchrotrons, when a short seminar, of say 45 minutes, was supposed to be more than sufficient for describing the RF system of the synchrotrons of Caltech, Cornell, and Frascati. Now things are different, and it may require many specialized talks to give a complete description of a single RF system.

For everything there is a proper time. Now RF seems to dominate the scenario of particle accelerators, and a demand for more current leads to new demands for the RF system.

I do now know what lies ahead in the future.

We heard of new structures and new materials; superconductivity, cold and warm, seems very promising.

I think that it would be a very nice idea to check the progress made here, in Erice, in a couple of years from now.

Mario Puglisi

PARTICIPANTS

BARTALUCCI, Sergio	CERN-PS, 1211 Geneva 23, Switzerland
BIZZARRI, Ubaldo	ENEA-CRE, Cas. Postale 65, 00044 Frascati, Italy
BÖHNE, Dieter	Gesellschaft für Schwerionenforschung, Postfach 110541, 6100 Darmstadt, Germany
BONI, Roberto	INFN, Laboratori Nazionali di Frascati, Cas. Postale 13, 00044 Frascati, Italy
CAPPI, Roberto	CERN-PS, 1211 Geneva 23, Switzerland
CARPINELLI, Massimo	Dipartimento di Fisica dell'Università, and INFN, Sezione di Pisa, Via Livornese, 56010 S. Piero a Grado, Italy
CHALOUPKA, Heinz	Universität Wuppertal, Fachbereich 13, Gauss Strasse 20, 5600 Wuppertal 1, Germany
DOHAN, Donald A.	TRIUMF, 4004 Wesbrook Mail, Vancouver, British Columbia, Canada
FASCETTI, Mario	ENEA-CRE, Cas. Postale 65, 00044 Frascati, Italy
FERRUCCI Luca	INFN, Sezione di Milano, Via Celoria 16, 20133 Milano, Italy
FITZE, Hansruedi	SIN, 5324 Villigen, Switzerland
GALLUCCIO, Francesca	INFN, Sezione di Napoli, Mostra d'Oltremare, 80125 Napoli, Italy
GRIFFIN, James E.	FNAL, MS-341, P.O.Box 500, Batavia, Ill.60510, USA
KRAFFT, Geoffrey	CEBAF, 12070 Jefferson Avenue, Newport News, Virginia, USA
KUMMER, Jörg	Institut für Angewandte Physik der Universität, Robert Mayer Strasse 2-4, 6000 Frankfurt am Main 1, Germany

LANZ, Paul   SIN,
                            5324 Villigen, Switzerland

LAWSON, Wesley G.   Energy Research Building, University,
                            College Park, Maryland 20742, USA

LENGELER, Herbert   CERN-E.F.,
                            1211 Geneva 23, Switzerland

LOMBARDI, Augusto   CERN-PS,
                            1211 Geneva 23, Switzerland

MASSAROTTI, Antonio   Dipartimento di Fisica dell'Università,
                            and INFN, Sezione di Trieste,
                            Via A.Valerio 2, 34127 Trieste, Italy

MATRONE, Antonio   ANSALDO Ricerche,
                            C.so Perrone 25, 16100 Genova, Italy

MESSINA, Giovanni   ENEA-CRE,
                            Cas. Postale 65, 00044 Frascati, Italy

MING WANG, Jiunn   CERN-LEP,
                            1211 Geneva 23, Switzerland

PAGANI, Carlo   Dipartimento di Fisica dell'Università,
                            and INFN, Sezione di Milano,
                            Via Celoria 16, 20133 Milano, Italy

PALMIERI, Vincenzo   INFN, Laboratori Nazionali di Legnaro,
                            Via Romea 4, 35020 Legnaro, Italy

PARODI, Renzo   INFN, Sezione di Genova,
                            Via Dodecaneso 33, 16146 Genova, Italy

PASOTTI, Cristina   Dipartimento di Fisica dell'Università,
                            and INFN, Sezione di Pavia,
                            Via Bassi 6, 27100 Pavia, Italy

PELLEGRINI, Claudio   BNL-NSLS,
                            Upton, Long Island, NY 11973, USA

PICASSO, Emilio   CERN-LEP,
                            1211 Geneva 23, Switzerland

PIEL, Helmut   Bergische Universität, Gesamthochschule
                            Wuppertal, Fachbereich Physik,
                            Gauss Strasse 20, 5600 Wuppertal 1, Germany

PILAT, Fulvia   CERN-LEP,
                            1211 Geneva 23, Switzerland

PISENT, Andrea   Dipartimento di Fisica dell'Università,
                            and INFN, Sezione di Padova,
                            Via Marzolo 8, 35100 Padova, Italy

PORCELLATO, Anna   INFN, Laboratori Nazionali di Legnaro,
                            Via Romea 4, 35020 Legnaro, Italy

PUGLISI, Mario   Dipartimento di Fisica dell'Università,
                               and INFN, Sezione di Pavia,
                               Via A.Bassi 6, 27100 Pavia, Italy

RAKA, Eugene   BNL-AGS, Dept. 911-B,
                               UPTON, Long Island, NY 11973-5000, USA

REES, Graham   Rutherford Appleton Laboratory, Building R2,
                               Chilton, Didcot, Oxon OX11 0QX, England

REISER, Martin   University of Maryland, Laboratory for Plasma
                               and Fusion Energy Studies,
                               College Park Campus, Maryland 20742, USA

SANTORO, Antonella   Dipartimento di Fisica dell'Università,
                               and INFN, Sezione di Pavia,
                               Via A.Bassi 6, 27100 Pavia, Italy

SCHAFFER, Georg   Kernforschungszentrum, Bau 416,
                               Postfach 3640, 7500 Karlsruhe, Germany

SCHEMPP, Alwin   Institut für Angewandte Physik der
                               Universität, Robert Mayer Strasse 2-4,
                               6000 Frankfurt am Main 1, Germany

SCHNELL, Wolfgang   CERN-LEP,
                               1211 Geneva 23, Switzerland

SERAFINI, Luca   INFN, Sezione di Milano,
                               Via Celoria 16, 20133 Milano, Italy

TIVERON, Bruno   INFN, Laboratori Nazionali di Legnaro,
                               Via Romea 4, 35020 Legnaro, Italy

TORELLI, Gabriele   Dipartimento di Fisica dell'Università,
                               and INFN, Sezione di Pisa,
                               Via Livornese, 56010 S. Piero a Grado, Italy

TRAN, Duc-Tien   C.G.R. MEV, BP 34,
                               78530 Buc, France

VACCARO, Vittorio G.   Dipartimento di Fisica dell'Università,
                               and INFN, Sezione di Napoli,
                               Mostra d'Oltremare, 80125 Napoli, Italy

VIGNATI, Angelo   ENEA-CRE,
                               Cas. Postale 65, 00044 Frascati, Italy

VRETENAR, Maurizio   CERN-PS,
                               1211 Geneva 23, Switzerland

WALDNER, Flavio   Istituto di Fisica dell'Università di Udine,
                               and INFN, Sezione di Trieste,
                               Via Larga 36, 33100 Udine, Italy

INDEX

Accelerating gap, 138, 207
Accelerating structures, 56
Alvarez structure, 38, 198
Amplifiers, 30
Amplitude, 21, 23

Beam loading, 196, 207, 210
Beamstrahlung, 69, 76
Bending magnetic field, 69
Biperiodic structure, 197
Bode's diagram, 25
Brillouin diagram, 45
Bunch, 72, 231
Buncher, 67

Child's law, 57
Circulator, 298
Collider, 13, 15
Constant gradient structure, 217
Coupled resonators, 38, 146, 219
Coupling, 211

Dipole weak-field, 62,72
Directional coupler, 279
Disk and washer structure, 193,199
Disk loaded structure, 193
Dispersion relation, 44, 191
Duty cicle, 47

Efficiency, 53, 83, 206, 228
Electron gun, 111
Emittance, 18, 89
Energy gain, 83
Energy spread, 63
Energy stored, 83
Equation of motion, 182

FDM, FEM-codes, 200
Feedback, 23, 249, 259

FEL, 58, 145
Fill time, 52
Filter, 25
Floquet's theorem, 46, 190
Focusing, 63

Gradient, 54
Group velocity, 46, 190
Gyroklystron, 57, 107

Higher order mode, 231, 254

Injector, 67
Isolator, 298

Kickers, 21, 133
Kilpatrik limit, 43, 66
Klystron, 43, 270, 275

LEP, 16, 240
Longitudinal electric field, 146
Longitudinal motion, 181, 252, 254
Luminosity, 4, 65

Modulation, 274
Modulator, 109
Mono-periodic structure, 189
Multipacting, 164

Normalized emittance, 89

Oscillating mode, 147, 182
Oscillating damping, 216

Parasitic mode, 205, 214
Periodic structure, 193
Phase, 23, 32
Phase locked loop, 35, 36
Phase space, 185

Phase velocity, 46, 52
Pinch effect, 75
Pulse compression, 57

Q-value, 41, 148, 226

Revolution frequency, 133
RF break-down, 219
RFQ, 39, 48
Robinson equation, 247, 252, 262

Shunt impedance, 41, 228
Side-coupled cavities, 197
Skin depth, 148, 151
Sparking, 43
Stability, 206
Standing-wave structure, 38, 47

Stop band, 45
Stored energy, 47, 52
Superconducting cavities, 41, 225

Transfer-function, 24
Transit time, 141
Transverse oscillation, 62, 271
Travelling wave structure, 38,51,81
Tuning, 294

Vacuum window, 288
Van de Graaf, 37
V.C.O., 32
VSWR, 279

Wake-field, 86, 99, 206
Waveguide, 279